DATE DUE

The Conscious Universe

The Conscious Universe

The Scientific Truth of Psychic Phenomena

DEAN I. RADIN, PH.D.

HarperEdge
An Imprint of HarperSanFrancisco

 For information address HarperCollins Publishers, 10 East 53rd Street, New York, NY 10022.

Harper*Edge* Web Site: http://www.harpercollins.com/harperedge

FIRST EDITION

Library of Congress Cataloging-in-Publication Data
Radin, Dean I.
The conscious universe : the scientific truth of psychic phenomena / Dean I. Radin.
p. cm.
Includes bibliographical references and index.
ISBN 0-06-251502-0 (cloth)
ISBN 0-06-251526-8 (pbk.)
1. Parapsychology. 2. Parapsychology—Case studies. I. Title.
BF1031.R18 1997 133—dc21 97-8602

06 07 08 09 ❖ RRD H 20 19 18 17 16 15 14 13 12 11

To my parents, Jerome Radin and Hilda Radin,
and to my brother, Len Radin

Contents

Acknowledgments

I am indebted to numerous friends and colleagues who encouraged me to follow my instincts and take the road less traveled. David Waltz and Klaus Witz supported my interests in graduate school. Later, I was inspired by the words and deeds of Stanley Krippner, Charles Tart, Helmut Schmidt, and William Braud. Hal Puthoff and Edwin May were my role models at SRI International. Robert Jahn and Brenda Dunne helped make Princeton University a thrilling place to work. Robert Morris and Deborah Delanoy were good friends and colleagues at the University of Edinburgh, Scotland. And Alan Salisbury and Stuart Brodsky were visionary leaders at Contel Technology Center. I sincerely thank them all.

I also thank Jessica Utts, Roger Nelson, Jerry Solfvin, Marilyn Schlitz, and Dick Bierman for many stimulating discussions that helped shape the tone and content of this book; Donald Baepler, for his unwavering support of my lab at the University of Nevada, Las Vegas; Robert Bigelow, for his vision and support of scientific exploration; and Jannine Rebman and the students at UNLV who made valuable contributions to our research.

I gratefully acknowledge the organizations and foundations that have provided funding to sustain our research. We have received grants from the Bigelow Foundation (Las Vegas, Nevada), the Parapsychology Foundation (New York City), the Institut für Grenzgebiete der Psychologie und Psychohygiene (Freiburg, Germany), the Society for Psychical Research (London, England), and the Fundação Bial (Porto, Portugal).

I thank my literary agent, Sandra Martin, and my editor, Eamon Dolan, for their expertise in shepherding these words through the maze of the publishing world. Finally, I thank my good friend Susie and my little poodle dog, Holly, for forcing me occasionally to do something other than work.

Preface

"Nonsense!" barked the man in the pinstriped suit. "There isn't a shred of evidence for psychic phenomena!" The clacking sound of the rails punctuated his blunt dismissal.

His companion, a young woman with luminous eyes and an immense halo of hair, was unimpressed. "Harry," she said, glaring at him, "the evidence is staring you in the face."

When I had boarded the commuter train a few minutes earlier, I was looking forward to an uneventful trip. But as the train started to move, two latecomers rushed in and took the seats next to me. Their argument had clearly been percolating for some time.

Harry was an advertisement for Brooks Brothers, complete with attaché case and *Wall Street Journal* tucked under one arm. She was dressed in saffron and carried a well-worn book bag.

"In my meditation last night," she said, pouting, "I received a message from Zeron."

Harry rolled his eyes and, voice dripping with sarcasm, said, "Would that be the Zeron from the planet Pluto or the Zeron from Atlantis?"

"Oh, the one from Atlantis, of course. You know the Plutonians aren't telepathic! We communed mentally through his dolphin friends. He said my psychic abilities would improve if I got my aura cleaned."

Harry's smirk at life's stupidity had permanently creased his forehead with an angry gash, but this last remark caused a vein to leap forward. Exasperated, he caught my eye, leaned over, and said in a stage whisper, "Shirley's gone off the deep end with all that New Age crap." I uttered a noncommittal grunt, not wishing to get sucked into what appeared to be a long-standing disagreement.

But I did not have the luxury of remaining neutral, for Shirley overheard the remark and righteously replied, "If you just *listened* to Zeron for once, you wouldn't be such a skeptic. His words are pure truth!"

"More like pure bull," he grumbled. "There isn't a shred of evidence for ESP, telepathy, or any of that hokum. Not one shred."

She protested: "If you *feel* it, that's proof enough. You just live in your head too much."

Sensing a concession, Harry bellowed, "Your belief about ESP doesn't mean it's true! It just says that you *believe* it's true. If science hasn't proved it, then it isn't true! It's just superstitious, mythological, folkloric, mumbo-jumbo, mystical *crap*."

I couldn't stand this anymore, so I said, "Excuse me, but I couldn't help but overhear your conversation. Actually there is quite a bit of scientific evidence for psychic phenomena. They really do exist."

Shirley smiled beatifically, pressed her palms together, and said "Bless you" with a bow. At the same time, Harry's expression snapped into such a stupendous grimace, with one eye squeezed tight and the other twitching like a guppy out of water, that I was a little concerned that his head might explode. I quickly added, "On the other hand, regardless of how persuasive your personal psychic experiences may be, science has shown time and again that personal *beliefs* are often mistaken."

After my little speech, both of my new acquaintances adopted scowls for different reasons. Shirley's face wavered between awe and bewilderment, while Harry narrowed his one functioning eye and said suspiciously, "What makes you think *you* know anything?"

I sighed, realizing that I had just made a mistake. From past experience, I knew that it would take about six hours of discussion about science, history, psychology, and physics just to reach the starting ground of "educated opinion."

I wanted to explain to Harry and Shirley that what many people think they know about psychic phenomena "ain't necessarily so." I wanted to describe how scientists have essentially proven that psi exists, using the same well-accepted experimental methods familiar to scientists in many disciplines. I also wanted to explain why hardly anyone knew this yet. But no one likes a lecture, so instead I wished I just had a book I could hand to them that would explain all this for me.

This is that book.

The Conscious Universe

Introduction

The psyche's attachment to the brain, i.e., its space-time limitation, is no longer as self-evident and incontrovertible as we have hitherto been led to believe. . . . It is not only permissible to doubt the absolute validity of space-time perception; it is, in view of the available facts, even imperative to do so.

CARL JUNG, *PSYCHOLOGY AND THE OCCULT*

In science, the acceptance of new ideas follows a predictable, four-stage sequence. In Stage 1, skeptics confidently proclaim that the idea is impossible because it violates the Laws of Science. This stage can last for years or for centuries, depending on how much the idea challenges conventional wisdom. In Stage 2, skeptics reluctantly concede that the idea is possible but that it is not very interesting and the claimed effects are extremely weak. Stage 3 begins when the mainstream realizes not only that the idea is important but that its effects are much stronger and more pervasive than previously imagined. Stage 4 is achieved when the same critics who previously disavowed any interest in the idea begin to proclaim that they thought of it first. Eventually, no one remembers that the idea was once considered a dangerous heresy.

The idea discussed in this book is in the midst of the most important and the most difficult of the four transitions—from Stage 1 into Stage 2. While the idea itself is ancient, it has taken more than a century to demonstrate it conclusively in accordance with rigorous, scientific standards. This demonstration has accelerated Stage 2 acceptance, and Stage 3 can already be glimpsed on the horizon.

The Idea

The idea is that those compelling, perplexing, and sometimes profound human experiences known as "psychic phenomena" are real. This will come as no surprise to most of the world's population, because the majority already believes in psychic phenomena. But over the past few years, something new has propelled us beyond old debates over personal beliefs. The reality of psychic phenomena is now no longer based solely upon faith, or wishful thinking, or absorbing anecdotes. It is not even based upon the results of a few scientific experiments. Instead, we know that these phenomena exist because of new ways of evaluating massive amounts of scientific evidence collected over a century by scores of researchers.

Psychic or "psi" phenomena fall into two general categories. The first involves perceiving objects or events beyond the range of the ordinary senses. The second is mentally causing action at a distance. In both categories, it seems that *intention,* the mind's will, can do things that—according to prevailing scientific theories—it isn't supposed to be able to do. We wish to know what is happening to loved ones, and somehow, sometimes, that information is available even over large distances. We wish to speed the recovery of a loved one's illness, and somehow that person gets better quicker, even at a distance. Mind willing, many interesting things appear to be possible.

Understanding such experiences requires an expanded view of human consciousness. Is the mind merely a mechanistic, information-processing bundle of neurons? Is it a "computer made of meat" as some cognitive scientists and neuroscientists believe? Or is it something more? The evidence suggests that while many aspects of mental functioning are undoubtedly related to brain structure and electrochemical activity,[1] there is also something else happening, something very interesting.

This Is for Real?

In discussions of the reality of psi phenomena, especially from the scientific perspective, one question always hovers in the background: You mean this is for real? In the midst of all the nonsense and excessive silliness proclaimed in the name of psychic phenomena, the misinformed use of the term "parapsychology" by self-proclaimed "paranormal investigators," the perennial laughingstock of magicians and conjurers . . . this is for real?

The short answer is, Yes.

A more elaborate answer is, Psi has been shown to exist in thousands of experiments. There are disagreements over how to interpret the evidence, but the fact is that virtually all scientists who have studied the evidence, *including the hard-nosed skeptics,* now agree that something interesting is going on that merits serious scientific attention. Later we'll discuss why very few scientists and science journalists are aware of this dramatic shift in informed opinion.

Shifting Opinions

The most important indication of a shift from Stage 1 to Stage 2 can be seen in the gradually changing attitudes of prominent skeptics. In a 1995 book saturated with piercing skepticism, the late Carl Sagan of Cornell University maintained his lifelong mission of educating the public about science, in this case by debunking popular hysteria over alien abductions, channelers, faith healers, the "face" on Mars, and practically everything else found in the New Age section of most bookstores. Then, in one paragraph among 450 pages, we find an astonishing admission:

> At the time of writing there are three claims in the ESP field which, in my opinion, deserve serious study: (1) that by thought alone humans can (barely) affect random number generators in computers; (2) that people under mild sensory deprivation can receive thoughts or images "projected" at them; and (3) that young children sometimes report the details of a previous life, which upon checking turn out to be accurate and which they could not have known about in any other way than reincarnation.[2]

Other signs of shifting opinions are cropping up with increasing frequency in the scientific literature. Starting in the 1980s, well-known scientific journals like *Foundations of Physics, American Psychologist,* and *Statistical Science* published articles favorably reviewing the scientific evidence for psychic phenomena.[3] The *Proceedings of the IEEE,* the flagship journal of the Institute for Electronic and Electrical Engineers, has published major debates on psi research.[4] Invited articles have appeared in the prestigious journal *Behavioral and Brain Sciences.*[5] A favorable article on telepathy research appeared in 1994 in *Psychological Bulletin,* one of the top-ranked journals in academic psychology.[6] And an article presenting a theoretical model for precognition appeared in 1994 in *Physical Review,* a prominent physics journal.[7]

In the 1990s alone, seminars on psi research were part of the regular programs at the annual conferences of the American Association for the Advancement of Science, the American Psychological Association, and the American Statistical Association. Invited lectures on the status of psi research were presented for diplomats at the United Nations, for academics at Harvard University, and for scientists at Bell Laboratories.

The Pentagon has not overlooked these activities.

From 1981 to 1995, five different U.S. government–sponsored scientific review committees were given the task of examining the evidence for psi effects. The reviews were prompted by concerns that if psi was genuine, it might be important for national security reasons. We would have to assume that foreign governments would exploit psi if they could.

Reports were prepared by the Congressional Research Service, the Army Research Institute, the National Research Council, the Office of Technology Assessment, and the American Institutes for Research (the latter commissioned by the Central Intelligence Agency). While disagreeing over fine points of interpretation, all five reviews concluded that the experimental evidence for certain forms of psychic phenomena merited serious scientific study.

For example, in 1981 the Congressional Research Service concluded that "Recent experiments in remote viewing and other studies in parapsychology suggest that there exists an 'interconnectiveness' of the human mind with other minds and with matter. This interconnectiveness would appear to be functional in nature and amplified by intent and emotion."[8] The report concluded with suggestions of possible applications for health care, investigative work, and "the ability of the human mind to obtain information as an important factor in successful decision making by executives."

In 1985 a report prepared for the Army Research Institute concluded that "The bottom line is that the data reviewed in [this] report constitute genuine scientific anomalies for which no one has an adequate explanation or set of explanations. . . . If they are what they appear to be, their theoretical (and, eventually, their practical) implications are enormous."[9]

In 1987 the National Research Council reviewed parapsychology (the scientific discipline that studies psi) at the request of the U.S. Army. The committee recommended that the army monitor parapsychological research being conducted in the former Soviet Union and in the United States, suggested that the army consider funding specific experiments, and most significantly, admitted that it could not propose plausible alternatives to the "psi hypothesis" for some classes of psi experiments. Dr. Ray Hyman, a psychology professor at the University of Oregon and a longtime skeptic of psi phenomena, was chairman of the National Research Council's review committee on parapsychology. He stated in a 1988 interview with the *Chronicle of Higher Education* that "Parapsychologists should be rejoicing. This was the first government committee that said their work should be taken seriously."[10]

In early 1989 the Office of Technology Assessment issued a report of a workshop on the status of parapsychology. The end of the report stated that "It is clear that parapsychology continues to face strong resistance from the scientific establishment. The question is—how can the field improve its chances of obtaining a fair hearing across a broader spectrum of the scientific community, so that emotionality does not impede objective assessment of the experimental results? Whether the final result of such an assessment is positive, negative, or something in between, the field appears to merit such consideration."[11]

In 1995 the American Institutes for Research reviewed formerly classified government-sponsored psi research for the CIA at the request of the U.S. Congress. Statistician Jessica Utts of the University of California, Davis, one of the two principal reviewers, concluded that "The statistical results of the studies examined are far beyond what is expected by chance. Arguments that these results could be due to methodological flaws in the experiments are soundly refuted. Effects of similar magnitude to those found in government-sponsored research . . . have been replicated at a number of laboratories across the world. Such consistency cannot be readily explained by claims of flaws or fraud. . . . It is recommended that future experiments focus on understanding how this phenomenon works, and on how to make it as useful as possible. There is little benefit to continuing experiments designed to offer proof."[12]

Surprisingly, the other principal reviewer, skeptic Ray Hyman, agreed: "The statistical departures from chance appear to be too large and consistent to attribute to statistical flukes of any sort. . . . I tend to agree with Professor Utts that real effects are occurring in these experiments. *Something* other than chance departures from the null hypothesis has occurred in these experiments."[13]

These opinions are even being reflected in the staid realm of college textbooks. One of the most popular books in the history of college publishing is *Introduction to Psychology,* by Richard L. Atkinson and three coauthors. A portion of the preface in the 1990 edition of this textbook reads: "Readers should take note of a new section in Chapter 6 entitled 'Psi Phenomena.' We have discussed parapsychology in previous editions but have been very critical of the research and skeptical of the claims made in the field. And although we still have strong reservations about most of the research in parapsychology, we find the recent work on telepathy worthy of careful consideration."[14]

The popular "serious" media have not overlooked this opinion shift. The May 1993 issue of *New Scientist,* a popular British science magazine, carried a five-page cover story on telepathy research. It opened with the lines, "Psychic research has long been written off as the stuff of cranks and frauds. But there's now one telepathy experiment that leaves even the sceptics scratching their heads."[15] And in the last few years, *Newsweek,* the *New York Times Magazine, Psychology Today,* ABC's *Nightline,* national news programs, and television and print media around the world have begun to moderate previously held Stage 1 opinions. They're now beginning to publish and broadcast Stage 2–type stories that take scientific psi research seriously.[16]

If all this is true, then a thousand other questions immediately bubble up. Why hasn't everyone heard about this on the nightly news?[17] Why is this topic so controversial? Who has psi? How does it work? What are its implications and applications? These are all good questions, and this book will attempt to answer them through four general themes: *Motivation, Evidence, Understanding,* and *Implications.*

Theme 1: Motivation

Why should anyone take psychic phenomena seriously? The answer rests on the strength of the scientific evidence, which stands on its own merits.

But to appreciate fully *why* the scientific case is so persuasive, and why any scientific controversy exists at all, we have to take a slightly circuitous route.

That route will first consider the *language* used to discuss psi, since much of the confusion about this topic comes from misunderstood and misapplied words (chapter 1). This is followed by examples of common human *experiences* that provide hints about the existence and nature of psi phenomena (chapter 2). We will then consider the topic of *replication*, where we will learn what counts as valid scientific evidence (chapter 3). And we'll end with *meta-analysis*, where we will see how replication is measured and why it is so important (chapter 4).

In sum, the motivations underlying this scientific exploration can be found in mythology, folktales, religious doctrines, and innumerable personal anecdotes. While sufficient to catch everyone's attention, stories and personal experiences do not provide the hard, trustworthy evidence that causes scientists to accept confidently that a claimed effect is what it appears to be. Stories, after all, invariably reflect subjective beliefs and faith, which may or may not be true.

Beginning in the 1880s and accumulating ever since, a new form of scientifically valid evidence appeared—empirical data produced in controlled, experimental studies. While not as exciting as folklore and anecdotes, from the scientific perspective these data were more meaningful because they were produced according to well-accepted scientific procedures. Scores of scientists from around the world had quietly contributed these studies.

Today, with more than a hundred years of research on this topic, an *immense* amount of scientific evidence has been accumulated. Contrary to the assertions of some skeptics, the question is not *whether* there is any scientific evidence, but "What does a proper evaluation of the evidence reveal?" and "Has positive evidence been independently replicated?"

As we'll see, the question of replicability—can independent, competent investigators obtain approximately the same results in repeated experiments?—is fundamental to making the scientific case for psi.

Theme 2: Evidence

Theme 2 discusses the main categories of psi experiments and the evidence that the effects seen in these experiments are genuinely replicable. The evidence is based on analysis of more than a thousand experiments investigating various forms of telepathy, clairvoyance, precognition, psychic healing, and psychokinesis (presented in chapters 5 through 9). The evidence for these basic phenomena is so well established that most psi researchers today no longer conduct "proof-oriented" experiments. Instead, they focus largely on "process-oriented" questions like, What influences psi performance? and How does it work?

Also presented are experiments exploring how psi interacts with more mundane aspects of human experience, such as unusual physical effects associated with the "mass mind" of groups of people (chapter 10), psi effects in casino gambling and lottery games (chapter 11), and applications of psi (chapter 12).

Theme 3: Understanding

The wealth of scientific evidence discussed in theme 2 will show that some psi phenomena exist, and that they are probably expressed in more ways than anyone had previously thought. The vast majority of the information used to make this case has been publicly available for years. One might expect then that the growing scientific evidence for genuine psi would have raised great curiosity. Funding would flow, and researchers around the world would be attempting to replicate these effects. After all, the implications of genuine psi are profoundly important for both theoretical and practical reasons. But this has not yet been the case. Few scientists are aware that any scientifically valid case can be made for psi, and fewer still realize that the cumulative evidence is highly persuasive.

In theme 3 we consider why this is so. One reason is that the information discussed here has been suppressed and ridiculed by a relatively small group of highly skeptical philosophers and scientists (chapter 13). Are the skeptics right, and all the scientists reporting successful psi experiments over the past century were simply delusional or incompetent? Or is there another explanation for the skepticism?

We will see that because scientists are also human, the process of evaluating scientific claims is not as pristinely rational or logical as the general public believes (chapter 14). The tendency to adopt a fixed set of beliefs and defend them to the death is incompatible with science, which is essentially a loose confederation of evolving theories in many domains. Unfortunately, this tendency has driven some scientists to continue to defend outmoded, inaccurate worldviews. The tendency is also seen in the behavior of belligerent skeptics who loudly proclaim that widespread belief in psi reflects a decline in the public's critical thinking ability. One hopes that such skeptics would occasionally apply a little skepticism to their own positions, but history amply demonstrates that science progresses mainly by funerals, not by reason and logic alone.

Understanding why the public has generally accepted the existence of psi and why science has generally rejected it requires an examination of the *origins* of science (chapter 15). In exploring this clash of beliefs, we will discover that the scientific controversy has had very little to do with the evidence itself, and very much to do with the psychology, sociology, and history of science.

Discussions about underlying assumptions in science rarely surface in skeptical debates over psi, because this topic involves deeply held, often unexamined beliefs about the nature of the world. It is much easier to imagine a potential flaw in one experiment, and use that flaw to cast doubt on an entire class of experiments, than it is to consider the overall results of a thousand similar studies. A related issue is how science deals with *anomalies,* those extraordinary "damn facts" that challenge mainstream theories.[18] As we look at the nature and value of anomalies, and how scientists react to them, we will also explore the role that prejudice, in the literal sense of "prejudging," has played in controlling what is presumed to be scientifically valid. Other issues, like how scientific disciplines rarely talk to one another, and the historical abyss between science and religion, make it abundantly clear that if psychic experiences were any other form of curious natural phenomena, they would have been adopted long ago by the scientific mainstream on the basis of the evidence alone.

Beyond the themes of motivation, evidence, and understanding, resides the question, So what? Why should anyone care if psi is real or not?

Theme 4: Implications

The eventual scientific acceptance of psychic phenomena is inevitable. The origins of acceptance are already brewing through the persuasive weight of the laboratory evidence. Converging theoretical developments from many disciplines are offering glimpses at ways of understanding how psi works (chapter 16). There are explorations of psi effects by major industrial labs, evaluation of claims of psychic healing by the Office of Alternative Medicine of the National Institutes for Health, and articles about psi research appearing in the "serious" media.

As acceptance grows, the implications of psi will become more apparent. But we already know that these phenomena present profound challenges to many aspects of science, philosophy, and religion (chapter 17). These challenges will nudge scientists to reconsider basic assumptions about space, time, mind, and matter. Philosophers will rekindle the perennial debates over the role of consciousness in the physical world. Theologians will reconsider the concept of divine intervention, as some phenomena previously considered to be miracles will probably become subject to scientific understanding.

These reconsiderations are long overdue. An exclusive focus on what might be called "the outer world" has led to a grievous split between the private world of human experience and the public world as described by science. In particular, science has provided little understanding of profoundly important human concepts like *hope* and *meaning.* The split between the objective and the subjective has in the past been dismissed as a nonproblem, or as a problem belonging to religion and not to science.

But this split has also led to major technological blunders, and a rising popular antagonism toward science. This is a pity, because scientific methods are exceptionally powerful tools for overcoming personal biases and building workable models of the "truth." There is every reason to expect that the same methods that gave us a better understanding of galaxies and genes will also shed light on experiences described by mystics throughout history.

Now let's explore a little more closely what we're talking about. What is psi?

THEME I

MOTIVATION

What is psi? What does it mean to study the scientific evidence for psi? What counts as scientific evidence? How do we evaluate that evidence?

To answer these questions, we'll begin by considering what is meant by psi, to help distinguish it from the wild, wacky world of the paranormal. We'll reflect on how some doubts about psi can be traced to confusions over related words like "supernatural," and we'll consider what science is and how it fits into the study of psi.

Next, we'll read some case studies that provide the motivation for studying whether what seems to be happening in psi experiences is really happening. Can the real-life anecdotes about psi be confirmed under controlled conditions? Then we'll cover two very important topics—replication and meta-analysis—that will allow us to make sense of the scientific evidence presented in theme 2.

CHAPTER 1

What Is Psi?

Many errors, of a truth, consist merely in the application of the wrong names of things.

BARUCH SPINOZA

Since primeval times, people have spoken of strange and sometimes profoundly meaningful personal experiences. Such experiences have been reported by the majority of the world's population and across all cultures. In modern times, they're still reported by most people, including the majority of college professors. These experiences, called "psychic" or psi, suggest the presence of deep, invisible interconnections among people, and between objects and people. The most curious aspect of psi experiences is that they seem to transcend the usual boundaries of time and space.

For over a century, these very same experiences have been systematically dismissed as impossible, or ridiculed as delusionary, by a small group of influential academics and journalists who have assumed that existing scientific theories are inviolate and complete. This has created a paradox. Many people believe in psi because of their experiences, and yet the defenders of the status quo have insisted that this belief is unjustified.

Paradoxes are extremely important because they point out logical contradictions in assumptions. The first cousins of paradoxes are anomalies, those unexplained oddities that crop up now and again in science. Like paradoxes, anomalies are useful for revealing possible gaps in prevailing theories. Sometimes the gaps and contradictions are resolved peacefully and the old theories are shown to accommodate the oddities after all. But that is not always the case, so paradoxes and anomalies are not much liked by scientists who have built their careers on conventional theories. Anomalies present annoying challenges to established ways of thinking, and because theories tend to take on a life of their own, no theory is going to lie down and die without putting up a strenuous fight.

Though anomalies may be seen as nuisances, the history of science shows that each anomaly carries a seed of potential revolution. If the seed can withstand the herbicides of repeated scrutiny, skepticism, and prejudice, it may germinate. It may then provoke a major breakthrough that reshapes the scientific landscape, allowing new technological and sociological concepts to bloom into a fresh vision of "common sense."

A long-held, commonsense assumption is that the worlds of the subjective and the objective are distinct, with absolutely no overlap. Subjective is "here, in the head," and objective is "there, out in the world." Psi phenomena suggest that the strict subjective-objective dichotomy may instead be part of a continuous spectrum, and that the usual assumptions about space and time are probably too restrictive.

The anomalies fall into three general categories: ESP (extrasensory perception), PK (psychokinesis, or mind-matter interaction), and phenomena suggestive of survival after bodily death, including near-death experiences, apparitions, and reincarnation (see the following definitions and figure 1.1). Most scientists who study psi today expect that further research will eventually explain these anomalies in scientific terms. It isn't clear, though, whether they can be fully understood without significant, possibly revolutionary, expansions of the current state of scientific knowledge.

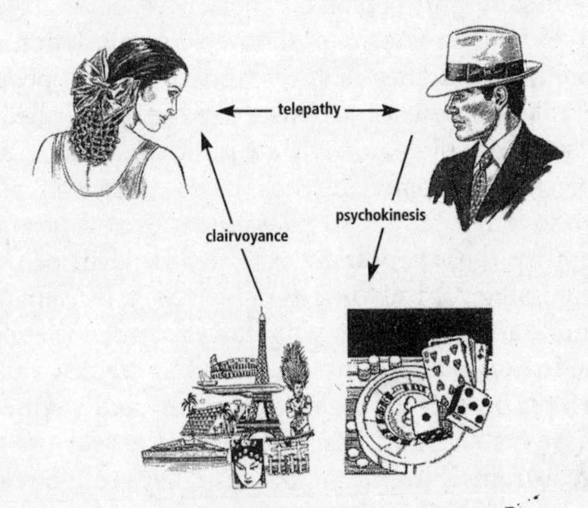

Figure 1.1. The flow of information in telepathy, clairvoyance, and psychokinesis.

What's in a Name?

In popular usage, psychic phenomena may be defined as follows:

telepathy Information exchanged between two or more minds, without the use of the ordinary senses.

clairvoyance Information received from a distance, beyond the reach of the ordinary senses. A French term meaning "clear-seeing." Also called "remote viewing."

psychokinesis Mental interaction with animate or inanimate matter. Experiments suggest that it is more accurate to think of psychokinesis as information flowing from mind to matter, rather than as the application of mental forces or powers. Also called "mind-matter interaction," "PK," and sometimes, "telekinesis."

precognition Information perceived about future events, where the information could not be inferred by ordinary means. Variations include "premonition," a foreboding of an unfavorable future event, and "presentiment," a sensing of a future emotion.

ESP Extrasensory perception, a term popularized by J. B. Rhine in the 1930s. It refers to information perceived by telepathy, clairvoyance, or precognition.

psi A letter of the Greek alphabet (ψ) used as a neutral term for all ESP-type and psychokinetic phenomena.

Related Phenomena

OBE Out-of-body experience; an experience of feeling separated from the body. Usually accompanied by visual perceptions reminiscent of clairvoyance.

NDE Near-death experience; an experience sometimes reported by those who are revived from nearly dying. Often refers to a core experience that includes feelings of peace, OBE, seeing lights, and certain other phenomena. Related to psi primarily through the OBE experience.

reincarnation The concept of dying and being reborn into a new life. The strongest evidence for this ancient idea comes from children, some of whom recollect verifiable details of previous lives. Related to psi by similarities to clairvoyance and telepathy.

haunting Recurrent phenomena reported to occur in particular locations, including sightings of apparitions, strange sounds, movement of objects, and other anomalous physical and perceptual effects. Related to psi by similarities to psychokinesis and clairvoyance.

poltergeist Large-scale psychokinetic phenomena previously attributed to spirits but now associated with a living person, frequently an adolescent. From the German for "noisy spirit."

Mistaking the Map for the Territory

Though the terms listed above are in common usage, scientists who study psi try to think about these phenomena in neutrally descriptive terms. This is because popular labels such as "telepathy" carry strong, unstated connotations that cause us to think we understand more than we actually do. As psycholinguists often point out, it's very easy to mistake the *name* of the thing for the thing itself. And when we are not clear about what "the thing" is, mistaking the map for the territory can lead to enormous confusions.

Some names also carry hidden theoretical assumptions. For example, some people have imagined that telepathy may literally be a transfer of mental signals from one mind to another. This commonly evokes the image of "mental radio," which has been proposed by various people over the years, including the author Upton Sinclair, who wrote a famous book by that title.

The concept of "mental radio" naturally suggests that telepathy is based on something like electromagnetic signaling. Brain-wave signals, however, are exceptionally weak, and in cases of telepathy where the "receiver" and "sender" are many miles apart, it is difficult to imagine that anything could detect the infinitesimally tiny signals "broadcast" from the sender. Still, because psi does not fit easily into conventional theories, researchers have repeatedly put the "electromagnetic" theories to the test. The results show that when telepathic receivers are isolated by heavy-duty electromagnetic and magnetic shielding (specially constructed rooms with steel and copper walls), or by extreme distance, they are still able to obtain information from a sender without using the ordinary senses.

So we know that telepathy doesn't work like conventional electromagnetic signaling. And yet, because the metaphor provides a powerful way of thinking about telepathy, many people still imagine that telepathy "works" through some form of mental radio.

Besides the problems that can arise from taking labels too literally, the strength of the evidence for various categories of psi varies widely. Simply labeling an effect without qualification tends to give the false impression that all these phenomena stand on equally firm scientific ground, and this is not the case.

Keep in mind that the names and concepts used to describe psi say more about the situations in which the phenomena are observed than about any fundamental properties of the phenomena themselves. This is always true in science but is often glossed over for the sake of simplicity. Depending on what we wish to measure, a photon can be either a wave or a particle. We may call it one thing or the other, but that does not change what it "really" is: something that is neither a wave nor a particle, but apparently both at once.

In addition, in scientific practice many of the basic terms for psi effects are accompanied by strings of qualifiers such as "apparent," "putative," and "ostensible." This is because many claims supposedly involving psi may not be caused by psi, but by normal psychological or misinterpreted physical factors. Here we avoid the repetitive use of qualifiers because they can become monotonous. But it is useful to remember that science deals with hypotheses, theories, and models, and not with absolutes. Every scientific concept carries some qualification.

What Are We Talking About?

Psi research continues to be controversial partly because of confusion about the term "paranormal." The common view of the paranormal, especially as reflected in the popular media, is of anything bizarre, occult, or mysterious. In this view, ESP, telepathy, and precognition are lumped together with "bleeding" statues, alien abductions, and five-headed toads.

Other terms commonly used to refer to all things strange include supernatural, psi, psychic, parapsychological, mystical, esoteric, occult, and for some unfathomable reason, "PSI," pronounced letter by letter, *p, s, i,* as though that meant something. (It doesn't in this context.)

The indiscriminate mixing of these terms has led to vast misunderstandings. There really is a difference between the scientific study of psi phenomena and, say, the belief that Elvis has reincarnated into a forty-pound zucchini that bears a striking resemblance to the late King of Rock and Roll. To clarify precisely what is meant by the phrase "scientific study of psi phenomena" and to prepare for the concept of replication in science, we must briefly consider five concepts: paranormal, supernatural, mystical, science, and the scientific method.

This review may seem a bit tedious, especially when compared with the fun stories about psychic experiences coming up in the next chapter. Surely we can skip all this worrying about words. Possibly, but consider that all a book can offer is a bunch of words, so a clear understanding of some key words now will become progressively more important later. Think of it like brushing your teeth. You don't really *want* to brush your teeth every single day, but if you don't, somewhere down the line you won't have anything left to brush. No brushing, no teeth. No words, no understanding. Simple.

Paranormal

Webster's Third New International Dictionary defines *paranormal* as "beyond the range of scientifically known phenomena." Note that this definition does not specify psychic phenomena per se, so paranormal can be used to refer to any unexplained, but potentially explainable, phenomenon. Also note that the definition uses the phrase "scientifically known," which itself raises a rather complicated issue involving the scientific method and the nature of evidence and proof in science. For now, let us take paranormal to mean something like "beyond the range of phenomena presently accepted by most scientists."

Many subjects now considered perfectly legitimate areas of scientific inquiry, including hypnosis, dreams, hallucinations, and subliminal perception, were relegated to the wackiest fringes of the paranormal in the late nineteenth century. A few hundred years before that, topics like physics, astronomy, and chemistry were so far out that those who merely dabbled in them risked accusations of heresy, or worse.

This simply points out that science, like most other things, is part of an evolutionary process: odd events considered paranormal eventually become normal after satisfactory scientific explanations are developed. In this sense—although some scientists would probably shudder at the analogy—virtually *all* cutting-edge, basic research can be viewed as the systematic practice of probing and explaining the paranormal.

Curiously, many effects that science cannot explain are generally not regarded as paranormal. In psychology, for example, there are some remarkable but completely unexplained phenomena such as photographic memory (the ability to remember images in perfect detail), lightning calculation in autistic savants (the ability to perform mental arithmetic with astonishing speed and accuracy), extraordinary musical aptitude in prodigies who seem to spring from the womb ready for Carnegie Hall, and so on.

Perhaps the most widely accepted, yet totally baffling phenomenon is conscious awareness itself, but this too is not regarded as paranormal. Thus, in general usage "paranormal" has taken on a connotation of eerie, bizarre, or ominous in addition to its dictionary meaning. As Marcello Truzzi, a sociologist at Eastern Michigan University, says:

> The term paranormal was created to designate phenomena considered natural—not supernatural—and which eventually should find scientific explanation but thus far have escaped such explanations. . . . Unfortunately, many critics of the paranormal continue to equate anything purportedly paranormal with the supernatural. This is particularly ironic since those who truly believe in the supernatural (such as the Roman Catholic church when it speaks of miracles) have long understood that a paranormal explanation precludes a supernatural one.[1]

Supernatural

Supernatural has several meanings; the usual is "miraculous; ascribed to agencies or powers above or beyond nature; divine." Because science is commonly regarded as a method of studying the natural world, a supernatural phenomenon is by this definition unexplainable by, and therefore totally incompatible with, science.

Today, a few religious traditions continue to maintain that psi is supernatural and therefore not amenable to scientific study. But a few hundred years ago virtually *all* natural phenomena were thought to be manifestations of supernatural agencies and spirits. Through years of systematic investigation, many of these phenomena are now understood in quite ordinary terms. Thus, it is entirely reasonable to expect that so-called miracles are simply indicators of our present ignorance. Any such events may be more properly labeled first as paranormal, then as normal once we have developed an acceptable scientific explanation. As astronaut Edgar Mitchell put it: "There are no unnatural or supernatural phenomena, only very large gaps in our knowledge of what is natural, particularly regarding relatively rare occurrences."[2]

Mystical

Mystical refers to the direct perception of reality; knowledge derived directly rather than indirectly. In many respects, mysticism is surprisingly similar to science in that it is a systematic method of exploring the nature of the world. Science concentrates on outer, objective phenomena, and mysticism concentrates on inner, subjective phenomena. It is interesting that numerous scientists, scholars, and sages over the years have revealed deep, underlying similarities between the goals, practices, and findings of science and mysticism. Some of the most famous scientists wrote in terms that are practically indistinguishable from the writings of mystics.

Science

Science may be defined as a well-accepted body of facts and a method of obtaining those facts. Scientists are quick to disagree, however, over what "well-accepted" means, what "facts" mean, what "methods" mean, what "mean" means, and even sometimes what "and" means. As a result, the definition of science depends to a large extent on whom you ask. We are not too far off the mark by repeating the pithy phrase "science is what scientists do." In any case, most scientists would probably agree that what made science great was the *scientific method.* So what's this method, and why is it so great?

If scientists cannot easily agree on what science is, then it seems unlikely that they can agree on something more complex like "the" scientific method. Psychologists Robert Rosenthal of Harvard University and Ralph Rosnow of Temple University maintain that "scientific method" is difficult to define because "The term 'scientific method' is itself surrounded by controversy, and is a misnomer to boot, since there are many recognized and legitimate methods of science."[3]

A common element among most varieties of scientific method is the use of controlled and disciplined *observation*. However, observation alone is insufficient. As philosopher Jerome Black wrote, "Neither observation, nor generalization, nor the hypothetic-deductive use of assumptions, nor the use of instruments, nor mathematical construction—nor all of them together—can be regarded as essential to science."[4]

Many other scientists and philosophers have agreed that simple definitions are too restrictive to capture the essence of the scientific method. Attempts to clarify the definition have ranged from the witty ("The scientist has no other method than doing his damnedest")[5] to the anarchistic ("Success in science occurs only because scientists break every methodological rule and adopt the motto 'anything goes'").[6] But this is not very enlightening. The specialness of the scientific method can be illustrated more effectively by comparing it with earlier, prescientific methods of pursuing knowledge. As L. L. Whyte explained, "About 1600 Kepler and Galileo simultaneously and independently formulated the principle that the laws of nature are to be discovered by measurement, and applied this principle in their own work. Where Aristotle had classified, Kepler and Galileo sought to measure."[7]

In addition to careful observations and measurements, a fundamental strength of the scientific method is its reliance on public, consensus agreement that the measurements are in fact correct. This differs dramatically from earlier approaches to knowledge, such as the logical arguments favored by philosophers, or the dogmatic acceptance of scripture demanded by religious authorities.

The idea of public agreement about measurements has led to the strong requirement in science (at least in the experimental sciences) that phenomena must be independently and repeatedly measurable to allow this consensus to form. In other words, the idea of repeatability, or replication, has become roughly equivalent to a test for *stability*.

If a phenomenon is highly unstable, we can't be sure whether we are measuring a real effect, some other effect, or just random variations. With this sort of confusion, no consensus can be reached and the existence of the effect in question remains in doubt. Scientists in the seventeenth century had not yet developed methods of clearly distinguishing between real effects and chance, so they were forced to bypass many interesting physical, biological, and psychological phenomena—in fact, almost everything studied in the sciences today.

Fortunately, some physical and astronomical effects were stable enough (or were precisely periodic) that early attempts at measurement were successful. Without such stable effects, science as we know it would have failed miserably and we would still be arguing as in Aristotle's time. Such philosophical debates typically went something like: Yes, it is so. *No, it is not so.* Yes it is. *No it isn't.* 'Tis! *'Tisn't!* As the philosopher Bertrand Russell remarked, "This may seem odd, but that is not my fault."[8]

Now, before we study in more detail what "stable" means in scientific terms, let's examine some commonly reported psi experiments to see why this topic is so intriguing.

CHAPTER 2

Experience

Is it not rather what we expect in men, that they should have numerous strands of experience lying side by side and never compare them with each other?

GEORGE ELIOT (1819–1880)

Spontaneous human experiences provide the core motivation for studying psychic phenomena. While there is always some distortion and exaggeration when recalling unusual incidents, sufficient similarity appears in reports from different people, in tens of thousands of case studies around the world, to form basic categories for the phenomena. The scientific challenge is to take these raw experiences and try to figure out what they mean. Are they what they appear to be, genuine phenomena that transcend the usual boundaries of space and time, or are they better understood as conventional psychological and physical effects?

To provide an experiential base for the experiments discussed later, here are some examples of commonly reported human experiences.

Feeling at a Distance

The prototypical case of feeling at a distance, published by author Bernard Gittelson,[1] concerns an event reported by the nineteenth-century landscape painter Arthur Severn and his wife, Joan. According to Mrs. Severn:

> I woke up with a start, feeling I had had a hard blow on my mouth, and with a distinct sense that I had been cut and was bleeding under my upper lip, and seized my pocket-handkerchief and held it (in a little pushed lump) to the part, as I sat up in bed, and after a few seconds, when I removed it, I was astonished not to see any blood, and only then realized it was impossible anything could have struck me there, as I lay

> fast asleep in bed, and so I thought it was only a dream!—but I looked at my watch, and saw it was seven, and finding Arthur (my husband) was not in the room, I concluded (rightly) that he must have gone out on the lake for an early sail, as it was so fine.
>
> I then fell asleep. At breakfast (half-past nine) Arthur came in rather late, and I noticed he rather purposely sat further away from me than usual, and every now and then put his pocket-handkerchief furtively up to his lip, in the very way I had done. I said, "Arthur, why are you doing that?" and added a little anxiously, "I know you have hurt yourself! but I'll tell you why afterwards."
>
> He said, "Well, when I was sailing, a sudden squall came, throwing the tiller suddenly round, and it struck me a bad blow in the mouth, under the upper lip, and it has been bleeding a good deal and won't stop." I said then, "Have you any idea what o'clock it was when it happened?" and he answered, "It must have been about seven." I then told what had happened to me, much to his surprise, and all who were with us at breakfast.

In a modern version of the same type of experience, Fred, an executive in a high-technology company, recounted the following story:

> In the middle of the night, out of a deep sleep, Fred suddenly jerked upright into a sitting position. He clutched his chest, gasping for breath. His wife, abruptly awakened by her husband's sudden movement, anxiously asked, "What's wrong?" A few moments later, when Fred was able to breathe normally again, he told his wife that he was all right, but he had a feeling that something terrible had happened. They glanced at the clock: 2:05 A.M.
>
> Fifteen minutes later, as they settled back to sleep, the phone rang. Fred's father was on the line. "I have bad news," he said. "Your mother just had a heart attack. We were sleeping, when she suddenly sat bolt upright, clutched her chest, and . . . she passed away." Fred was shocked. "When did this happen?" he asked. "About fifteen minutes ago, just after 2:00 A.M.," replied his father.[2]

Is it really possible to feel someone else's experience at a distance? Is there evidence for genuine telepathy, where all normal sensory cues have been eliminated and we know that the experience was not exaggerated or misremembered? We will see later that the answer is yes—at least some cases of telepathy do include the perception of feelings and possibly the experiences of another person.

Seeing at a Distance

Like many spontaneous psychic experiences, clairvoyance is often associated with dangerous situations, life crises, and other moments of high need or motivation. For example, Beverley Nichols, a reporter for the Canadian Broadcasting Corporation, tells of an incident in 1963, when he was in England describing a procession of Queen Elizabeth in her royal coach.

> Without any warning I had a sharp feeling of discomfort, almost of nausea, accompanied by an acute headache. The picture of the Queen and her cavalcade vanished as swiftly as if it had been blacked out in a theatrical performance, to be replaced by an equally vivid picture of President Kennedy driving in an open car, flanked by his escort of motorcyclists with their snarling exhausts. And as though it were being dictated to me, I began to describe the scene.[3]

When the procession was over a few minutes later, Nichols was leaving for a drink when a stranger rushed up to him and said, "President Kennedy has been assassinated. Six minutes ago."

The next example is more mundane, except for the fact that it was part of a government program funded in 1974 at Stanford Research Institute (SRI), a Menlo Park, California, think tank. The Central Intelligence Agency wanted to know if clairvoyance could be used to "see" distant, strategically important locations. The narrator of this tale is physicist Russell Targ, who used an experimental method with a "remote viewer" named Pat Price, a retired police commissioner from Burbank, California. Price was given only longitude and latitude coordinates of an unknown site somewhere on the other side of the world, and was asked to describe what he "saw" there.

> On July 10 of 1974, one of our contract monitors came to SRI with a new task for us to do. . . . [This] contract monitor, a physicist from the CIA, had brought us coordinates from what he described as a "Soviet site of great interest to the analysts." They wanted any information we could give them, and they were eager to find out if we could describe a target ten-thousand miles away, with only coordinates to work from.
>
> Armed with a slip of paper bearing the coordinates, Price and I climbed to the second floor of SRI's Radio Physics building and locked ourselves into a small electrically shielded room which we had been using for our experiments. . . . As always, I began our little ritual of starting the tape recorder, giving the time and date, and describing who we were and what we were doing. I then read the coordinates.
>
> As was Pat's custom, he polished his spectacles, leaned back in his chair and closed his eyes. He was silent for about a minute . . . then began his description: "I am lying on my back on the roof of a two or three story brick building. It's a sunny day. The sun feels good. There's the most amazing thing. There's a giant gantry crane moving back and forth over my head. . . . As I drift up in the air and look down, it seems to be riding on a track with one rail on each side of the building. I've never seen anything like that." Pat then made a little sketch of the layout of the

buildings, and the crane, which he labeled as a "gantry." Later on, he again drew the crane [shown in figure 2.1].

After several days we completed the remote viewing. We were astonished when we were told [later] that the site was the super-secret Soviet atomic bomb laboratory at Semipalatinsk, where they were also testing particle beam weapons. . . . The accuracy of Price's drawing is the sort of thing that I, as a physicist, would never have believed, if I had not seen it for myself. The drawing in [figure 2.2] was made by the CIA from satellite photography of the Semipalatinsk facility.[4]

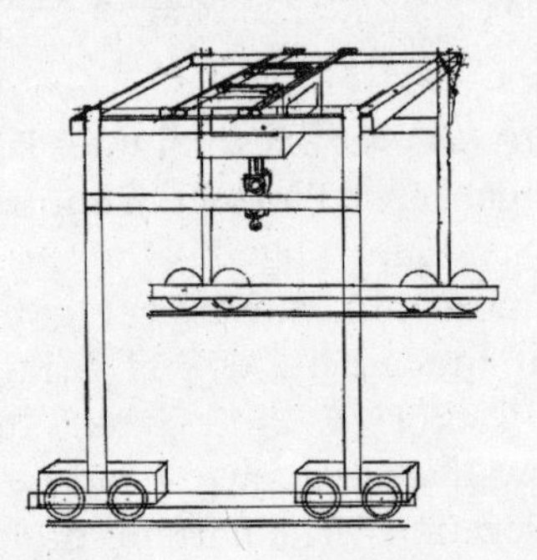

Figure 2.1. Drawing of the gantry crane by Pat Price.
From *Journal of Scientific Exploration.*

Figure 2.2. Drawing of the Semipalatinsk site by a CIA artist.[5]
From *Journal of Scientific Exploration.*

This was a remarkably accurate case, one of hundreds of experiments conducted by the SRI researchers. But is there any reason to believe that clairvoyance is a genuine ability that most people can use if they put their mind to it? After 110 years of experimental research, the answer is clear: yes, it appears that various forms of extended perception are authentic and are probably distributed among the general population like any other talent, such as musical or sports ability.

Mind Over Matter

"I can't work there!" screamed Gail. She had just returned from her first day as a psychiatric nurse in the lockup ward of a state mental hospital for the criminally insane. A sensitive, emotionally expressive woman, Gail was in shock after trying to cope all day with psychotic patients on a dangerous ward. She kept repeating that she couldn't work there.

Because she was so upset, her roommate, Dan, suggested that she try something she found relaxing, like sewing. Gail reluctantly agreed and turned on the machine. A few seconds later, a foot-long blue flame shot out of the control switch and the sewing machine abruptly died. Gail jumped up angrily, saying, "This doesn't work!" She stomped over to the stove to heat up a teapot. She turned the knob and nothing happened. The gas oven was completely silent. There was no snapping sound of the electric pilot light, no hissing sound of gas—nothing.

Still wanting some tea, Gail put a mug of water in the microwave oven, set the timer, and hit the start button. Nothing happened. The oven was dead. As Dan went over to examine the microwave oven, Gail stormed out of the kitchen, muttering angrily how nothing seemed to work anymore. She started her tape recorder in the living room. A few seconds later she shrieked, "What the . . . ?" Dan ran into the living room just in time to see the cassette player chewing up tape and spitting it out onto the floor.

By now Gail was becoming skittish about touching anything. Dan tried to calm her by saying, "Don't be silly, these are coincidences. Play a record instead." She put a record on the turntable and turned it on; it started to move, then they both heard a loud snap and the turntable stopped dead. Gail said angrily, "See, it doesn't work!"

At this point, Dan suggested that perhaps she should just go to bed, as he couldn't afford any more repair bills even if the strange events *were* just coincidences.

A week later, Gail had had enough and quit her exceedingly stressful job. In the meantime, Dan had procrastinated about getting the appliances repaired, and when things settled down, he tried them all again to confirm that they were broken. Sure enough, the sewing machine, turntable, and cassette recorder were still quite dead, and when Dan opened them up, he found that several of the electrical components had been fused inside each of the appliances. But to his amazement, the gas and the microwave ovens worked perfectly. There was no sign that anything had ever been wrong with them.[6]

Is there any reason to believe that psychokinesis, perhaps amplified by high anxiety or stress, may affect the behavior of machines, actually causing them to fail? The answer is yes, psychokinesis may be responsible for the failure of some machines, especially sensitive electronic machines. Later, we'll discuss how more than seventy researchers around the world have been studying mind-matter interactions in the laboratory for more than fifty years.

Intuitive Hunches

You are driving down a nearly deserted highway on a dry, clear day. A single car ahead of you is going a little slower than you like to drive, so you check your rearview mirror and prepare to pass. As you start to accelerate, you inexplicably get a bad feeling, so you immediately take your foot off the accelerator.

As soon as your foot leaves the accelerator, the car in front of you suddenly swerves into the passing lane without warning, its front tire having just blown out. If you had continued to accelerate, you would have been in a serious, high-speed accident. Was the "bad feeling" that saved you merely a coincidence, or was it something more?

Intuition may be thought of as being aware of something without knowing how you are aware of it. Oftentimes, an intuitive feeling is an educated guess, a combination of past experience and astute judgment. In this sense, intuition may be thought of as the mental analog of a gymnast's well-trained body: with talent and training, the mind and the body can both do exceptionally difficult maneuvers on their own, without conscious direction.

In the case of the car-driving example, perhaps you unconsciously observed a slight wobble in the car ahead of you. This might have presaged a tire blowout, which filtered its way to your conscious mind as a caution, and this is what caused the "bad feeling." But there are some intuitive hunches that cannot be explained so easily. For example, a colleague named Alex recalled the following dramatic experience:

> Preparing for a hunting trip later in the month, Alex was cleaning a double-action, six-shot revolver.[7] For safety's sake, he normally kept five bullets in the revolver, with the hammer resting on the sixth, empty chamber. He carefully removed the five bullets, cleaned the gun thoroughly, then began to put the bullets back in the pistol. When he arrived at the fifth and final bullet, he unexpectedly got a bad feeling, a distinct sense of dread that had something to do with that bullet.
>
> Alex worried about this odd feeling, because nothing like it had ever happened to him before. He decided to trust his intuition, so he put the bullet aside and positioned the pistol's hammer as usual over the sixth chamber. The chamber next to it, which normally held the fifth bullet, now was also empty.
>
> Two weeks later, Alex was at the hunting lodge with his fiancée and her parents. That evening, an ugly argument broke out between the par-

> ents over their impending divorce. Alex tried to calm them down, but the father, in a violent rage, grabbed Alex's gun, which was in a drawer, and pointed it at his wife. Alex tried to stop the impending disaster by jumping between the gun and the woman, but he was too late—the trigger had already been pulled. For a horrifying split second, Alex knew that he was about to get shot at point-blank range. But instead of a sudden, blazing death, the pistol went "click." The cylinder had revolved to an empty chamber—the very chamber that *would have contained the fifth bullet* if Alex had not set it aside two weeks before.[8]

In this case, Alex's intuition saved his life. To this day, he keeps that fifth bullet in a safe-deposit box, because, as he puts it, "It's said that everyone has one bullet with their name on it. I'm one of the few who knows exactly where that bullet is, and I'm never letting it out of my possession."

Alex's hunch cannot be attributed to unconscious sensory information, or even to an educated guess, because the details, exact timing, and even the people involved in Alex's hunting trip were not known when he was cleaning his gun. And he had never been involved in such a deadly altercation before, or since.

So where did the bad feeling come from? One possibility is that the information comes from our own future experience. We call such time-displaced perceptions "precognition" for pre-knowledge of the future, or "presentiment" for pre-feeling. While precognitions are fairly common experiences, especially in the form of intuitive hunches and dreams, many scientists consider true precognition to be impossible, mainly because it suggests that causation is not as simple as we thought. It raises the perplexing possibility that causation sometimes "flows backward." This is deeply troubling to some scientists because most scientific models assume that cause and effect "flow" in only one direction. Despite the disquieting implications, the results of a half-century of experimental tests, described later, indicate that some forms of precognition do exist.

The Feeling of Being Stared At

A commonly reported form of distant mental influence on living organisms is "the feeling of being stared at," which is closely related historically to the notion of the evil eye. Considerable folklore endorses the idea that gazing at someone carries special powers, favors, or influence. Folklore aside, contemporary opinion polls consistently confirm that the feeling of being stared at is known in all cultures.

The classic episode is as follows: A woman is having lunch alone at a diner when she slowly starts to feel agitated. She initially thinks it may be the caffeine from one too many cups of coffee, but the hair on the back of her neck is beginning to tingle uncomfortably. Soon she just can't shake the creepy feeling that someone is watching her. She gets the sense that the "someone" is behind her, so she turns to look, and sure enough, an intense young man is staring directly at her.

A similar tale is described in this paragraph from a short story by Sir Arthur Conan Doyle, the creator of Sherlock Holmes:

> At breakfast this morning I suddenly had that vague feeling of uneasiness which comes over some people when closely stared at, and, quickly looking up, I met his eyes bent upon me with an intensity which amounted to ferocity, though their expression instantly softened as he made some conventional remark upon the weather.[9]

Can a starer's intense focus affect the human nervous system? As with the study of intuition, to answer this we first have to ensure that the "staree" cannot be cued through normal sensory means, or through any form of unconscious information that might have alerted the staree to the presence of the "starer."

The "feeling of being stared at" has been studied in the laboratory for nearly a century. In the most recent versions of these experiments, the starer and staree are isolated from each other, and the starer watches the staree over closed-circuit video to maintain strict sensory separation. Results of such experiments, discussed later, reveal that focusing on another person from a distance does in fact affect the nervous system of the stared-at person. People in laboratory tests are rarely consciously aware of these physiological changes, suggesting that we may be influenced by others far more than we know.

Distant Mental Healing

A related topic is prayer, spiritual healing, or any form of what might be called "distant mental healing." In a prototype case, a friend is scheduled for exploratory surgery the next morning to remove a tumor. She asks you to pray for her fast recovery. You pray intensely that night, and the next morning you receive a phone call from her. She reports that the preparatory X rays show that the tumor has disappeared! After it is confirmed that no trace remains of the tumor that was there the day before, the surgery is canceled, much to the consternation of the doctors and the amazement of your friend. Were your prayers answered?

Author Bernard Gittelson reports the following similar story:

> Physician Rex Gardner of Sunderland District Hospital in Britain has investigated several inexplicable cases of apparent healing through prayer. In one, a woman asked the parishioners of her church to pray for her recovery from a severely ulcerous varicose condition in her leg. Her doctor

had told her that even if she were healed, the scar would require skin grafting; but the ulcer healed completely the day after the prayer meeting, and no grafting was required.

In his report, Gardner says that the story is "so bizarre that I would not have included it had I not been one of the doctors who examined the patient's leg at the next monthly prayer meeting and had all the people who had been present not been available for interrogation."[10]

More generally, is there any evidence that thinking about people at a distance, directing either calm, loving thoughts or aggressive, malevolent thoughts, actually affects their physiology? Four decades of laboratory experiments, discussed later, reveal that the answer is quite clearly yes.

When a Billion Minds Think Alike

Substantial laboratory evidence suggests that when a mind directs its focus toward a distant object, that object will change its behavior. Given this, the question naturally arises as to whether the effects caused by one person in the laboratory may "scale up" to include real-world events involving millions or billions of people.

In a series of recent studies with extraordinary implications, our laboratory, along with colleagues at Princeton University, the University of Edinburgh in Scotland, and the University of Amsterdam in the Netherlands, has begun to examine this question. The results indicate that when many minds think alike, the state of the world does indeed change.

For example, on October 3, 1995, at 10:00 A.M. PST, a remarkable incident took place. A sizable fraction of the globe's population, perhaps a billion minds, paid rapt attention to a single event—the announcement of the verdict in the murder trial of former football star O. J. Simpson.

Our research suggests that when groups of people concentrate on a single event, this mental focusing actually affects the physical world in unexpected ways. Later, we discuss how we went about studying "field-consciousness effects" during the O. J. Simpson verdict, the broadcasts of the 1995 and 1996 Academy Awards, and the Opening Ceremonies of the Centennial Olympic Games. We also discuss independent replications of these surprising "mass mind" effects by colleagues at other universities.

Psi in the World at Large

Compounding evidence suggesting that psi events occur in the world at large immediately raises the question of whether psi also occurs in casinos. Presumably, most of the people playing casino games *want* to win. Their wishes directed toward the games closely resemble the mental intention studied in laboratory-based psi experiments. While casinos always win in the long run because the odds are stacked in favor of the house, on rare days they can lose money if several big slot-machine jackpots, or big wins on the table games, are hit on the same day. Do casino profits fluctuate in predictable ways? Are some people luckier than usual on certain days?

Because the University of Nevada, Las Vegas is within a mile of the highest density of casinos in the world, through personal contacts our laboratory obtained the unusual opportunity to examine the daily win-loss ratios for both table games and slot machines in a casino. We found evidence of predictable variations in casino profits that are also related to factors associated with psi ability. This suggests that psi does indeed manifest in casinos. Later, we'll discuss what we found, along with corroborating findings about possible psi-mediated fluctuations in the payout rates of state and national lottery games.

Limitations of Stories

While the anecdotes in this chapter and the following chapters are intriguing, as stories they are not sufficiently persuasive to convince most scientists that psi is real. But what about a hundred similar stories? Or a thousand? Surely tens of thousands of such stories would cause one to pause and seriously wonder if psi is real. Still, because psychologists have shown that memory is much more fallible than most people think, and because eyewitness testimony is easily distorted, simply collecting more anecdotal stories will not settle the scientific question about psi. To do this, we need to consider two topics in some detail: What counts as scientific evidence? And how do we measure this evidence? Let's begin with the concept of *replication*.

CHAPTER 3

Replication

I am not interested in the ephemeral—such subjects as the adulteries of dentists. I am interested in those things that repeat and repeat and repeat in the lives of the millions.

THORNTON WILDER

Imagine that we paid certain people millions of dollars a year to do something. We'd revere them as heroes, reminisce enthusiastically about their accomplishments, and attempt to emulate them. Imagine that what these people did to earn unparalleled public esteem was something they could do only about one-third of the time, in spite of genetic superiority, native talent, and decades of daily practice. And imagine that their one-third "hit rate" was considered an outstanding accomplishment.

We call such people baseball players.

A prime example, Mickey Mantle, was at bat 8,102 times from 1951 to 1968. He managed to get 2,415 base hits, for a lifetime average hit rate of 29.8 percent. Mantle's performance varied widely over the years, ranging from a high of nearly 37 percent in 1957 to a low of 24 percent in 1968. One might think that baseball fans would be upset because Mantle did not even get a base hit about two-thirds of the time. But because baseball is known to involve highly skilled human performance, we accept relatively low hit rates and variations in performance, even among the best in the world.

Measuring Performance

Figure 3.1 shows Mantle's yearly hit rates from 1951 to 1968, and his lifetime average. The hit rates are shown as dots called "point estimates," and the vertical bars are called "95 percent confidence intervals." These two concepts are very important to understand at the outset because they will be used again later to show why we can place high confidence in the results of psi experiments.

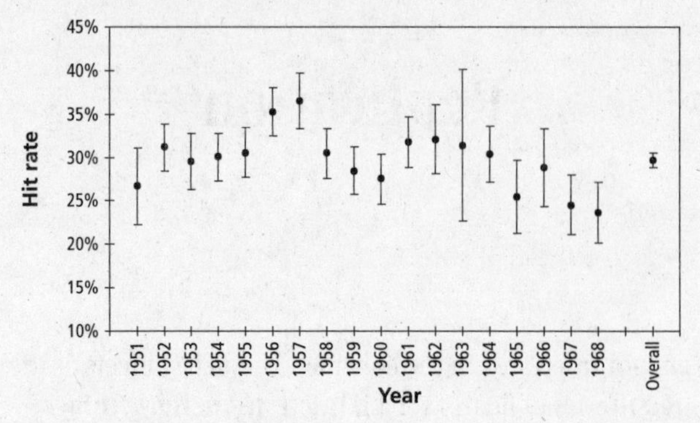

Figure 3.1. Mickey Mantle's yearly average hit rates and overall hit rate, with 95 percent confidence intervals shown.

The "point estimate" is the average hit rate, calculated from the number of times at bat per year and the number of hits that year. It is called a "point" estimate because an average is a single value, which implies high precision. The problem is, we know that from one year to the next the number of times that Mantle was at bat differed. So if one year he was at bat only ten times and got eight hits, his hit rate that year would be 80 percent. The next year he might be at bat three hundred times and get one hundred hits for a 33 percent hit rate.

If we paid attention only to the point estimate, we might think that Mantle's degree of skill changed enormously from one year to the next. But as soon as we know that he was at bat only ten times in one case, and three hundred in the other, we understand that an 80 percent estimate, despite being an *exact* 80 percent, is not a very good measure of his long-term performance.

So besides the best estimate of performance—the point estimate—we need something else to indicate the *degree of confidence* we can have in that point estimate. And here is where the "95 percent confidence interval" comes in. In general, a small confidence interval around a point estimate means that the point estimate was based on lots of trials, or that the repeated observations were very similar to one another, or both.[1] A large confidence interval means that the point estimate was based on only a few trials, or the repeated observations were very different from one another, or both. The prefix "95 percent" means that we can be 95 percent sure that the true, long-term skill level is somewhere in that range.

In figure 3.1 we see that Mantle's performance varied over the years, both in terms of point estimates and in our confidence in those point estimates. In 1960, for example, Mantle's hit rate was 27 percent, and we have fairly good confidence that this was a pretty accurate reflection of his skill level because the 95 percent confidence interval ranges from about 24 percent to 30 percent. We can tell from this that he was at bat lots of times, or that his performance was highly consistent, or both. By contrast, in 1963, at first glance it looks like he performed much better than in 1960 because his point estimate was 32 percent. But by paying attention to the 95 percent confidence interval, we see that he was not at bat many times, or his performance varied wildly, or both, and we can only have confidence that his performance was somewhere between 22 percent and 40 percent.

Now here is another important point: because the two 95 percent confidence intervals for Mantle's hit rates in 1960 and 1963 overlap each other, we cannot say with confidence that the two hit rates actually differed from each other. In other words, it is quite true that he hit 27 percent in 1960 and 32 percent in 1963 on average, but we have *low confidence* that this 5 percent difference (32 percent minus 27 percent) reflected a genuine difference in Mantle's true performance level. The 5 percent may be attributable to purely chance fluctuations. A statistician would say that in this case we do not have confidence that the two measurements are significantly different from each other. Saying that two values are "significant" means that, by convention, we have at least 95 percent confidence that two observed values reflect real underlying differences. This is equivalent to saying that the odds against chance (that the two values are really the same) are twenty to one or greater.

Now we are prepared to introduce another important concept. Notice that at the right side of figure 3.1, Mantle's "overall" hit rate is shown as 29 percent, with a very small confidence interval. This tells us with 95 percent confidence that his actual lifetime hit rate was somewhere between 27 percent and 31 percent. By considering all the data, we can be sure about Mantle's true skill level within a range of 3 percent. Compare this with the 5 percent to 10 percent ranges that we had for any given year. In sum, *combining data increases our confidence in the true value of a measurement.*

Psi and Baseball

What does all this talk about baseball have to do with psi performance? The link is this: All forms of human performance vary widely from one moment to the next, even among highly skilled players like Mickey Mantle. We could not predict when Mantle would get a hit, and we certainly could not predict when one of his hits might become a home run. But, of course, this does not mean that he did not get hits or that home runs do not exist. What we can do is form point estimates to measure performance at different times and in different contexts and then calculate confidence intervals to determine how sure we can be about those performances.

This is precisely what we will do later as we review psi performance over many experiments. We will discover that by combining thousands of people's performances over hundreds of experiments, we can obtain very high levels of confidence about the existence of psi.

At Bat: Joe Sixpack

So far we have seen how to obtain high confidence in a measurement. How do we use that measurement to see if psi occurred in an experiment?

While the particular definitions of psi performance are different for each kind of experimental design, the basic test ideas are the same: we compare performance when psi was presumably used against what would happen if psi weren't used.

To illustrate, imagine that one day we decide to give a college baseball player—let's call him Joe Sixpack—a thrill by putting him in our major league starting lineup. We let him play a couple of games to measure his batting average, and we observe that Joe's hit rate was about 25 percent. Joe accomplished the amazing feat of getting a hit about one in four times at bat. We can now compare Joe Sixpack to Mickey Mantle to see if Mickey really deserved all that acclaim.

Figure 3.2 shows the point estimates and 95 percent confidence intervals for hit rates measured for Joe and Mickey, and for comparison, also for Hank Aaron. We see that Joe's performance was nowhere near as good as either Mickey's or Hank's. His 95 percent confidence interval didn't even come close to overlapping Mickey's and Hank's confidence intervals. So we can say with confidence that there is a *significant difference* between Joe's true skill level and Mickey's or Hank's skill level. We also see that the 95 percent confidence intervals for Mickey and Hank *do* overlap, so for them we cannot say with confidence that their skills really differed.

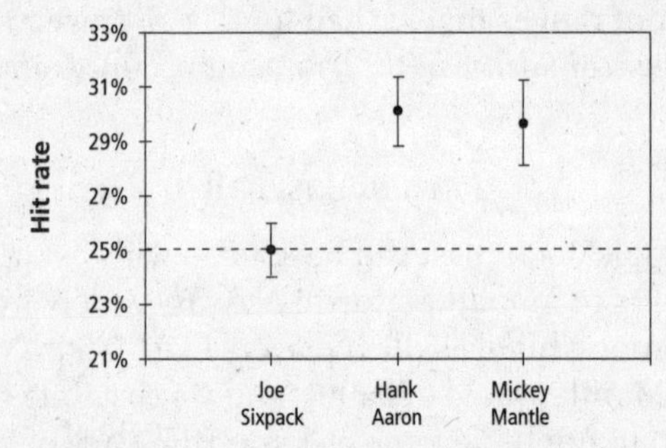

Figure 3.2. Hit rate point estimates and 95 percent confidence intervals for Mickey Mantle, Hank Aaron, and a fictional average player.

Back to the Laboratory

Now imagine that we perform a psi experiment in the laboratory. We select as our test subjects volunteers who claim no special abilities whatsoever. Unlike in baseball, in most lab experiments some "hits" can occur without any cause, just by chance. So, say that the *chance* hit rate in this experiment is 25 percent, which is common in experiments requiring the participant to use some form of ESP to select one correct target out of a group of four targets.

Now we run, say, a hundred such volunteers through the laboratory procedure, and instead of seeing the chance expected "hit rate" of 25 percent, we observe a hit rate of 34 percent, an increase over chance of 9 percent. Rather than getting a hit about one in four times "at bat," we would be in Mickey Mantle's class, getting a hit about one in three times at bat.

If we find this result sufficiently intriguing, we might run the same experiment several more times, each time with a new batch of one hundred volunteers. If we continue to see the 9 percent advantage over chance, we will gain confidence that the first experiment was not just a fluke, but a fairly good measure of the average volunteer's performance. If we also had high confidence in the experimental design, meaning that we could be sure no hints were given to the subjects, accidentally or intentionally, then we would have produced a series of psi experiments that showed significant evidence for psi.

Measuring the results of a psi experiment is analogous to our baseball example because in both cases we are interested in comparing two conditions: high versus low skill, and observed effects versus chance expectation. With repeated measurements, we can create highly reliable estimates of performance and judge whether the results obtained in the two conditions are the same or different.[2]

Variations

Variations in performance make repeated trials necessary when testing human skills, and explain the need for replication in the experimental sciences in general. Indeed, replication[3] is universally accepted as one of the most important criteria for scientifically establishing the existence of a claimed phenomenon. Unfortunately, as in the baseball example, it's not always possible to repeat and authenticate every event at will. This is especially true in the life sciences, where the objects of study are "open systems" that react and change as a result of the experiment itself. Under these conditions, obtaining successful replications based on just a few events is the exception rather than the rule.

In a psi experiment, when someone gets a hit by selecting the one correct target out of four possibilities, this is taken as evidence of one of two things: ESP or chance. On the basis of a single trial, we have no way of knowing for sure if the hit was a lucky guess or involved some ESP. Similarly, in baseball, when a batter gets a hit, we usually don't have a clue whether that particular hit involved skill or dumb luck. It is only on the basis of long-term averages that baseball skill or ESP can be measured with confidence.

Shifting gears from skilled physical performance, which is directly observable, to mental performance like mathematics skill or ESP, which we can only infer, we find that the process of replication suddenly becomes more complicated.

Reproducibility

When we're dealing with claims for a subtle mental skill like ESP, the inability to reliably produce experimental effects *on demand* makes it difficult to establish the nature or even the reality of the phenomenon. For years this has been the single biggest stumbling block for parapsychology. But this is not exclusively a problem for ESP; it is widespread for most of the really interesting problems in psychology, sociology, and medicine.

While there's little question that many normal psychological phenomena, such as conscious awareness and creativity, are "real," attempting to capture those phenomena with laboratory techniques can be exceptionally frustrating. Psychologist Seymour Epstein summarized the situation for conventional psychology in *American Psychologist,* the flagship journal of the American Psychological Association:

> Psychological research is rapidly approaching a crisis as the result of extremely inefficient procedures for establishing replicable generalizations. The traditional solution of attempting to obtain a high degree of control in the laboratory is often ineffective because much human behavior is so sensitive to incidental sources of stimulation that adequate control cannot be achieved. . . . Not only are experimental findings often difficult to replicate when there are the slightest alterations in conditions, but even attempts at exact replication frequently fail.[4]

Earlier, psychologists J. D. Bozarth and R. R. Roberts had surveyed 1,334 articles in standard psychology journals and found that only eight articles, or 0.7 percent of the published studies, involved replications of previous work.[5] This means that practically no one ever bothers to repeat previous studies. Considering the high scientific value placed on the ability to repeatedly demonstrate an effect, this is most perplexing, because it suggests that, in general, psychologists are willing to accept the reality of a claimed effect based on a *single study,* even though many psychological effects are known to be exceptionally difficult to repeat. How can this be? How do we know that a successful study published in a psychology journal wasn't just a fluke, or a mistake, or fraudulent?

Sociologist Harry Collins conducted an extensive study of replication in science, in light of its importance in establishing the reality of a phenomenon. He came to the surprising conclusion that not only are positive replications exceptionally rare in science, but

> Experiments hardly ever work the first time; indeed, they hardly ever work at all. Thus, any sensible experimenter ought to expect that most of what he or she does in the way of practical activity will be trial and (mostly) error. It will comprise not proper experiments, but one preliminary run after another.[6]

Also contributing to the "crisis" of replication in the life sciences is the fact that in most disciplines the scientific reward system places a high premium on original work and much less value on replications of previous work. Indeed, many professional journals have editorial guidelines that preclude the publication of "mere" replications.[7] As a result, when an occasional replication is conducted, the design of the study tends to be significantly altered from the original to provide the researcher with an opportunity to discover something new as well as to confirm previous findings. Moreover, given the time and expense involved in conducting any well-designed, rigorous experiment, repeating previous work exactly is usually considered a waste of resources.

A Paradox

So now we run headlong into a paradox: science places high value on replication of claimed effects, and yet for the most interesting phenomena, replications are surprisingly difficult to find. We will see later that the situation is not quite as dismal for psi research, because psi is so curious and represents such a huge challenge to scientific assumptions that hundreds of investigators over the years have conducted thousands of replication studies.

An engaging analogy to this paradox was noted by former undersecretary of the army Norman Augustine in his analysis of the defense contracting business:

> Were one to examine the relationship between the amount of testing that is required of a newly developed item and the complexity of that item, it might not be unreasonable to expect that the less complex the product the less testing it requires. . . .
>
> [However,] the correlation is not direct but rather is inverse. . . . Thus, one finds that the amount of testing needed decreases as an item becomes more complex. . . . Relatively simple unguided artillery projectiles somehow demand literally thousands of test rounds, whereas a new intercontinental ballistic missile needs only a few handfuls of test flights to demonstrate its adequacy.[8]

In other words, simple, cheap, not-very-interesting but easy-to-demonstrate effects are replicated to exhaustion, whereas the really interesting, complex, and difficult-to-demonstrate effects are hardly ever replicated. We are all familiar with experiments designed to illustrate the force of gravity and the "blind spot" in human vision. These have been replicated untold thousands of times; the replications provide a sense of comfort that if we do such and such, we'll get this or that familiar response.

Because of their repeatable stability, these effects are considered real, reasonable, and self-evident. If this same degree of stability cannot be easily demonstrated in a high-school physics or psychology class for a more unusual effect, like psi, then the reality of the effect is suspect.

Why Replication Can Be Difficult

Psi effects do not fall into the class of *easily* replicated effects. There are eight typical reasons why replication is difficult to achieve: (1) the phenomenon may not be replicable; (2) the written experimental procedures may be incomplete, or the skills needed to perform the replication may not be well understood; (3) the effect under study may change over time or react to the experimental procedure; (4) investigators may inadvertently affect the results of their experiments; (5) experiments sometimes fail for sociological reasons; (6) there are psychological reasons that prevent replications from being easy to conduct; (7) the statistical aspects of replication are much more confusing than most people think; and (8) complications in experimental design affect some replications.

Since all these points help explain why psi phenomena have traditionally been so difficult to establish, let's examine each point in detail.

Nonreplicable Phenomena

Some phenomena simply cannot be reproduced at will. Observation of a spontaneous or rare event such as a supernova, a meteor, or ball lightning cannot be repeated on demand. As a result, such phenomena are often difficult to establish as "real." Even fairly well documented effects like ball lightning are still considered doubtful by some scientists because descriptions of the effect seem to defy known scientific principles. Ball lightning appears to be a highly energized plasma, generally spherical in shape and about the size of a basketball. It exhibits peculiar, unpredictable effects, such as floating around a room, displaying apparently "willful" behavior at times, and sometimes causing great destruction when it discharges.

Countless other reports have resisted scientific study, including sightings of unidentified flying objects (UFOs) and mysterious animals (e.g., the Loch Ness Monster and Bigfoot) and the formation of crop circles. Directly reflecting the importance of replication in science, such spontaneous events are usually classified as paranormal, hallucinations, or hoaxes. It is interesting to consider, however, that people who think they have witnessed a UFO, or a ghost, or a Bigfoot, tend to become convinced of such phenomena because of their experience, whereas those who have not seen them find it difficult to believe. As we shall see later, the old saying "I'll believe it when I see it" is only half the story. It's equally true that "I'll see it when I believe it."

Some nonreplicable phenomena are not particularly rare but are uncontrollable. Today's weather forecast is a notoriously poor predictor of the weather next week because of the incredible complexity of the global environment.

Another type of nonreplicable phenomenon is the "nonphenomenon." In late 1987 some scientists reported finding higher than expected levels of oxygen in a sample of air trapped in a piece of million-year-old amber. The report was appealing because it offered an explanation for why dinosaurs were able to grow to such huge proportions and flourish (huge animals need lots of oxygen). Another group of scientists tried to repeat the experiment with new samples of amber and found normal levels of oxygen. Because the original "inflated oxygen effect" was not replicated, it was considered more likely that the original claims were due to contamination, or to a measurement error, or to an incorrect set of assumptions. The "nonphenomenon" is how some skeptics have (in the past) categorized claims of psi phenomena.

Incomplete Knowledge

Some experiments are difficult to replicate because the nuances of experimental procedures can be difficult to capture in words. Psychologist Michael Polanyi introduced the concept of *tacit knowledge* to refer to information that cannot easily be communicated verbally.[9] Hunches and intuitions are examples of tacit knowledge; they are often things learned from direct experience and training. In the words of psychologists Robert Rosenthal and Ralph Rosnow:

> . . . The scientist who fails to replicate a device or an experiment, whether in behavioral or natural science, may conclude that the claimed phenomenon is not replicable—which, of course, is one possibility. It is also possible that the scientist did not carry out the study "properly" because he or she did not have the benefit of tacit knowledge.[10]

The effects of tacit knowledge are easily seen when following cookbook recipes. Anyone who has tried to follow a recipe without knowing in advance how the dish is supposed to turn out knows only too well that printed instructions are woefully incomplete. Even if each written step is procedurally complete, the "tricks of the trade" that make the difference between a gourmet's delight and something the dog won't touch are difficult to learn without, say, apprenticing to a master chef for several years.

Indeed, many skills require more than pure intellectual knowledge. Playing a musical instrument, carpentry, and gymnastics, for example, are skills that demand long hours of kinesthetic practice beyond the intellectual mastery of the task. Kinesthetic knowledge is nearly impossible to convey with words, as anyone knows who has tried to describe how to ride a bicycle.

Because laboratory procedures are roughly similar to cookbook recipes, it's not surprising that laboratories have their equivalents to gourmet chefs. Some people have "golden hands" and are legendary for their ability to make things happen in the laboratory. Others are disasters waiting to happen. Tacit knowledge in laboratory procedures seems to play an especially crucial role when psychological behavior is being studied, because the interpersonal behaviors of experimenters and participants have to be taken into account.

In psi research, the problem is further compounded. For example, in a clairvoyance experiment, the researcher has to carry out strict, double-blind procedures to prevent any normally sensed information about the target from reaching the participant. And he or she has to ensure that the participant cannot discover information about the target by fraudulent means. Moreover, natural biases on the part of the participants and the experimenters have to be controlled to prevent revealing clues about the target, and so on. These and many other design aspects must be carefully monitored by the main investigator and the assistant experimenters.

Stochastic and Reactive Effects

Replication may be difficult to achieve if the phenomenon under study is inherently stochastic, that is, if it changes with time. Moreover, the phenomenon may react to the experimental situation, altering its characteristics because of the experiment. These are particularly sticky problems in the behavioral and social sciences, for it is virtually impossible to guarantee that an individual tested once will be exactly the same when tested later. In fact, when dealing with living organisms, we cannot realistically expect strict stability of behavior over time. Researchers have developed various experimental designs that attempt to counteract this problem of large fluctuations in behavior.

Replication is equally problematic in medical research, for the effects of a drug as well as the symptoms of a disease change with time, confounding the observed course of the illness. Was the cure accelerated or held back by the introduction of the test drug? Often the answer can only be inferred based on what happens on average to a group of test patients compared to a group of control patients.

Even attempts to keep experimenters and test participants completely blind to the experimental manipulations do not always address the stochastic and reactive elements of the phenomena under study. Besides the possibility that an effect may change over time, some phenomena may be *inherently* statistical; that is, they may exist only as probabilities or tendencies to occur.

Experimenter Effects

In a classic book entitled *Pitfalls in Human Research,* psychologist Theodore X. Barber discusses ten ways in which behavioral research can go wrong.[11] These include such things as the "investigator paradigm effect," in which the investigator's conceptual framework biases the way an experiment is conducted and interpreted, and the "experimenter personal attributes effect," where variables such as age, sex, and friendliness interact with the test participants' responses. A third pitfall is the "experimenter unintentional expectancy effect"; that is, the experimenter's prior expectations can influence the outcome of an experiment.

Researchers' expectations and prior beliefs affect how their experiments are conducted, how the data are interpreted, and how other investigators' research is judged. This topic, discussed in chapter 14, is relevant to understanding the criticisms of psi experiments and how the evidence for psi phenomena has often been misinterpreted.

Sociological Factors

Science is a social activity, and replication involves certain sociological factors. Any scientist making unexpected claims must interest other researchers in trying to replicate those effects. Otherwise, the scientist becomes a lone wolf howling into the wind. Lone wolves can howl all they like, but they're generally not taken very seriously.

The lone-wolf tactic was used in a 1995 *Newsweek* cover story on the paranormal. That story will be examined in more detail later to illustrate how the media often distort the facts about psi research. Here we are just concerned with a statement in the article that independent researchers have not been able to reproduce the psychokinesis (mind-matter interaction) effects reported by Professor Robert Jahn's laboratory at Princeton University. The exact phrase used in the article was "Other labs, using Jahn's machine, have not obtained his results."[12]

If true, this is an important criticism, because it implies that a lone researcher's results might be due to error or fraud. The statement, however, is pure fiction. As we will see in chapter 8, Jahn's research has been replicated by more than seventy researchers worldwide, both before and after Jahn produced the main body of his work. Jahn is not a lone wolf, but the social consequences of the implication are powerful persuaders of popular opinion.

To distinguish the lone wolves from the properly socialized wolf packs, skeptics have invented labels like "pseudoscience" and "pathological science." Terms such as these have been used to refer to psi research in important scientific journals like *Science* and *Nature*. Such labels suggest that the lone wolf cannot be trusted, or is sloppy or incompetent.

Labels like "pseudoscience" are valued rhetorical tactics of extreme skeptics because they help reduce the pain of cognitive dissonance (that is, the pain caused by getting stuck in the mental loop: "the evidence looks good, but it can't be real, but I can't find the problem, but . . ."). Unfortunately, the fear of attracting such labels also makes it difficult for more conventionally minded scientists to take psi experiments seriously. They worry about being tainted by the pseudoscience label, and they may also worry that their reputations will suffer if their colleagues suspect their interests.

An excellent illustration of this is the case of the distinguished physicist John Wheeler. Wheeler is deeply interested in the problem of the observer and the observed in quantum theory. He has waxed poetic about the implications of quantum theory, writing, for example, that "There may be no such thing as the 'glittering central mechanism of the universe' to be seen behind a glass wall at the end of the trail. Not machinery but magic may be the better description of the treasure that is waiting."[13]

One might think that Wheeler would be sympathetic to psi research. After all, psi research is also concerned with trying to understand the peculiar interactions between the subjective and the objective. And yet, as part of his published contribution to an American Association for the Advancement of Science symposium on the role of consciousness in the physical world, Wheeler felt compelled to add a postscript entitled "Drive the pseudos out of the workshop of science":

> The author would be less than frank if he did not confess he wanted to withdraw from this symposium when—too late—he learned that so-called extrasensory perception . . . would be taken up in one of the papers. How can anyone be happy at an accompaniment of pretentious pseudoscience who wants to discuss real issues about real observations in real science?[14]

Like Wheeler, very few scientists would be happy about being associated in any way with the dreaded label "pseudoscience." Besides incurring the fear and loathing of the mainstream, those conducting research on topics stamped "pseudoscience" may find that funding sources mysteriously dry up, journals refuse to publish their research, and opportunities for academic tenure vanish.

The difficulty of getting scientists to attempt to replicate, or even pay attention to, psi experiments is related to what Thomas Gold of Cornell University has called the "herd effect." This is the tendency for scientists (or any people, for that matter) to cluster together in groups where only certain ideas or techniques are acceptable. A scientific herd forms for essentially the same reason that sheep form a herd—to protect individuals. It is very risky for one's career to stand apart from the herd, given the rapidly diminishing likelihood that one can continue to practice science outside the herd. Without exception, scientists who conduct psi research are high risk-takers, because the academic world lets them know very quickly that "we don't take kindly to strangers in these here parts."

Psychological Factors

It is well known that most scientists are "theory-driven" rather than "data-driven." This means that scientists are uncomfortable with "facts" unless some theory can explain them. Parapsychological "facts" are uncomfortable because there are no well-accepted explanations for why the facts should exist. This does not mean that no scientific theories of psychic phenomena exist; actually, there are dozens. It is the *adequacy* of the theories that is in question.

Being theory-driven also means that scientists fail to see data that contradict their theoretical expectations. This does not mean that they fail to *understand* the data, but rather that they have a strong tendency literally *not to perceive* the offending data. As discussed in some detail in chapter 14, a substantial body of conventional psychological research supports this strong consequence. Witnessing this effect in action is truly astonishing. It is like trying to get a dog to look at something that you know he will find interesting. "There it is! Look at the evidence there!" *Where? I don't see anything.* "There I say. Look where I'm pointing, not at my hand!" *Nope, I don't see anything.*

Another consequence of being theory-driven is reflected in the well-worn phrase "Extraordinary claims require extraordinary evidence." This is probably a good rule of thumb, but the problem is that different people have very different ideas about what "extraordinary" means. In fields with unusual claims and weak theories, like parapsychology, cold fusion, or homeopathy, enormous amounts of evidence are required. In fields with equally unusual claims but with strong theoretical backing, like the idea of nonlocal correlations in quantum mechanics, a relatively tiny amount of experimental evidence is required.

The consequence for experimental replication is that the *degree* of replication, the *ease* of replication, and even the *requirement* for replication are all influenced by the presence and strength of existing theory. This translates into the psychological condition of *expectation.*

Another psychological factor is that the quality of an experiment is in the mind of the beholder. When people with different degrees of personal belief in a hypothesis are asked to assess the quality of an *identical* set of studies, the people who believe in the hypothesis see that set of experiments as obviously competent, whereas the people who do not believe the hypothesis see the same set of experiments as obviously flawed. The consequence for psi research is that confirmed skeptics always feel justified in rejecting positive replications because they always see the methodology as flawed, whether it is actually flawed or not!

Still another psychological factor is the finding that preexisting beliefs persist regardless of new evidence to the contrary. For example, some scientists have claimed that they found no evidence of psychic effects in their experiments, and yet when their own data were reanalyzed later, clear evidence was found. One case is Professor John E. Coover, from Stanford University, who was one of the first academic scientists to conduct ESP tests in the early twentieth century. In his tests of people's ability to guess the suit of playing cards, he found a 30.1 percent hit rate, whereas 25 percent would be expected by chance. His conclusion was that

> Since [the result] lies within the field of chance deviation, although the probability of its occurrence by chance is fairly low, it cannot be accepted as a decisive indication of some cause beyond chance which operated in favor of success in guessing.[15]

And yet, researchers who reexamined Coover's publications many years later found that these experiments actually produced positive evidence for ESP with odds against chance of 160 to 1.[16]

Another example is Dr. James Kennedy, a skeptical psychologist who tried to replicate J. B. Rhine's work with ESP cards in the late 1930s. Kennedy's design involved 204 subjects, and he described his results as "entirely negative." The actual results, in a series testing for telepathy, were odds against chance of ten million to one.[17]

A third example is psychologist Dr. Susan Blackmore, a skeptic of psi phenomena who reported nineteen psi experiments in her doctoral dissertation, five of which achieved statistical significance (that is, each of these studies resulted in odds against chance of twenty to one or greater). Dr. Blackmore has repeatedly claimed that because she obtained mostly negative results in her own psi research, and was unable to replicate her occasional successful experiments, she was driven by her experiences to become a skeptic. Yet the odds against chance of obtaining five successful experiments out of the nineteen she conducted for her doctoral work are five hundred to one![18]

Still another example is Professor Ray Hyman, a long-term critic of parapsychology, who in a review of telepathy tests rejected thirteen out of twenty-four experiments on the grounds that they were statistically nonsignificant. And yet, when the data from those thirteen "nonsignificant" replications were combined into one grand experiment, they produced an overall result that *was* statistically significant.[19]

What these examples show is that beliefs strongly affect the perceived success or failure of replications. The lesson is that just as we should be skeptical of strong proponents of psychic phenomena who claim to get consistently positive results, we should also be skeptical of those opponents who claim that they consistently see only negative results.

Statistical Factors

Take an arbitrary experiment with, say, fifty subjects. Let's assume that this experiment produced a statistically significant result (again, consider this to mean a *successful* result with odds against chance of at least twenty to one). Now, you plan to run an exact replication of this experiment, again using fifty subjects. The question is, what is the probability that if you run the exact same experiment, with fifty new subjects, you'll get another successful outcome?

When experienced experimental psychologists and professional statisticians were asked this question, they gave answers ranging between 80 and 90 percent. That is, they thought there was an 80 percent to 90 percent chance that the replication study would be successful.[20]

It turns out that the correct answer is only about 50 percent. In other words, in a faithful replication of a previously successful study, using exactly the same number of subjects, you will obtain another successful result only half the time (this is a consequence of something called the "power" of a statistical test). This may seem odd, because it certainly *seems* that if you do an exact repetition of a successful study, the replication ought to be successful as well. But it is not so.

The reason is that experiments involving human beings never turn out *exactly* the same way twice, and the statistical implications of evaluating experiment results are not always obvious even to experienced experimentalists and statisticians. One of the consequences of understanding the nature of replication is an awareness that skeptics who demand extremely high rates of repeatability for psi experiments simply do not understand the statistics of replication.

Another statistical issue has been discussed by University of California statistician Jessica Utts.[21] She uses the example of a genetic-engineering experiment to show why we cannot have high confidence in an experiment if we run too few trials. Suppose, she asks, that we find that seventy out of one hundred births that were genetically designed to produce boys actually resulted in boys. This is a success rate of 70 percent instead of the 51 percent average rate expected by normal population birth rates.

This experiment with a 70 percent success rate would result in odds of ten thousand to one that the genetic-engineering method was better than the chance expected rate. This would convince most scientists that the genetic-engineering method was effective.

Now suppose that a skeptic came along and tried to replicate this experiment, but only checked ten births. To his surprise, he found seven boys, also for a 70 percent success rate. The problem is that because the smaller experiment had less statistical power (remember, fewer trials provide less confidence in the point estimate), this would result in odds against chance of only five to one. Since by convention we need odds against chance of at least twenty to one to claim a statistically significant result, the skeptic could loudly proclaim that the replication was a failure. Scientists have lost research grants based on such faulty proclamations.

In other words, if we focused only on whether the odds against chance in an experiment were greater than twenty to one, the second experiment would be considered a failure even though it found exactly the same male birth rate of 70 percent as the first study!

Experimental Design Factors

Some skeptics have claimed that as experimental quality improves, the evidence for psychic effects will decrease. The implication is that if we run perfect, high-quality experiments, we will never get systematic evidence for psi effects. This is a potentially valid criticism, but it has been tested (as we will see in later chapters) and does not hold up under scrutiny.

No experiments are absolutely immune from design flaws, and not all flaws are created equal. One valid type of flaw is an artifact that can be demonstrated to *cause* the observed results. This is sometimes called the "smoking gun" flaw. Another type of experimental flaw is a "plausible alternative" to the hypothesis. This is something that can plausibly account for the observed results, like the smell of burnt gunpowder, even if a smoking gun itself cannot be found.

But not all proposed flaws are valid. One invalid flaw is the so-called "dirty test tube." This is essentially an assertion that the experiment was not perfect, and therefore some *hypothetical* "dirt in the test tube" might account for the observed effect. Alleging the presence of such a flaw is an invalid criticism because it is "nonfalsifiable." In other words, we can't *test* whether some unspecified dirt somewhere might have made an important difference in the experiment. The real issue in criticisms of experimental methodology is whether claimed flaws *in fact* caused the observed results. If they did not, then claims for dirt in the test tube are inconsequential, whether or not the dirt ever existed in the first place.

How Many Replications Are Needed?

Now that we have a better understanding of the need for replications, and we understand some of the difficulties involved in conducting replications, let's say we do a psi experiment and get a positive result. Would that one experiment convince anyone that psi is real? This depends to a large degree on the claim. For example, say a group of confirmed skeptics did an experiment with a yogi who claimed he could levitate. And say the experiment was shown on live TV, and the yogi levitated as claimed, and the skeptics were clearly shocked because they didn't believe such a thing was possible. Some people might become convinced that levitation was real. One assumes, of course, that the experimental procedures precluded fraud as an explanation. But even then, because prior beliefs drive what we can see, it is likely that the skeptics would not believe their eyes and would end up discounting the levitation as a trick.

In the early days of psi experiments, skeptics insisted that "it would be sufficient . . . to convince them of ESP if a parapsychologist could perform successfully a single 'fraud-proof' experiment."[22] But even the skeptics soon saw that this was a mistake, because no experiment is perfect, and a flaw in a single critical study might be overlooked. In addition, experimental results in the empirical sciences are often reported in terms of *probabilities*, or as odds against chance for one or another hypothesis. Thus, if a single experiment obtained odds against chance of a thousand to one, that seems pretty impressive, but such results could have occurred, by definition, purely by chance one in a thousand times. Most scientists would not be willing to change their beliefs about the nature of the world on the basis of a single experiment that produced odds against chance of a thousand to one. Or would they?

Sometimes, when an effect is predicted on the basis of a well-respected theory, or when the people reporting the effect are prominent scientists, or when the claim is not too remote from accepted scientific knowledge, then just one or two successful studies can convince scientists that a claimed effect is real. A striking example is the evidence upon which the "omega-minus" particle was accepted in physics. The omega-minus particle was considered to be "found" on the basis of only *two events* out of a total of nearly 200,000 experimental trials. In other words, an event with an extremely poor replication rate—observed only once in a hundred thousand times—was still considered sufficient to convince most physicists that the particle was real.[23]

Returning to the world of psi phenomena, what if we conducted a *second* experiment and it too produced odds against chance of, say, one thousand to one? It is unlikely that two such experimental results, both with high odds, could be obtained by chance alone. Perhaps some scientists might begin to take notice. What about three successful experiments? A dozen? Where do we draw the line?

The presumption that an experiment must work every single time without fail is clearly too strong a requirement for any phenomenon involving human performance. We do not expect baseball legends to bat a thousand, so why should we require even better performance for psychics? Softening this requirement, however, opens the question of *degrees* of replication. This is a troublesome issue, for the degree of replication required to establish an effect is closely related to how strange that effect is supposed to be. And the "strangeness" of an effect is directly related to how much the observed phenomenon deviates from theory. This is equivalent to saying how far it deviates from our prior expectation, and this in turn raises the issue of how to *measure* degrees of replication, and what counts as evidence.

Skeptical British psychologist Mark Hansel proposed the following recipe for "how much evidence" would be required to convince him that something interesting was going on:

> If a result is significant at the .01 level [that is, odds against chance of one hundred to one] and this result is not due to chance but to information reaching the subject [by psi], it may be expected that by making two further sets of trials the antichance odds of one hundred to one will be increased to about a million to one, thus enabling the effects of ESP—or whatever is responsible for the original result—to manifest itself to such an extent that there will be little doubt that the result is not due to chance.[24]

Thus, Hansel claimed that if an experiment producing odds against chance of one hundred to one could be repeated three times with the same or greater odds in each test, he would be satisfied that the result was not chance. Hansel would be pleased to know that this has been achieved dozens of times, in numerous categories of psi experiments. This is why informed skeptics today agree that chance is no longer a viable explanation for the results obtained in psi experiments.

How do we know that psi experiments have been replicated? This leads to the idea of meta-analysis—the analysis of analyses.

CHAPTER 4

Meta-analysis

[Meta-analysis] is going to revolutionize how the sciences . . . handle data. And it's going to be the way many arguments will be ended.

THOMAS CHALMERS

Because independent replication is the key to producing acceptable scientific evidence, we need ways of measuring how much replication has taken place. The technique most widely accepted today is called meta-analysis—the analysis of analyses.

Conceptually, meta-analysis is simple. Taking the example of baseball, we have high confidence that Mickey Mantle was an unusually good baseball player because his lifetime batting average was about 30 percent. Our confidence is not based on his performance in a single game, but on the long-term average of thousands of times at bat, and thousands of hits. By comparison, in a single psi experiment, one person's outstanding performance would hardly give us confidence that the performance was more than, say, an interesting coincidence. The same person's consistently successful performance over thousands of tests would be much more convincing.

Let's say we wanted to judge whether Michael Jordan of the Chicago Bulls or Clyde Drexler of the Houston Rockets was a better basketball player, in terms of points scored per game. Figure 4.1 shows the points these two players scored in games played from November 3, 1995, to April 18, 1996. To judge who was the better player, let's say we sent our trusty analyst, Joe Sixpack, to see some of the games and keep track of how many points the two players scored. If Joe attended only the seven games when Drexler happened to score more points than Jordan (for example, on December 2, 1995, Drexler scored forty-one points and Jordan thirty-seven points), Joe's reasonable conclusion would be that Drexler was a better player than Jordan.

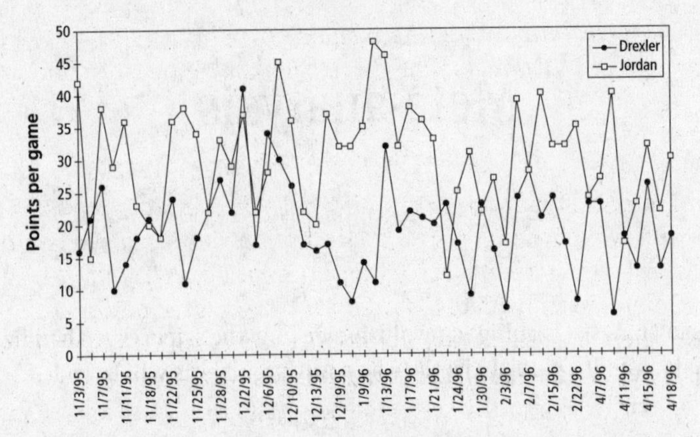

Figure 4.1. Points scored per game by Michael Jordan and Clyde Drexler, in games played from November 3, 1995, to April 18, 1996.

We would be skeptical of Joe's conclusion, but only because when we saw the two players' performance over many games, we would see that Jordan almost always scored higher than Drexler. In fact, when we considered all the games played from November 3 to April 18, we would immediately see that Jordan's average was clearly better than Drexler's. This shows how repeated measurements increase our confidence in the evidence we use to form judgments, whether we are interested in basketball players or psi performance.

We may also be interested, not in how any *particular* player performs, but in how an entire *team* performs. For example, a combined batting average for a baseball team would give us some indication of how the team performs in general. In a psi experiment, the equivalent would be to combine the performances of many people, each of whom participated in multiple experimental trials. Combining results in this way enables us to judge psi performance among groups.

Science is primarily interested in generalizations and lawful *tendencies* rather than unique events. This is because part of the goal of science is to be able to understand and describe phenomena accurately enough to predict future events. By definition, single-shot or totally spontaneous events are unpredictable.

What meta-analysis allows us to do is ask even higher-level questions, such as how an entire baseball *league* performs. Here we examine the results not from just one player, or from a group of players on a given team, but from groups of players on groups of teams. Now we are interested in baseball performance in general, independent of a particular team or player.

For psi experiments, we can ask questions not only about how an individual performed, or how a group of individuals performed in a given experiment, but how people perform *in general* across many experiments. Asking "meta" questions allows us to develop higher confidence about *per-*

formance in general without getting bogged down in the specifics of individuals, or groups of individuals. The more data, the higher the confidence.

Reviewing Research

Measuring replication rates across different experiments requires that research be reviewed in some fashion. Research reviews can be classified into four types. A type 1 review simply identifies and discusses recent developments in a field, usually focusing on a few exemplar experiments. Such reviews are often found in popular-science magazines such as *Scientific American*. They are also commonly used in skeptical reviews of psi research because one or two carefully selected exemplars can provide easy targets to pick apart.

The type 2 review uses a few research results to highlight or illustrate a new theory or to propose a new theoretical framework for understanding a phenomenon. Again, the review is not designed to be comprehensive but only to illustrate a general theme.

Type 3 reviews organize and synthesize knowledge from various areas of research. Such narrative reviews are not comprehensive, because the entire pool of combined studies from many disciplines is typically too large to consider individually. So again, a few exemplars of the "best" studies are used to illustrate the point of the synthesis.

Type 4 is the integrative review or meta-analysis, which is a structured technique for exhaustively analyzing a complete body of experiments. It draws generalizations from a set of observations about each experiment.[1]

Integration

Meta-analysis has been described as "a method of statistical analysis wherein the units of analysis are the results of independent studies, rather than the responses of individual subjects."[2] In a single experiment, the raw data points are typically the participants' individual responses. In meta-analysis, the raw data points are the results of separate *experiments*.

The basic ideas of meta-analysis have been around since the 1930s, but over the last few decades the techniques have been refined and clarified. Today meta-analyses have exploded in popularity because the behavioral, social, and medical sciences were all in the same boat: they needed a method of formally determining whether the highly variable effects measured in their experiments were replicable.[3]

Because meta-analysis is used to combine data from a group of similar experiments, the technique involves some reevaluation of the original data. In some cases, originally reported results are recast into statistics that are amenable to making a grand combination. The next step is coding and quantifying the experimental procedures, including factors such as the type of controls, where and when the reports were published, the number of test participants, and so on.

Then, these results are examined to see if there are any clear patterns among the studies. As mentioned before, skeptics have long maintained that better-controlled studies would show smaller effects, implying that all psi effects are accidents due to poorly controlled experiments. Meta-analysis allows such an assertion to be tested by examining how study quality is related to the actual results of each experiment.

Accuracy

Meta-analysis provides a much more accurate assessment of a body of research than the traditional descriptive or narrative literature review. In studying the idea that meta-analysis is more accurate than a single analysis, psychologists Cooper and Rosenthal discovered that reviewers who rely on the traditional narrative literature review can reach conclusions that are *completely contrary* to what the data actually reveal by use of meta-analysis.[4] This is because after reading a literature review, which typically describes each study in only a paragraph or two, we are left with lots of disconnected details that are difficult to integrate into meaningful statements. With meta-analysis, there is a single, quantitative measurement of the effect of interest.

Some critics of meta-analysis (critics are everywhere) have argued that these integrative techniques can be biased or oversimplified.[5] Such criticisms are answered by noting that meta-analysis requires explicit details of how the analysis was performed, thus allowing independent analysts to confirm the evaluation. Also, when we use *all* the relevant studies in the analysis rather than just the "good" studies, most of the problems related to reviewer bias are prevented.

Apples and Oranges

Critics also have argued that because meta-analysis combines results from a wide variety of studies, it is actually a way of mixing "apples" and "oranges."[6] Is it valid to generalize about effects of interest by combining studies employing different experimenters, different experimental designs, and different subjects?

The answer is yes, it is valid to combine apples and oranges if we wish to discover what is general about both, namely, something about *fruit*. When a series of psi experiments are combined, the apples and oranges are represented by the slight differences among the studies, but the common effect in all studies—the fruit—is psi.

Another criticism of meta-analysis is that authors tend to report only studies with significant results and leave the nonsignificant studies unpublished. This is called the *file-drawer problem,* referring to reports of unsuccessful studies that may be languishing forgotten in the back of researchers' file cabinets.[7] If the size of this hidden file drawer is large, it tends to inflate the estimate of an overall effect. If researchers publish only their successful studies, we will come to the unavoidable conclusion that all studies are successful. And this may or may not be true. We discuss the file-drawer problem in more detail later, including ways of measuring its impact on the meta-analytic outcome.

Take an Aspirin

Figure 4.2 reproduces the results of a meta-analysis of twenty-five medical studies investigating whether aspirin helps reduce heart attacks. The analysis was first reported in the *British Medical Journal* in 1988. The result of the analysis was widely described in the news media as a medical breakthrough, and in 1990 a contributor to the prominent journal *Science* used it as an example of how to perform a meta-analysis.[8]

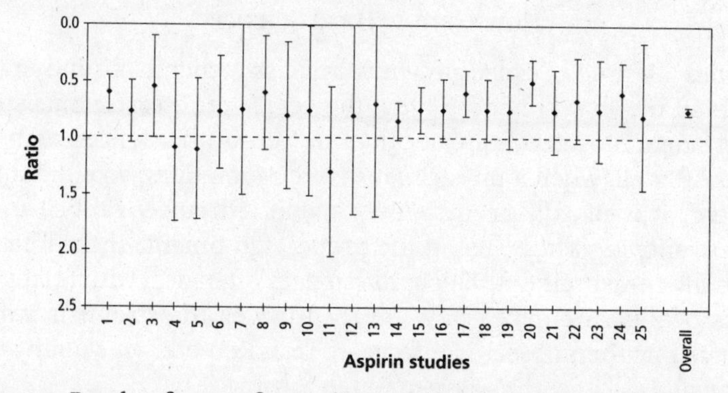

Figure 4.2. Results of twenty-five studies examining whether aspirin reduces heart attacks. Only five of the twenty-five studies were individually "successful," but overall—shown as the right-most point estimate—there is no doubt that aspirin really does have therapeutic value.

In the figure, the point estimate showing the result of each study is the ratio between the effect of active treatments versus no treatments, and the confidence intervals are 99 percent. Thus a value of 1.0 in this figure means that the treatment was no better than the control in reducing heart attacks. A value of *less* than 1.0 means that the treatment was *better* than the control (because the treatment resulted in fewer heart attacks).

The main point for the present discussion is the observation that only five of the twenty-five studies rejected chance with 99 percent confidence. That is, considered individually, the majority (80 percent) of the studies are "failures" because their confidence intervals included 1.0. A reviewer who was skeptical of aspirin's ability to reduce heart attacks might examine these individual studies and go away unimpressed, confident that there was no evidence that aspirin has any clear therapeutic value.

But notice that when the results of all the studies are *combined*, the overall result (shown at the right end of the graph) is above the 1.0 chance line, at about 0.75, and the tiny error bars clearly exclude chance. Thus, even though the effect is uncertain when considered in individual experiments, it was widely advertised (and rightly so) that taking aspirin really does make a significant difference.

In other words, the aspirin effect was declared to be "real" based on the combined results of all studies. This is exactly what meta-analysis has done for psi experiments. Considered individually, some psi experiments have been successful but the effects did not appear to be easily repeatable. This uncertainty—along with a lack of theories predicting psi effects—has fueled the skeptics' doubt for over a century. But when studies are combined, there is no doubt that the psi effects are real.

How Hard Is Hard Science?

It is often assumed, especially by physicists, chemists, and other researchers in the "hard sciences," that the degree of research consistency in the hard sciences is much greater than in the softer sciences such as psychology. After all, when a physicist measures something, like the weight of a steel bar, it is usually an extremely stable, extremely precise measurement. Anyone, anywhere, using the proper equipment, should be able to replicate the measurement. The hard sciences often deal with relatively tiny amounts of measurement error, which provides the scientists with high confidence that their observed effects are real. Remember, stability of measurement allows consensus agreement to form.

By contrast, when a psychologist tries to measure some aspect of human behavior, or a sociologist tries to measure some aspect of society, the measuring tools—questionnaires, surveys, and psychophysiological measurements—require statistics to make sense of the resulting data. This is because random variation, the "noise" in living systems, tends to be huge.

Thus, differences in accuracy and precision of measurement are the primary basis for the distinctions drawn between the hard and the so-called soft sciences. But now there is reason to believe that the assumed differences have been vastly exaggerated. Using meta-analytic techniques, psychologist Larry Hedges of the University of Chicago discovered a surprising result: some experiments in the soft sciences are as replicable as those in the hard sciences! According to Hedges:

> Research results in the social and behavioral sciences are often conceded to be less replicable than research results in the physical sciences. . . .

> Comparison [of the consistency of research results in physics and the social sciences] suggests that the results of physical experiments may not be strikingly more consistent than those of social or behavior experiments. The data suggest that even the results of physical experiments may not be cumulative in the absolute sense by statistical criteria.[9]

By "cumulative," Hedges meant the degree to which measurements among replicated experiments tend to agree. Hedges studied the issue of *how much* consistency should be expected in replicated experiments. Although there is no technically precise answer to this question, Hedges argued that one way to judge an acceptable degree of empirical cumulativeness is to compare the observed effects in the behavioral sciences with what is observed in the hard sciences. In particular, Hedges examined empirical replication rates in a discipline known for its quality of research and elegance of theory—particle physics. He concluded that

> social science research may not be overwhelmingly less cumulative than research in the physical sciences. In fact, the evidence shows several parallels in the reviews of social and physical science domains. Experimental results are not always consistent by statistical criteria. About 45% of the reviews in both domains exhibited statistically significant disagreements when no studies were omitted from the results.[10]

This is a remarkable conclusion, for it states that even in particle physics, one of the most rigorous, well-funded, and hardest of the hard sciences, the *actual* replication rates are comparable to the replication rates observed in the soft, pliable world of the behavioral sciences.

To illustrate, let's consider a publication of the Particle Data Group (PDG) of the American Physical Society. This is an international, multiuniversity team that reviews the experimental evidence for fundamental properties of elementary particles—properties like mass and lifetimes. On the basis of published experimental evidence, the PDG determines the best estimates for these properties.[11]

What is important for this discussion is that in these reviews, some data are used and some are discarded. Among the reasons listed for discarding data are "The results involve some assumptions we do not wish to incorporate" and "The measurement is clearly inconsistent with other results which appear to be highly reliable."[12] In other words, data are discarded to reduce "outliers" that are thought to be flawed in some way. As new data are added to the old, the precision of the estimate increases. However, as the PDG writes:

> Some cases of rather wild fluctuation are shown; this usually represents the introduction of significant new data or the discarding of some older data. Older data are sometimes discarded in favor of more modern data if

> it is felt that the newer data had fewer systematic errors. . . . By and large, a full scan of our history plots shows a rather dull progression towards greater precision at a central value completely consistent with the first data point shown.[13]

Rather dull, that is, *only if* the outliers are removed. As Hedges found, up to 45 percent of the data are discarded to achieve these results. What would happen if we used Hedges's measure of the highest degree of research consistency one can expect and compared that with the research consistency found in certain psi experiments? And what if that comparison showed similar results?

As discussed in the following chapters, these comparisons *have* been made, and we are forced to conclude that when psi research is judged by the same standards as any other scientific discipline, then the results are *as consistent* as those observed in the hardest of the hard sciences!

To borrow the words of Coleridge:

> What if you slept?
> And what if in your sleep, you dreamed?
> And what if in your dream you went to heaven and there
>   plucked a strange and beautiful flower?
> And what if, when you woke, you had the flower in your hand?
> Ah! What then?[14]

Well, then we would have to reconsider whether we were really dreaming after all. Perhaps the tens of thousands of psi experiences reported throughout the centuries really do mean that psi exists. And maybe there is something of real scientific interest going on.

Now that we understand a bit about the need for replication and the purpose of meta-analysis, we are prepared to explore the second major theme in this book: *Evidence.*

THEME 2

EVIDENCE

The theme of the first part of the book was *motivation*. Why investigate psi in the first place? We've seen that the primary motivation is the fact that people have frequently reported strange experiences that don't fit within the established scientific worldview. These experiences have been recorded throughout history, suggesting that we are dealing with something fundamental to human nature. What are these experiences, and how can we begin to understand them?

To address these questions, we've sharpened our focus from the popular concept of "the paranormal" to a class of unexpected information and energy exchanges called psi. We've learned that in spite of centuries of folklore and interesting anecdotes, providing persuasive *scientific* evidence for psi requires independently replicated, controlled experiments. After reviewing the purpose and nature of replication, and ways of measuring replication, we are now ready to see if the holy grail has been achieved.

The theme in this part of the book is the *evidence* for psi. We'll look at four general categories of psi experiments. We'll also explore two new experimental categories: "field-consciousness" effects and psi influences in casino and lottery games. And we'll end with a review of how psi is already being put to practical use. Let's begin with the evidence for one of the most commonly reported psi experiences—telepathy.

CHAPTER 5

Telepathy

Before I picked up the phone, I *knew* it was you.

On Monday, for no apparent reason I found myself thinking about an old friend from college. We hadn't corresponded in years and I had lost track of where she lived. Then out of the blue, I received a letter from her. I phoned her and it turns out that on Monday, when I was thinking about her, she was looking through our college yearbook and decided to contact me!

I was with my husband at the Hollywood Bowl when complete scenes from the movie *King Solomon's Mines* flashed before me, and I knew my husband was thinking about the movie. Without realizing what I was saying, I turned to my husband and said, "I saw *King Solomon's Mines* too." He was shocked. "How did you know I was thinking about that?"[1]

The experience of direct communication between two minds has been reported so frequently throughout history that it eventually gained its own name: telepathy. Coined in 1882 by the British scholar Frederic W. H. Myers, a founding member of the London-based Society for Psychical Research, the word *telepathy* means "feeling at a distance."[2]

In almost all cases, the reason such communications were reported is because they were meaningful to the experiencer. We often place telephone calls to each other, we often receive letters from old friends, and we sometimes *seem* to know what others are thinking. But when we have strong feelings about such events, and we know we didn't use the ordinary senses to get this information, *and* the information was verified in due course, this may be a reflection of genuine telepathy.

Is it possible that such episodes can be explained as educated guesses, or as misinterpreted coincidences? Yes. In many cases what appears to be telepathy probably *is* due to psychological factors such as selective memory, wishful thinking, misremembered events, or subliminal cues. Regardless of how convincing, interesting, or personally moving any given story may be, we know that evidence relying solely on eyewitness testimony or memory is notoriously inaccurate, and there are simply too many normal explanations for such experiences. So, to provide a scientifically valid answer to the question "How do we know that genuine telepathy really exists?" we cannot rely entirely on anecdotal reports.

Many popular books about psychic phenomena attempt to prove their case by citing dozens of "well-documented" cases and asking the reader to conclude that the psychic effect *must* be true. But even after reading thousands of meticulously recorded case studies, all that we realistically end up with is "face validity" for the claimed effect. That is, we've gained motivation to believe that *something* intriguing is going on, but we do not know much beyond that. To provide scientific evidence requires demonstrating the same sort of effects under well-controlled conditions. If it can be shown that information really does get from one person to another, even after we control for the effects of belief, motivation, memory, and sensory input, then we know that in *some* of the case studies what seemed to be telepathy may actually have been telepathy.

Thoughts About Thought Control

Before going any further, let's consider the popular notions of thought control and mind reading. Parapsychologists are constantly being sought out by agitated people who complain that they're being inundated by the thoughts of one or more distant persons, or worse, that their minds are being invaded by others. While such experiences are undoubtedly very disturbing, they are usually not related to telepathy but indicate the need for professional psychotherapy.

This is not to say that such nefarious intrusions are impossible in principle. We know with high confidence that some forms of distant mental influence on the human nervous system are probably true, as discussed in chapter 9. On the other hand, a gigantic gulf separates claims that "the FBI and the CIA are controlling my mind" and the small, generally unconscious effects observed in the laboratory. While our minds are undoubtedly "controlled" to some extent by constant exposure to certain ideas (like advertisements compelling us to buy something we don't need), most people who experience *uncontrollable* disturbing thoughts should check with a psychotherapist before calling a parapsychologist.[3]

The fact is that effective telepathic control of a randomly selected individual, if such a thing existed, would be a powerful manipulative tool. It wouldn't be wasted on influencing people who are not in positions of economic or political power. One can imagine that exercising a little mind control over, say, the leader of a terrorist group would be a far cheaper, more humane, and more potent means of enforcing a change in attitude than putting soldiers in harm's way. Of course, abuse of such power is a frightening prospect, which is one of the reasons that it is useful to discuss openly what is known about telepathy and psi in general.

Early Case Studies

The modern history of the study of telepathy is fairly straightforward, though the brief version presented here sidesteps many clever experimental variations. The first studies of telepathy were based on collections of spontaneous experiences. Systematic collections began in 1886 with the publication of the seminal work *Phantasms of the Living,* by the British scholars Edmund Gurney, Frederic Myers, and Frank Podmore.[4] Although most of the cases reported in that classic work were contributed by people in Britain, the authors also surveyed some cases in the United States. Here's an example, reported by a Dr. Walter Bruce of Micanopy, Florida.[5]

> On Thursday, the 27th of December last [1884], I returned from Gainesville . . . to my orange grove, near Micanopy. I have only a small plank house of three rooms at my grove, where I spend most of my time when the grove is being cultivated. There was no one in the house but myself at the time, and being somewhat fatigued with my ride, I retired to my bed very tired, probably 6 o'clock; and, as I am frequently in the habit of doing, I lit my lamp on a stand by the bed for the purpose of reading. After reading a short time, I began to feel a little drowsy, put out the light, and soon fell asleep.
>
> Quite early in the night I was awakened. I could not have been asleep very long, I am sure. I felt as if I had been aroused intentionally, and at first thought someone was breaking into the house. I looked from where I lay into the other two rooms (the doors of both being open) and at once recognized where I was, and that there was no ground for the burglar theory; there being nothing in the house to make it worth a burglar's time to come after.
>
> I then turned on my side to go to sleep again, and immediately felt a consciousness of a presence in the room, and singular to state, it was not the consciousness of a live person, but of a spiritual presence. This may

provoke a smile, but I can only tell you the facts as they occurred to me. I do not know how to better describe my sensations than by simply stating that I felt a consciousness of a spiritual presence. This may have been part of the dream, for I felt as if I were dozing off again to sleep; but it was unlike any dream I ever had. I felt also at the same time a strong feeling of superstitious dread, as if something strange and fearful were about to happen.

I was soon asleep again or unconscious, at any rate, to my surroundings. Then I saw two men engaged in a slight scuffle; one fell fatally wounded—the other immediately disappeared. I did not see the gash in the wounded man's throat, but knew that his throat was cut. I did not recognize him, either, as my brother-in-law. I saw him lying with his hands under him, his head turned slightly to the left, his feet close together.

I could not, from the position in which I stood, see but a small portion of his face; his coat, collar, hair or something partly obscured it. I looked at him the second time a little closer to see if I could make out who it was. I was aware it was someone I knew, but still could not recognize him. I turned, and then saw my wife sitting not far from him. She told me she could not leave until he was attended to. (I had got a letter a few days previously from my wife, telling me she would leave in a day or two, and was expecting every day a letter or telegram, telling me when to meet her at the depot.)

My attention was struck by the surroundings of the dead man. He appeared to be lying on an elevated platform of some kind, surrounded by chairs, benches, and desks, reminding me somewhat of a schoolroom. Outside of the room in which he was lying was a crowd of people, mostly females some of whom I thought I knew. Here my dream terminated.

I awoke again about midnight; got up and went to the door to see if there were any prospects of rain; returned to my bed again, and lay there until nearly daylight before falling asleep again. I thought of my dream and was strongly impressed by it. All strange, superstitious feelings had passed off.

It was not until a week or 10 days after this that I got a letter from my wife, giving me an account of her brother's death. Her letter, which was written the day after his death, was mis-sent. The account she gave me of his death tallies most remarkably with my dream. Her brother was with a wedding party at the depot at Markham station, Fauquier County, Va. He went into a store nearby to see a young man who kept a bar-room near the depot, with whom he had some words. He turned and left the man, and walked out of the store. The bar-room keeper followed him out, and without further words deliberately cut his throat.

It was a most brutal and unprovoked murder. My brother-in-law had on his overcoat, with the collar turned up. The knife went through the collar and clear to the bone. He was carried into the store and laid on the counter, near a desk and show case. He swooned from loss of blood soon after being cut. The cutting occurred early Thursday night, December 27th. He did not die, however, until almost daylight, Saturday morning.

What makes this case particularly striking is a dream reported at the same time by his sister-in-law, Mrs. Stubbing, who was visiting her cousin in Kentucky. Mrs. Stubbing independently wrote this account of her dream:

> I saw two persons—one with his throat cut. I could not tell who it was, though I knew it was somebody that I knew, and as soon as I heard of my brother's death, I said at once that I knew it was he that I had seen murdered in my dream; and though I did not hear how my brother died, I told my cousin, whom I was staying with, that I knew he had been murdered. This dream took place on Thursday or Friday night, I do not remember which. I saw the exact spot where he was murdered, and just as it happened.

Early Experiments

One of the first experimental studies of telepathy was reported by the British physicist Sir William Barrett, who in 1883 conducted "thought-transference" tests between distant hypnotized subjects.[6] Several years later, Sir Oliver Lodge, a British physicist renowned for his pioneering work in radio receivers, published studies involving a pair of young women who claimed to have telepathic abilities. Both Barrett's and Lodge's experiments were reportedly successful and encouraged other scientists to investigate telepathy.

In 1917, psychologist John E. Coover from Stanford University conducted telepathy tests using a deck of forty regular playing cards. Coover separated the telepathic "sender" and "receiver" in adjoining rooms and remained in the room with the sender. He eventually ran 105 students as receivers and 97 as senders, collecting ten thousand individual trials. The receivers were able to guess the identity of the cards being "sent" to them with odds against chance of 160 to 1. Coover's published opinions about his findings were more pessimistic than his data actually revealed, possibly because of disapproving pressure from his peers at Stanford.

About the same time that Coover was conducting his experiments, Leonard Troland of the Psychology Department at Harvard University was conducting telepathy tests using a test machine that automatically selected a target card, recorded it, and also recorded the subjects' responses. His results, based on 605 trials, indicated that the subjects were actually *avoiding* the correct targets with odds against chance of fourteen to one. Ten years later, George Estabrooks, then a graduate student in psychology at Harvard University, conducted telepathy tests of students segregated in adjoining rooms. The results of three series of experiments were highly significant, producing odds against chance of millions to one. Estabrooks's fourth study isolated the subjects in distant rooms, and while the results were not as strong as in the first three studies, they were still successful in demonstrating telepathic links between pairs of students.

Mental Radio

In 1930 an influential book describing a series of telepathy tests was published by the Pulitzer Prize–winning author and social activist Upton Sinclair.[7] The book, *Mental Radio,* created a popular sensation because Sinclair was widely known as a no-nonsense realist. Sinclair's wife, Mary Craig Sinclair, had developed an interest in telepathy and had trained herself to perceive sketches drawn by someone else. Typically her husband drew the sketches, but on occasion other family members or her husband's secretary provided drawings.

Starting in 1928, the number of successful, direct "hits" produced by Mrs. Sinclair after one year of testing was judged to be 65 out of 290 picture-drawing sessions. A hit was counted if Mrs. Sinclair drew an obvious likeness of the target sketch. Some of the experiments took place with the person drawing the target sketch located many miles from Mrs. Sinclair. Sinclair and his wife noted that their telepathy tests could also have been "explained" as clairvoyance, and in some tests as precognition. They later tested these possibilities and confirmed that no "sender" was necessary to accurately describe the target sketch. Upton Sinclair asked his friend Albert Einstein to comment on the experiments. Einstein wrote the following, which was included as the preface to the book:

> I have read the book of Upton Sinclair with great interest and am convinced that the same deserves the most earnest consideration, not only of the laity, but also of the psychologists by profession. The results of the telepathic experiments carefully and plainly set forth in this book stand surely far beyond those which a nature investigator holds to be thinkable. On the other hand, it is out of the question in the case of so conscientious an observer and writer as Upton Sinclair that he is carrying on a conscious deception of the reading world; his good faith and dependability are not to be doubted.[8]

The second edition of *Mental Radio,* published in 1962, included a reprint of an article published in 1932 by Dr. Walter Franklin Prince, the research officer of the Boston Society for Psychical Research. Prince was impressed by Sinclair's book and wrote to him to see if he could get the original sketches and notes to conduct an independent analysis of the experiments. Sinclair agreed, and Prince exhaustively reanalyzed the data, trying to see if the telepathy tests might be reinterpreted in any "normal" terms. This included the possibilities that the results were due to pure chance, the "kindred ideas of relatives" (that is, that Sinclair and his wife knew each other so well that when he sketched a target photo, she could make an educated guess about what it was), conscious or subconscious fraud, and even "involuntary whispering."

Prince carefully tested each idea and was able to demonstrate that none was sufficient to explain the correspondences between the sketches and Mrs. Sinclair's responses. Taking the most conservative position, he concluded that telepathy had been demonstrated based on tests conducted between Mrs. Sinclair and her brother-in-law, some thirty miles apart, and on experiments that took place between separate rooms. Prince wrote:

> The results were so remarkable that they deserve to arrest the attention of every psychologist. The next seven experiments were made with agent and percipient in different rooms, shut off from each other by solid walls; and their results were very impressive.[9]

ESP Card Tests

Probably the best-known series of telepathy experiments were the ESP card tests pioneered by Professor Joseph Banks Rhine and his colleagues at Duke University from the late 1920s to 1965. Rhine developed a "forced-choice" technique, using a special deck of twenty-five cards consisting of five groups of five symbols (square, circle, wavy lines, star, and triangle). A person acting as the "sender" thoroughly shuffled the deck, selected the top card, and tried to mentally "send" that symbol to a remote person. Using a prearranged timing scheme, or a way to signal the sender to "transmit" the next card, the remote person eventually made twenty-five guesses, one for each card in the deck, and the resulting number of matches between the actual cards and the guesses was compared to chance expectation of five "hits."

For about sixty years, roughly from 1880 to 1940, ESP card experiments provided increasingly persuasive evidence for psi (we will examine this evidence in detail in the next chapter). The experiments were reported in over a hundred publications, and they involved thousands of participants who contributed more than four million individual trials. While some of the tests were originally designed to study telepathy, it was soon realized that most of the observed effects could also be "explained" by clairvoyance. That is, it was possible that senders were not required, and instead the receivers were using clairvoyance to directly perceive the cards.

To the present day, no one has come up with a persuasive experimental design that can unambiguously distinguish between telepathy and clairvoyance. Some have argued that "pure" telepathy might be tested by having the sender simply think about a purely mental target, and not write it down anywhere, and then see whether the receiver could describe that target. Unfortunately, that would not work for two reasons. First, it is well known that people are not good at selecting targets without unwittingly introducing their personal biases. This is a valid criticism that applies to many (but not all) of the experiments described by Upton Sinclair in *Mental Radio*. Magicians take advantage of these known response biases by "forcing" unsuspecting members of an audience to select a card or an object that the magician wants them to select. Because it is possible to make educated guesses about what someone is thinking, these biases muddy how "pure" telepathy tests should be interpreted.

Second, at some point the identity of even a "purely mental" target must become objectively known. For example, the target may be written down on paper to record the results of the trial. As soon as that occurs, the experiment suddenly shifts from a pure telepathy experiment into one that could also involve clairvoyance or precognition. As a result of this conceptual problem, for several decades most experimentalists concentrated on studying clairvoyance and precognition. The term "general ESP," or GESP, became popular to reflect the fact that it was (and still is) difficult to distinguish cleanly among the various forms of perceptual psi.

Dream Telepathy Experiments

In the 1960s, a growing number of researchers had become disenchanted with the forced-choice card tests pioneered by J. B. Rhine. Card tests provided rigorous testing conditions and allowed simple interpretations of the results, but after participants had tried to guess thousands of cards, they became bored with the exercise. A new generation of researchers wanted to develop experimental designs that both held the participants' interest and were closer to the "raw" psi experiences reported in spontaneous cases.

The result was "free-response" test designs that were similar to those used in the late nineteenth century by the first telepathy experimenters and that also resembled the picture-drawing studies reported by Sinclair in *Mental Radio*. In these tests, participants were encouraged to freely report their ongoing mental experiences. Researchers took these experiential reports and matched them against the actual psi targets, which were typically interesting photographs.

One of the most successful, systematic series of free-response telepathy studies was motivated by results of cross-cultural surveys showing that about half of all spontaneous psi experiences occur in the dream state.[10] From 1966 to 1972, researchers led by psychiatrist Montague Ullman and psychologist Stanley Krippner at Maimonides Medical Center in Brooklyn,

New York, devised a series of clever telepathy tests conducted in a dream-research laboratory. Their results suggested that if someone is asked to "send" mental images to a dreaming person, the dreamer will sometimes incorporate those images into the dream.

The dream telepathy studies spawned many experimental replications conducted over the six years of the Maimonides program. In those studies, a volunteer telepathic receiver—let's call her "Rose"—spent the night in the Maimonides dream lab. Rose met and talked with a lab experimenter—we'll call him "Sam"—who acted as the "sender." Rose also met the other experimenters taking part in the testing session that night.

When Rose was ready for sleep, she was ushered into an experimental chamber, which was both soundproof and electromagnetically shielded. An experimenter, "Earl," applied electrodes to her head in the usual way for monitoring brain waves (EEG) and eye movements. From that point on she had no further contact with Sam or Earl, or with any other experimenter, until the session was completed. In a room next to the experimental chamber, Earl monitored Rose's EEG and eye movements all night long. At the beginning of each period of rapid eye movement (REM), when Rose was probably dreaming, Earl notified Sam by pressing a buzzer.

In some of the Maimonides studies, Sam and Rose were located about thirty-two feet from each other, and in later studies ninety-eight feet, fourteen miles, and in one case, forty-five miles apart. Before Sam left for his remote site, a third experimenter gave him a sealed target picture that had been randomly selected from a pool of possible targets, usually a pool of eight or twelve pictures. A complex randomization method ensured that none of the experimenters, and of course none of the dreamers, knew the identity of the target.

Sam did not open the packet containing the target until he was isolated in the remote location. His only communication with the other experimenters was through a one-way buzzer, or through a planned sequence of telephone rings for longer-distance experiments. Whenever the experimenter in the dream lab signaled to Sam that Rose had entered a REM period, Sam concentrated on the target picture with the aim of sending it telepathically and influencing her dream.

Toward the end of each REM period, Rose was awakened by Earl (by an intercom announcement), who asked her to describe any dreams that she had just experienced. At the end of the night's sleep, Rose was again asked for her impressions about what the target picture might have been. Of course, to provide a valid test for telepathy (or any form of psi), Earl had to remain blind to the target throughout the session. Rose's responses were recorded and transcribed for later analysis by a group of independent judges.

The judges individually examined the transcript from a given session and compared it to the entire pool of pictures, one of which was the actual target used by Sam in that session. The judges were usually asked to provide a *ranking* for each picture. So for, say, a pool of eight pictures, the picture with the highest correspondence to the transcript would be ranked 1, and the picture with the least correspondence would be ranked 8. If the judges ranked the actual target picture in the top half of the pool, ranks 1 through 4, this was considered a "hit." Thus, if telepathy did not occur in dreams, over many repeated sessions the chance hit rate in this experiment would be expected to hover around one in two, or a 50 percent chance hit rate.

Example of a Correspondence

Here is an example of the kind of correspondence reported between a receiver's dream and the sender's target picture. In the session in question, the target photo was Max Beckmann's *Descent from the Cross*, a painting that depicts Christ being taken down from the cross.[11] Additional materials given to the sender to facilitate his involvement with the target idea were a small wooden crucifix, a Jesus doll, nails, and a red marker, along with the instructions, "Using these tacks, nail Christ to the cross" and "Using this marker, color his body with blood."

Two of the participant's dreams that night involved speeches by Winston Churchill and a native ceremonial sacrifice. Note the symbolic relevance of "church-hill" in the reported dream:

> In the Churchill thing there was a ceremonial thing going on, and in the native dream there was a type of ceremony going on . . . leading to whatever the ceremony would be to sacrificing two victims. . . . I would say the sacrifice feeling in the native dream . . . would be more like the primitive trying to destroy the civilized. . . . It believed in the god-authority . . . no god was speaking. It was the use of the fear of this, or the awe of god idea, that was to bring about the control.[12]

Results of Dream Studies

In journal articles published between 1966 and 1973, a total of 450 dream-telepathy sessions were reported. These experiments involved several design variations, including (*a*) designs in which receivers' dreams were monitored and recorded throughout the night and senders attempted to "send" the target pictures to the sleeper during each dreaming period; (*b*) designs in which senders mentally sent their targets the day *after* the dreams had been recorded, providing a precognitive twist to the original design; (*c*) designs in which the target was hidden and known to no one during the experiment; (*d*) designs in which senders sent the targets only at the *beginning* of the sleep period or sporadically; and (*e*) designs in which a single dream was used rather than a combination of dreams throughout the night.

Figure 5.1 shows the results in terms of the obtained hit-rate point estimate for each study and a 95 percent confidence interval. Note that the graph is centered on 50 percent because the method of judging in these studies resulted in a 50 percent hit rate purely by chance. The 95 percent confidence intervals in some cases extend above 100 percent, but these are not displayed because it is not possible to obtain a true hit rate greater than 100 percent. The wider confidence intervals, as in study 23, reflect the fact that there were fewer sessions in that study (in this case, only two), so in spite of the 100 percent hit rate (two hits in two trials), our confidence about this 100 percent is not very strong.

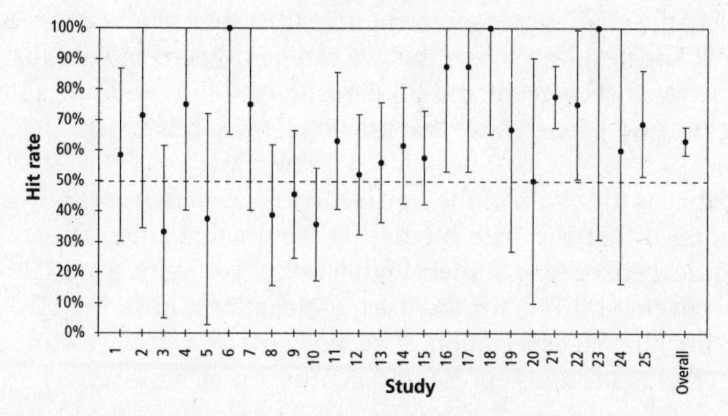

Figure 5.1. Summary of the dream-telepathy experimental results, with 95 percent confidence intervals and where 50 percent is chance expectation. The study numbers correspond to the studies listed in appendix A of the book *Dream Telepathy*. The combined hit rate for all trials is labeled "overall."

Notice that of the twenty-five studies displayed in figure 5.1, nineteen had positive outcomes. That is, they resulted in hit rates greater than 50 percent. This suggests that these experiments were successfully replicated. But also notice that in eighteen of the twenty-five studies the 95 percent confidence intervals included the chance level of 50 percent. For those studies we cannot confidently exclude the possibility that the real hit rate might have been chance. A critic reviewing these studies might therefore argue that this does not demonstrate a series of replicated studies, because in 72 percent (eighteen of twenty-five) of the experiments no "successful effects" were observed. But hold on, a major strength of meta-analysis is about to appear.

If we now consider the results of *all* the experiments combined, based on all 450 sessions, the overall hit rate is seen to be 63 percent (the rightmost point estimate in figure 5.1). The 95 percent confidence interval clearly excludes the chance expected hit rate of 50 percent. In fact, while it may not look like it from the graph, the odds against chance of getting a 63 percent hit rate in 450 sessions, where chance is 50 percent and the confidence interval is this small, is seventy-five million to one. In other words, while the majority of individual studies were not independently "successful," when taken as a whole the evidence is abundantly clear that something interesting occurred in dreams.

To illustrate that the "overall" confidence interval shown in figure 5.1 is more impressive than it at first appears to be, consider the 95 percent, 99 percent, and 99.999 percent confidence intervals shown in figure 5.2. The "95 percent" confidence interval is the same as that shown in figure 5.1. It indicates that we can be 95 percent sure that the average hit rate in the dream-telepathy studies was somewhere in the range of about 57 percent to 67 percent. The next line shows that we can be 99 percent sure that the hit rate was between 56 percent and 68 percent. And the third line shows that we can be 99.999 percent sure that the hit rate was between 52 percent and 72 percent.

Notice that as the width of the confidence interval increases only slightly, our confidence that the true hit rate is within that interval—and is *not chance*—quickly rises to extremely high levels of certainty. Thus, although a 63 percent overall hit rate for the dream-telepathy studies is "only" 13 percent over the chance expectation of 50 percent, we can have strong confidence that this 13 percent represents a genuine, nonchance effect.

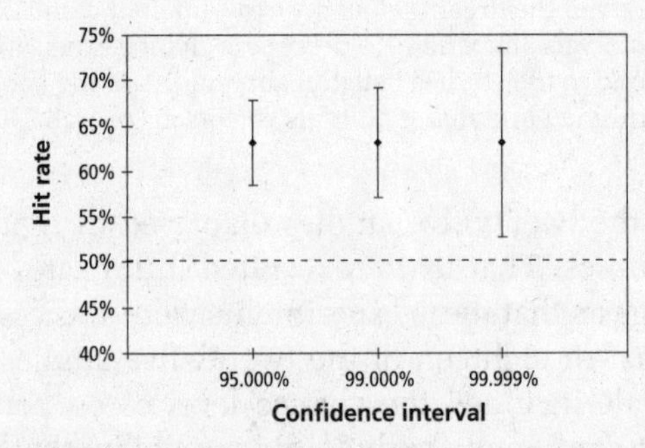

Figure 5.2. The 95 percent, 99 percent, and 99.999 percent confidence intervals for the dream-telepathy experiments. Our confidence increases from odds against chance of 20 to 1, then 100 to 1, and then 100,000 to 1 as the width of the confidence interval increases only slightly.

The point is that when we combine results of many similar studies to form the equivalent of a single, grand experiment conducted by many experimenters, from many locations, over many years, we also substantially increase our confidence in the outcome. Combining the results of the Maimonides dream-telepathy studies is only the beginning of a formal meta-analysis. These studies were described here mainly for historical reasons, and to demonstrate the value of examining replicated experiments to increase our confidence in the outcome.

Do odds of one in seventy-five million allow us to say that telepathy in the dream state was "proven"? No. All we know from the present overview is that chance can be soundly rejected as one of many possible explanations for the results observed in these studies. Left out of this analysis were other important factors such as assessing how many studies might have been conducted that were not published (the "file-drawer problem"), evaluating the quality of individual studies (because all experiments are not created equal), and assessing the degree of replication across different experiments and experimenters.

Ganzfeld Telepathy Experiments

As the dream-telepathy studies were winding down in the mid-1970s, parapsychologist Charles Honorton, one of the researchers on the Maimonides project, began a new series of telepathy experiments.[13] At almost the same time, William Braud, a psychologist at the University of Houston, and Adrian Parker, a psychologist at the University of Edinburgh, each independently developed ideas similar to Honorton's about how to develop a "psi conducive" state involving reduced sensory input.[14]

Honorton, Braud, and Parker had noticed that descriptions of mystical, meditative, and religious states often included anecdotes about psi experiences, and that the association between reduced mental noise and the spontaneous emergence of psi was noted long ago in the ancient religious texts of India, the *Vedas*. For example, in Patanjali's *Yoga Sutras,* one of the first textbooks on yoga dating back at least thirty-five hundred years, it is taken for granted that prolonged practice with deep meditation leads to a variety of *siddhis,* or psychic abilities.[15] States similar to deep meditation occur naturally during dreaming, prior to falling asleep, under hypnosis, with some drugs, and in sensory-isolation chambers. What these mental states have in common is an alert, receptive mind combined with reduced sensory input.

This suggests that when mental "noise" settles down, the mind may be able to attend more effectively to faint impressions, some of which may be psychic in origin. This makes sense from a point of view proposed by Henri Bergson, a French philosopher and Nobel laureate (for literature). In the early twentieth century, Bergson suggested that the human brain and nervous system function not only as a detector and a processor of sensory information but also as a filter. This filter preprocesses the overwhelming mass of sensory information that constantly bombards us, and it selectively presents to conscious awareness only those fragments of information that we consciously wish to attend to, or are likely to find useful for survival purposes.

This filtering mechanism is apparent when we suddenly hear our name spoken across a noisy, crowded room, or when we are absentmindedly driving a car and suddenly find ourselves stomping on the brakes before we are even aware of the child who has run out into the street. Unconscious mental processes like Bergson's filter have been studied in depth in subliminal perception and psychotherapy, and we now know through many experimental studies that we are continually processing enormous amounts of information, most of which does not reach conscious awareness.

Honorton, Braud, and Parker each decided to develop a telepathy experiment using a sensory deprivation technique called the "ganzfeld," a German word meaning "whole field."[16] The basic idea was that if a person was placed in a condition of sensory deprivation, the nervous system would soon become "starved" for new stimuli, and the likelihood of perceiving faint perceptions that are normally overwhelmed by ordinary sensory input would improve. The ganzfeld studies were an extension of the ideas underlying the dream-telepathy experiments, using a technique that provided a faster method of collecting data because it did not require the receiver to be dreaming.

The ganzfeld-telepathy experiments are particularly interesting in terms of providing acceptable scientific evidence for psi because the original concept was based upon theoretical *predictions* about the perceptual effects of reducing sensory noise. Moreover, researchers and skeptics had jointly agreed on specific *guidelines* for how these experiments should be conducted and evaluated, and the success of the technique has generated dozens of independent *replications*.[17]

Most of the ganzfeld experiments took advantage of lessons learned in past psi research, thereby avoiding many of the design problems discovered by early experimenters.[18] Also, the results of the ganzfeld studies have been discussed in several sophisticated, informed debates, so consideration of these studies has been elevated far beyond the usual rhetorical exchanges between proponents and skeptics of psi.

The Ganzfeld Method

The ganzfeld experiment has three phases: preparing the receiver and sender, sending the target, and judging the outcome. The advantage of this three-step method, refined over decades of critical scrutiny, is that it provides a clean separation of the sender, receiver, and experimenter, as well as an unambiguous way to measure success. The disadvantage is that the final outcome, which is simply a "hit" or a "miss," sacrifices the richness of the receiver's mental impressions for the sake of clarity. A single session takes two or three people about ninety minutes of effort, or about 4.5 person-hours. This is a substantial expenditure of human resources to collect a single data point, but it is more efficient than the 8 to 24 person-hours typically required to run a single dream-telepathy session.

Phase 1: Preparation

Let's say that "Rose" is the telepathic receiver in the experiment. She is placed into the ganzfeld state by sitting in a comfortable reclining chair, listening to continuous white noise played over headphones (like the static heard between radio stations), and wearing translucent hemispheres—usually halved table-tennis balls—over her eyes while a red light shines on her face. A ten-minute progressive relaxation audiotape is often played through headphones to help her relax.

At first, the soft, unpatterned sound and light of the ganzfeld environment are gently stimulating. But after a few minutes, because the nervous system primarily responds to changes, and the ganzfeld is specifically designed to present an *unchanging* sensory field, Rose achieves a state very similar to that reported under sensory-isolation conditions. As the brain becomes starved for new visual imagery and changing sounds, mild or sometimes vivid imagery is commonly experienced.

Before Rose is sealed into the ganzfeld chamber, the experimenter, "Earl," asks her to speak aloud any feelings or images that come to mind when the relaxation tape ends. She is told to continue speaking aloud until instructed to stop, about twenty minutes later. Earl then shuts the door to Rose's ganzfeld room and escorts the sender, "Sam," to a distant, securely isolated room.

Earl has previously asked an assistant to randomly select one "target pack" out of a large pool of such packs. Each pack contains four pictures, one of which the assistant randomly selected as the telepathic target for the session. All the target packs and target pictures are enclosed in opaque envelopes, with no indication on the outside of the envelopes as to their contents. This allows Earl to remain blind to the identity of the target. In a fully automated ganzfeld experiment, instead of asking an assistant to select the target, Earl would have a computer automatically select a video-based target pack and a video-clip target in that pack at random. Whether the experiment uses pictures or videos, the targets within a pack are carefully selected so that the four images are as different from one another as possible.

Phase 2: Sending

Earl hands Sam the target, still in its opaque envelope, and then seals him into the sender's chamber. In fully automated testing systems, a computer-controlled, closed-circuit video system presents the target to Sam over a video monitor. In the most sophisticated experiments today, all the interactions between Earl, Rose, and Sam are completely automated to ensure that the experimental procedures are followed exactly the same way in each session.

Sam now views the target and tries to mentally send it to Rose. In video-based experiments, Sam takes a short break while the videotape rewinds. Then, when it plays again, he views and "sends" the target, and so on, alternating between actively sending and relaxing for about twenty minutes. Sam is asked to try to become "immersed" in the target picture and to send Rose his full experience.

In some testing systems, during the sending phase both Earl and Sam can use headphones to listen in on everything Rose says. A one-way audio link runs from Rose's soundproof ganzfeld room to Earl's control room, and from there to Sam's remote, isolated room. The audio link is used for three reasons. One is to reassure Rose, who, being isolated in a soundproof chamber, may feel more comfortable knowing that someone out there is making sure she is still all right. The second reason is to create an audio recording of everything Rose says for future research purposes. And the third reason is to provide audio feedback to help Sam adjust his strategy in "mentally sending" his experience to Rose.

Phase 3: Judging

After fifteen to thirty minutes, depending on the experimental design, Earl informs both Sam and Rose that the sending phase is over. Rose removes her Ping-Pong ball eyeshades and turns off the red ganzfeld lamp. Earl turns off the white noise playing over Rose's headphones, then presents her with copies of the four targets, one of which Sam was trying to send. In automated systems, a computer automatically presents the four targets in random order to a video monitor in Rose's room. Earl and Sam can also view the same targets on video monitors in their respective rooms.

We might point out that in the automated-video ganzfeld system recently developed at the University of Edinburgh, Scotland, *two* video players are used.[19] The rationale for this additional feature is as follows: Video targets in these experiments are typically one minute in length, but the sending phase lasts from fifteen to twenty minutes. To keep Sam focused on the task, the target video clip must be played and replayed repeatedly. Critics have suggested that because the portion of the videotape containing the actual target image is replayed repeatedly during the sending phase, after a while that image might begin to look a little "noisy" or blurry. While videotape images cannot become scratched like film, over hundreds of repeated plays the magnetic tape can begin to degrade and video images can begin to lose their clarity.

Thus, if a single video player were used for both sending and judging, Rose might be able to notice that one of the four images—the actual target repeatedly replayed to Sam—was a bit noisier, and she might select that one. Of course, this is not a problem if the four images within a target pack are all used as targets about the same number of times, because then all the images would equally degrade over time. Nevertheless, critics insist on plugging every potential design loophole, so to avoid completely the possibility of a "noisier" image, at Edinburgh a second video player and a duplicate videotape are dedicated to the judging phase.

Returning to the judging process, remember that Earl does not know the identity of the actual target picture or video clip that Sam has tried to send. So Earl asks Rose to rank the four pictures 1 through 4, according to how well each matches her impressions during the ganzfeld stimulation period. After she ranks the targets, the experimental session is over, all parties reconvene, and Sam reveals the actual target.

A direct "hit" is assigned if Rose ranks the actual target number 1; otherwise the entire session is considered a "miss." By chance, this experiment should result in a hit every four sessions, for a 25 percent chance hit rate. A hit rate reliably greater than this would indicate that information about the sender's target picture somehow got to the receiver, even with rigorous, double-blind controls in place to prevent any form of sensory leakage or experimenter bias. And that, of course, is what the experiment is designed to answer: is information about one person's experience accessible to another, remote person without the use of the normal senses?

Mentation Examples

Some of the correspondences observed between the senders' targets and the receivers' impressions are remarkable. Here are three verbatim transcripts of receiver mentations recorded during actual ganzfeld experiments, along with descriptions of the targets. These examples are taken from experiments conducted by Charles Honorton and his colleagues using an automated ganzfeld testing system.[20]

The target: Salvador Dali's famous painting *Christ Crucified.*
The receiver's impressions:

> . . . I think of guides, like spirit guides, leading me and I come into like a court with a king. It's quite. . . . It's like heaven. The king is something like Jesus. Woman. Now I'm just sort of somersaulting through heaven. . . . Brooding. . . . Aztecs, the Sun God. . . . High priest. . . . Fear. . . . Graves. Woman. Prayer. . . . Funeral. . . . Dark. Death. . . . Souls. . . . Ten Commandments. Moses . . .

The target: A video clip of horses, from the film *The Lathe of Heaven.* The clip starts with an overhead view of five horses galloping in a snowstorm. The camera zooms in on the horses. The scene shifts to a close-up of a single horse trotting in a grassy meadow, first at normal speed, then in slow motion. The scene shifts again; the same horse trots slowly through empty city streets.

The receiver's impressions:

> I keep going to the mountains. . . . It's snowing. . . . Moving again, this time to the left, spinning to the left. . . . Spinning. Like on a carousel, horses. Horses on a carousel, a circus . . .

The target: A video clip of a collapsing suspension bridge taken from a 1940s newsreel. The film shows the bridge swaying back and forth and bending up and down. Light posts are swaying, suspension cables are dangling. The bridge finally collapses from the center and falls into the water.

The receiver's impressions:

> . . . Something, some vertical object bending or swaying, almost something swaying in the wind. . . . Some thin, vertical object, bending to the left . . . Some kind of ladder-like structure but it seems to be almost blowing in the wind. Almost like a ladder-like bridge over some kind of chasm that's waving in the wind. This is not vertical this is horizontal . . . A bridge, a drawbridge over something. It's like one of those old English type bridges that opens up from either side. The middle part opens up. I see it opening. It's opening. There was a flash of an old English stone bridge but then back to this one that's opening. The bridge is lifting, both sides now. Now both sides are straight up. Now it's closing again. It's closing, it's coming down, it's closed. Arc, images of arcs, arcs, bridges. Passageways, many arcs. Bridges with many arcs. . . .

The First Meta-Analysis

At the annual convention of the Parapsychological Association in 1982, Charles Honorton presented a paper summarizing the results of all known ganzfeld experiments to that date. He concluded that the experiments at that time provided sufficient evidence to demonstrate the existence of psi in the ganzfeld. Skeptical psychologist Ray Hyman disagreed, and decided to independently analyze the same studies.[21] This eventually led to two separate meta-analyses, one by Honorton and another by Hyman, both published in 1985. They agreed on some points and disagreed on others. Honorton was a dedicated researcher deeply involved in the research itself, and he tended to see psi in the data; Hyman was a lifelong confirmed skeptic, and he did not.[22]

At that time, ganzfeld experiments had appeared in thirty-four published reports by ten different researchers. These reports described a total of forty-two separate experiments. Of these, twenty-eight reported the actual hit rates that were obtained. The other studies simply declared the experiments successful or unsuccessful. Since this information is insufficient for conducting a numerically oriented meta-analysis, Hyman and Honorton concentrated their analyses on the twenty-eight studies that had reported actual hit rates. Of those twenty-eight, twenty-three had resulted in hit rates greater than chance expectation. This was an instant indicator that some degree of replication had been achieved, but when the actual hit rates for all twenty-eight studies were combined, the results were even more astounding than Hyman and Honorton had expected: odds against chance of ten billion to one. Clearly, the overall results were not just a fluke, and both researchers immediately agreed that *something* interesting was going on. But was it telepathy?

Independent Replications

At that time, investigators from ten different labs had conducted ganzfeld experiments. One of these labs, directed by British psychologist Carl Sargent, contributed nine of the studies, Honorton's lab contributed five, and the remaining laboratories each contributed one, two, or three studies. Thus, only two laboratories had conducted half the studies, and one of these labs was directed by Honorton, who, after all, was reporting the meta-analysis. Through differences in technique, experimental quality, or design artifacts, maybe Honorton's and Sargent's labs were able to get successful results, but no one else could replicate the effect. If so, this would understandably cast doubt on the results.

To address the concern about whether independent replications had been achieved, Honorton calculated the experimental outcomes for each laboratory separately. Significantly positive outcomes were reported by six of the ten labs, and the combined score across the ten laboratories still resulted in odds against chance of about a billion to one. This showed that no one lab was responsible for the positive results; they appeared across-the-board, even from labs reporting only a few experiments. To examine further the possibility that the two most prolific labs were responsible for the strong odds against chance, Honorton recalculated the results after *excluding* the studies that he and Sargent had reported. The resulting odds against chance were still ten thousand to one. Thus, the effect did not depend on just one or two labs; it had been successfully replicated by eight other laboratories.

Selective Reporting

Another factor that might account for the overall success of the ganzfeld studies was the editorial policy of professional journals, which tends to favor the publication of successful rather than unsuccessful studies. This is the "file-drawer" effect mentioned earlier. Parapsychologists were among the first to become sensitive to this problem, which affects all experimental domains. In 1975 the Parapsychological Association's officers adopted a policy opposing the selective reporting of positive outcomes.[23] As a result, both positive and negative findings have been reported at the Parapsychological Association's annual meetings and in its affiliated publications for over two decades.

Furthermore, a 1980 survey of parapsychologists by the skeptical British psychologist Susan Blackmore had confirmed that the file-drawer problem was not a serious issue for the ganzfeld meta-analysis. Blackmore uncovered nineteen completed but unpublished ganzfeld studies.[24] Of those nineteen, seven were independently successful with odds against chance of twenty to one or greater. Thus while *some* ganzfeld studies had not been published, Hyman and Honorton agreed that selective reporting was not an important issue in this database.

Still, because it is impossible to know how many other studies might have been in file drawers, it is common in meta-analyses to calculate how many unreported studies would be required to nullify the observed effects among the known studies.[25] For the twenty-eight direct-hit ganzfeld studies, this figure was 423 file-drawer experiments, a ratio of unreported-to-reported studies of approximately fifteen to one. Given the time and resources it takes to conduct a single ganzfeld session, let alone 423 hypothetical unreported experiments, it is not surprising that Hyman agreed with Honorton that the file-drawer issue could not plausibly account for the overall results of the psi ganzfeld database.[26] There were simply not enough experimenters around to have conducted those 423 studies.

Thus far, the proponent and the skeptic had agreed that the results could not be attributed to chance or to selective reporting practices. But perhaps, it was argued, the experiments were seriously *flawed* in some way. Perhaps these flaws accounted for the apparent success.

Design Flaws

Skeptics often contend that psi experiments are inadequately designed. They claim that the experimenters were sloppy about data collecting and recording, or that they failed to control against subject or experimenter fraud, or any number of other potential problems. These flaws, so the claim goes, produce false-positive results, and the more flawed the study, the more positive the results. Conversely, the better designed the study, the smaller the results—leading to the assertion that if perfectly designed experiments were conducted, they would show only null results (on average).

Meta-analysis provides a straightforward way of testing whether design flaws are systematically related to the results reported in a series of studies. To perform this test, judges assign a rating to each study indicating the degree to which certain design criteria are present or absent. If a criterion is absent in a study, the study is assigned a zero score for that criterion. If a criterion is present, the study gets a score of one for that criterion. After the presence or absence of each criterion has been determined by detailed study of the experimental report, the final quality rating is calculated by simply adding up the ones. The quality ratings for each experiment are then compared to the effects observed in the experiments.

Finding a large *negative* relationship between study quality and experimental results would support the critics' assertion. If this same analysis showed no systematic relationship between study quality and outcomes, it would suggest that despite the possible presence of flaws in some studies (and no experiment is perfect), those flaws did not account in any *systematic* way for the results of the studies. And thus design flaws would not be responsible for the observed outcomes.

In any form of psi research, a "fatal flaw" is a design feature or overlooked aspect of the experiment that allowed for explicit or inadvertent *sensory cueing*. This includes the absence of controls that may have allowed the telepathic receiver to deliberately or accidentally obtain information about the target picture through normal sensory means. Another potentially fatal flaw is *inadequate randomization* of the targets, because if the identity of the target can be inferred in any way, this can give the receiver a clue to its identity.

Sensory Leakage

Because the ganzfeld procedure uses a sensory-isolation environment, the possibility of sensory leakage during the telepathic "sending" portion of the session is already significantly diminished. After the sending period, however, when the receiver is attempting to match his or her experience to the correct target, if the experimenter interacting with the receiver knows the identity of the target, he or she could inadvertently bias the receiver's ratings. One study in the ganzfeld database contained this potentially fatal flaw, but rather than showing a wildly successful result, that study's participants actually performed slightly *below* chance expectation.

Another problem can occur if the pool of four targets given to the receiver for judging contains the *actual* physical target that the sender had used in a manual ganzfeld test. For example, say that the target was a photograph, and it was handled by the sender during the sending period. According to the "greasy finger" hypothesis, there might be cues like fingerprints or smudges on the actual target that could give the receiver a clue about which target was the real one and which were the decoys. Contemporary ganzfeld studies have eliminated this possibility by using duplicate target pictures or by presenting video targets, but some of the earlier studies did not.

Despite variations in study quality due to these and other factors, Hyman and Honorton both concluded that there was no systematic relationship between the security methods used to guard against sensory leakage and the study outcomes. Honorton proved his point by recalculating the overall results only for studies that had used duplicate target sets. He found that the results were still quite strong, with odds against chance of about 100,000 to 1.

At this point, the two meta-analysts agreed that the results were not due to chance, or to selective reporting, or to sensory leakage. But could poor randomization procedures somehow have allowed the receivers or the experimenters to figure out the identity of the targets?

Randomization

In psi experiments, the way a target is selected is important because if the participants can consciously or unconsciously guess what the targets are, and they are repeatedly guessing many targets in a row, as in an ESP card test, then their responses could look like psi when they are really educated guesses.

Say that an ordinary deck of playing cards was accidentally unbalanced to contain fewer clubs than there were supposed to be. With repeated guessing, and with feedback about the results of each trial, participants might be able to notice that clubs did not show up as often as expected. If they decided to slightly undercall the number of clubs in subsequent guesses, this could slightly inflate the number of successful hits they got on the remaining cards. Successful results in such a test would not indicate psi, but rather a clever (or unconscious) application of statistics.

In a ganzfeld study, however, the process of randomizing the targets is much less important because only one target is used per session, and most participants serve in only one session. So there is no possibility of learning any guessing strategies based on inadequate randomization. However, a critic could argue (and did) that if all the target pictures *within each target pool* were not selected uniformly over the course of the study, this could still produce inflated hit rates.

The reasoning goes like this: A person who has participated in the study tells a friend about her ganzfeld experience where the target was, say, a Santa Claus picture. Later, if the friend participated in the study, *and* he got the same target pool, *and* during the judging period he also selected the Santa Claus because of what his friend said, *and* the randomization procedure was poor, *and* Santa Claus was selected as the target again, then what looked like psi wasn't really psi after all, but a consequence of poor randomization.

A similar concern arises for the method of randomizing the *sequence* in which the experimenter presents the target and the three decoys to the receiver during the judging process. If, for example, the target is always presented *second* in the sequence of four, then again, a subject may tell a friend, and the friend, armed with knowledge about which of the four targets is the real one, could successfully select the real target without the use of psi.

Although these scenarios are implausible, skeptics have always insisted on nailing down even the most unlikely hypothetical flaws. And it was on this issue, the importance of randomization flaws, that Hyman and Honorton disagreed. Hyman claimed that he saw a significant relationship between randomization flaws and study outcomes, and Honorton did not. The sources of this disagreement can be traced to Honorton's and Hyman's differing definitions of "randomization flaws," to how the two analysts rated these flaws in the individual studies, and to how they statistically treated the quality ratings.

These sorts of complicated disagreements are not unexpected given the diametrically opposed convictions with which Honorton and Hyman began their analyses. When such discrepancies arise, it is useful to consider the opinions of outside reviewers who have the technical skills to assess the disagreements. In this case, ten psychologists and statisticians supplied commentaries alongside the Honorton-Hyman published debate that appeared in 1986. None of the commentators agreed with Hyman, while two statisticians and two psychologists not previously associated with this debate explicitly agreed with Honorton.[27]

In two separate analyses conducted later, Harvard University behavioral scientists Monica Harris and Robert Rosenthal (the latter a world-renowned expert in methodology and meta-analysis) used Hyman's own flaw ratings and failed to find any significant relationships between the supposed flaws and the study outcomes. They wrote, "Our analysis of the effects of flaws on study outcome lends no support to the hypothesis that ganzfeld research results are a significant function of the set of flaw variables."[28]

In other words, everyone agreed that the ganzfeld results were not due to chance, nor to selective reporting, nor to sensory leakage. And everyone, except one confirmed skeptic, also agreed that the results were not plausibly due to flaws in randomization procedures. The debate was now poised to take the climactic step from Stage 1, "It's impossible," to Stage 2, "Okay, so maybe it's real."

Effect Size

Some skeptics argue that even if psi effects observed in current experiments turn out to be replicable, and are not due to any known design problems, they are still too small to be of either theoretical or practical interest. This attempt to trivialize the effect is a red herring, of course, because *any* valid demonstration of a genuine psi effect is of outstanding importance. Just because the effect appears to be weak now says nothing about what it may become after improvements in experimental procedures and theoretical understanding.

In fact, a review of the history of science reveals that most discoveries are initially weak and sporadic. Only years later, after much hard work and technical refinement, do we begin to achieve the degree of reliability expected of a mature science. This argument aside, the psi ganzfeld effect turns out to be neither as weak nor as inconsequential as many had thought.

The easiest way to compare effects across the various ganzfeld experiments in the 1985 meta-analysis is to compare the hit rates among twenty-five of the forty-one experiments that used designs where the chance hit rate was 25 percent. Taking the number of trials and direct hits in those twenty-five studies from Honorton's 1985 publication, we can plot the results for each experiment. Figure 5.3 shows the hit-rate point estimates and 95 percent confidence intervals for each of the twenty-five studies. As indicated, the overall hit rate for the combined 762 sessions was 37 percent. This hit rate corresponds to odds against chance of about a trillion to one—even though the majority of the individual studies (fourteen of twenty-five) were not independently "successful" (their 95 percent confidence intervals included chance). This again demonstrates the value of combining all available studies as opposed to just a few selected experiments.

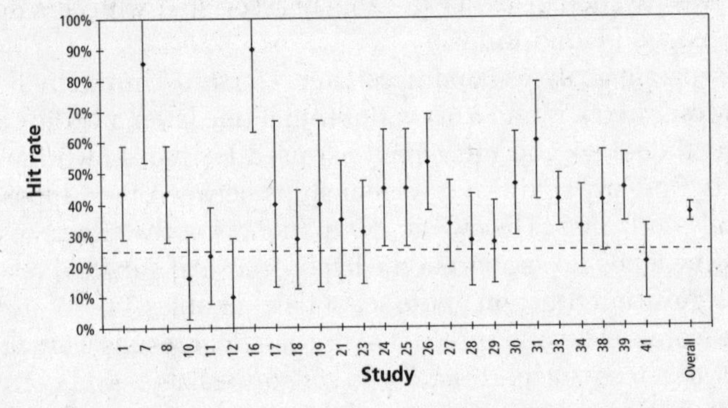

Figure 5.3. Point estimates and 95 percent confidence intervals of the 1985 ganzfeld meta-analysis. Study numbers correspond to the studies listed in Honorton's table A1.[29]

To show that the psi ganzfeld effect is larger than it first appears, let's compare it with the results of a widely publicized medical study investigating whether aspirin could prevent heart attacks (as discussed in chapter 4).[30] That study was discontinued after six years because it had become abundantly clear that the aspirin treatment was effective, and it was considered unethical to keep the control group on placebo medication. This was widely publicized as a major medical breakthrough, but despite its practical importance, the *magnitude* of the aspirin effect is extremely small. Taking aspirin reduces the probability of a heart attack by a mere 0.8 percent compared with not taking aspirin (that's eight-tenths of one percentage point). This effect is about ten times smaller than the psi ganzfeld effect observed in the 1985 meta-analysis.

The Joint Communiqué

After the 1985 meta-analyses were published, Hyman and Honorton agreed to write a joint communiqué. In the communiqué, which was published in 1986, they began by describing the points on which they agreed and disagreed:

> We agree that there is an overall significant effect in this data base that cannot reasonably be explained by selective reporting or multiple analysis. We continue to differ over the degree to which the effect constitutes evidence for psi, but we agree that the final verdict awaits the outcome of future experiments conducted by a broader range of investigators and according to more stringent standards.[31]

They then specified in detail the "more stringent standards" that future experiments would have to follow to provide evidence that satisfied the skeptics. Honorton was especially interested in getting Hyman to agree publicly to these criteria, as skeptics are notorious for changing the rules of the game after all previous objections have been met and new experiments continue to provide significant results.

The new standards, acceptable to both Honorton and Hyman, included such things as rigorous precautions against sensory leakage, extensive security procedures to prevent fraud, detailed descriptions of how the targets were selected, full documentation of all experimental procedures and equipment used, and complete specifications about what statistical tests were to be used to judge success. With a recipe agreed to by the leading ganzfeld psi researcher and the leading skeptic, the stage was set to see whether future ganzfeld studies would continue to show successful results. If they did, then the skeptics would be forced to admit that something interesting was going on.

The Autoganzfeld

Starting in 1983, Honorton and his colleagues initiated a new series of ganzfeld studies that were computer-controlled. Largely implemented by psychologist Rick Berger, the new automated ganzfeld system—called the "autoganzfeld"—was specifically planned to avoid the design problems that had been identified in the 1985 meta-analyses, and the experiments conducted with this system fully complied with the experimental recipe published in the 1986 joint communiqué.[32] Honorton's research program continued to collect ganzfeld data until September 1989, when a loss of funding forced the laboratory to shut down. The major innovations in the new studies were the use of computers to control most of the experimental procedures and the introduction of closed-circuit video to present short videotaped film clips and still pictures as ganzfeld targets.

The target pool in the autoganzfeld system consisted of eighty still pictures (called the "static targets") and eighty short audio-video segments (called the "dynamic targets"). These 160 targets were arranged in groups of four targets per set, for a total of twenty static and twenty dynamic target sets. The static targets included art prints and photographs, and the dynamic targets included short video clips taken from motion pictures, TV shows, and cartoons. All the targets were recorded on videotape.

Besides using a steel-walled, sound-proofed, and electromagnetically shielded room to isolate the receiver, the experimenters adopted computer-controlled procedures to help them ensure that the experimental procedures were not vulnerable to sensory leakage or to deliberate cheating. In addition, two professional magicians who specialized in the simulation of psi effects (called "mentalists" or "psychic entertainers") examined the autoganzfeld system and protocols to see if it was vulnerable to mentalist tricks or conjuring-type deceptions. One of the magicians was Ford Kross, an officer of the Psychic Entertainers Association. Kross provided the following written statement about the autoganzfeld setup:

> In my professional capacity as a mentalist, I have reviewed Psychophysical Research Laboratories' automated ganzfeld system and found it to provide excellent security against deception by subjects.[33]

The other magician was Cornell University psychologist Daryl Bem, who besides coauthoring a 1994 paper on the ganzfeld psi experiments with Honorton,[34] is also a professional mentalist and a member of the Psychic Entertainers Association.

Results

All together, 100 men and 140 women participated as receivers in 354 sessions during the six-year autoganzfeld research program.[35] The participants ranged in age from seventeen to seventy-four; and eight different experimenters, including Honorton, conducted the studies. The program included three preliminary and eight formal studies. Five of the formal studies employed only "novices"—participants who served as the receivers in just one session each. The remaining three formal studies used experienced participants.

The bottom line for the eleven series, consisting of a total of 354 sessions, was 122 direct hits, for a 34 percent hit rate. This compares favorably with the 1985 meta-analysis hit rate of 37 percent. Honorton's autoganzfeld results overall produced odds against chance of forty-five thousand to one.

Providing Proof of the Telepathy Pudding

In Hyman and Honorton's joint communiqué they wrote, "We agree that the final verdict awaits the outcome of future experiments conducted by a broader range of investigators and according to more stringent standards."[36] The autoganzfeld results published after the joint communiqué were statistically significant, and the hit rate (34 percent) was consistent with the results of the 1985 meta-analysis. One might expect that Hyman would concede that the psi ganzfeld effect had been demonstrated.

But it isn't so easy to disavow one's lifelong, dearly held convictions. In commenting on Honorton's triumphant autoganzfeld studies, Hyman offered the following quasi-concession:

> Honorton's experiments have produced intriguing results. If independent laboratories can produce similar results with the same relationships and with the same attention to rigorous methodology, then parapsychology may indeed have finally captured its elusive quarry.[37]

Hyman was not just being coy. The proof of the pudding in science really does reside in multiple, independent replications. So, did studies after the autoganzfeld studies continue to successfully replicate the psi ganzfeld effect?

Figure 5.4 summarizes all replication attempts as of early 1997.[38] As before, the graph shows the hit-rate point estimates and 95 percent confidence intervals. The left-most line records the results from the 1985 meta-analysis (indicated as "85 MA"), and the next line to the right shows the Psychophysical Research Laboratories (PRL) autoganzfeld results. The numbers in parentheses after each label refer to the number of ganzfeld sessions contributed by the various investigators. Thus, the 1985 meta-analysis hit rate was based on a total of 762 separate sessions.

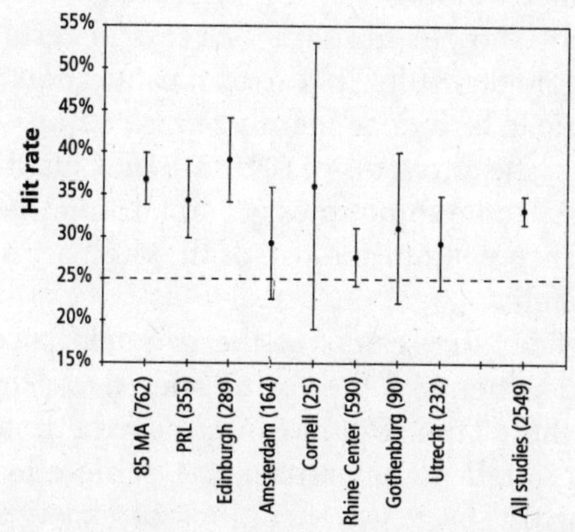

Figure 5.4. Results of all ganzfeld telepathy experiments as of early 1997.

The next replications were reported by psychologist Kathy Dalton and her colleagues at the Koestler Chair of Parapsychology, Department of Psychology, University of Edinburgh, Scotland. The Edinburgh experiments, conducted from 1993 through 1996 (and still ongoing), consisted of five published reports and 289 sessions using an improved, fully automated psi ganzfeld setup. It was based on Honorton's original autoganzfeld design and implemented in stages first by Honorton, then by psychologist Robin Taylor, then by me, and finally by Kathy Dalton.[39] Other replications have been reported by Professor Dick Bierman of the Department of Psychology at the University of Amsterdam; Professor Daryl Bem of Cornell University's Psychology Department; Dr. Richard Broughton and colleagues at the Rhine Research Center in Durham, North Carolina; Professor Adrian Parker and colleagues at the University of Gothenburg, Sweden; and doctoral student Rens Wezelman from the Institute for Parapsychology in Utrecht, Netherlands.[40]

While only the 1985 meta-analysis, the autoganzfeld study, and the Edinburgh study independently produced a hit rate with 95 percent confidence intervals beyond chance expectation, it is noteworthy that *each* of the six replication studies (after the autoganzfeld) resulted in point estimates greater than chance. The 95 percent confidence interval at the right end of the graph is the combined estimate based on all available ganzfeld sessions, consisting of a total of 2,549 sessions. The overall hit rate of 33.2 percent is unlikely with odds against chance beyond a million billion to one.

Ganzfeld Summary

From 1974 to 1997, some 2,549 ganzfeld sessions were reported in at least forty publications by researchers around the world. After a 1985 meta-analysis established an estimate of the expected hit rate, a six-year replication was conducted that satisfied skeptics' calls for improved procedures. That "autoganzfeld" experiment showed the same successful results. After publication of the autoganzfeld results in 1990, the question was whether the effects could continue to be independently replicated.

We now know that the answer is yes. We are fully justified in having very high confidence that people sometimes get small amounts of specific information from a distance without the use of the ordinary senses. Psi effects do occur in the ganzfeld.

Now jointly consider the results of the ganzfeld psi experiments, the dream-telepathy experiments of the 1960s and 1970s, the ESP cards tests from the 1880s to the 1940s, Upton Sinclair's experiments in 1929, and earlier studies on thought transference. The same effects have been repeated again and again, by new generations of experimenters, using increasingly rigorous methods. From the beginning, each new series of telepathy experiments was met with its share of skeptical attacks. These criticisms reduced mainstream scientific interest in the reported effects, but ironically they also refined the methods used in future experiments to the point that today's ganzfeld experiments stump the experts.

Features like computer controls, multiple video players, automatic random selection of targets, and so on, are not *required* to conduct a proper ganzfeld test. They were slowly added to the basic ganzfeld test-design to address one critical concern after another. As we have seen, virtually identical results have been observed in these experiments with or without all the elaborate precautions in place. Nevertheless, the fully automated ganzfeld tests were not a waste of money, because now we know that extremely unlikely possibilities, like the receiver noticing that some video targets may be slightly noisier than others, cannot explain away psi in the ganzfeld.

Long before this experimental evidence was available, Sigmund Freud, a staunch skeptic of supernatural and occult beliefs, was asked for his opinion of telepathy. He wrote:

> No doubt you would far prefer that I should hold fast to a moderate theism, and turn relentlessly against anything occult. But I am not concerned to seek any one's favor, and I must suggest to you that you should think more kindly of the objective possibility of thought-transference and therefore also of telepathy.[41]

If information can be exchanged between two minds, what about the more general case of between a mind and a distant object? In the next chapter, we examine the evidence for clairvoyance—perception at a distance.

CHAPTER 6

Perception at a Distance

Man also possesses a power by which he may see his friends and the circumstances by which they are surrounded, although such persons may be a thousand miles away from him at that time.

PARACELSUS (1493–1541)

In the preceding chapter, we first encountered the conceptual problem of how to distinguish cleanly between telepathy and clairvoyance. Although testing for "pure" telepathy remains an unsolved problem, methods for investigating clairvoyance—psi perception at a distance without a sender—are comparatively well understood. Starting in the late nineteenth century, two types of clairvoyance experiments have been widely replicated by dozens of researchers: studies using ESP cards and "remote-viewing" or picture-drawing experiments. In this chapter we'll briefly discuss the early experiments, then concentrate in more depth on modern remote-viewing studies.

The Phenomenon

Clairvoyance differs from telepathy in that no one "sends" the information that is received. That is, information is obtained from a distant or hidden location, beyond the ordinary bounds of space (and time, but that is discussed in the next chapter). While clairvoyance literally means "clear seeing," the actual psi perceptions can also resemble sound, called "clairaudience," or smell, touch, or taste, called "clairsentience." The popular term "extrasensory perception" (ESP), coined by J. B. Rhine in 1934 in a book by that title,[1] is synonymous with clairvoyance, as are the modern phrases "remote viewing" and "remote perception."

The classic spontaneous experience of clairvoyance involves a distant crisis, it displays features characteristic of "pure" telepathy, and it often occurs in nonordinary states of awareness, typically dreams. Here is a case as retold by author Bernard Gittelson:

> A woman on an Oregon farm was jolted awake one morning at 3:40 by the sound of people screaming. The sound quickly vanished, but she felt a smoky, unpleasant taste in her mouth. She woke her husband, and together they scoured the farm but found nothing irregular. That evening on a television newscast, they heard about a plant explosion that started a huge chemical fire which killed six people. The explosion had occurred at 3:40 A.M.[2]

Another example that blurs the distinction between clairvoyance and telepathy is recorded in the psychical research classic *Phantasms of the Living,* compiled by members of the Society for Psychical Research. The person reporting this story was a Mrs. Morris Griffith, who lived in North Wales, Britain, in 1884. She reported the following experience:

> On the night of Saturday, the 11th of March, 1871, I awoke in much alarm, having seen my eldest son, then at St. Paul de Loanda on the south-west coast of Africa, looking dreadfully ill and emaciated, and I heard his voice distinctly calling to me. I was so disturbed I could not sleep again, but every time I closed my eyes the appearance recurred, and his voice sounded distinctly, calling me "Mamma."
>
> I felt greatly depressed all through the next day, which was Sunday, but I did not mention it to my husband, as he was an invalid, and I feared to disturb him. We were in the habit of receiving weekly letters every Sunday from our youngest son, then in Ireland, and as none came that day, I attributed my great depression to that reason, glad to have some cause to assign to Mr. Griffith rather than the real one. Strange to say, he also suffered from intense low spirits all day, and we were both unable to take dinner, he rising from the table saying, "I don't care what it costs, I must have the boy back," alluding to his eldest son.
>
> I mentioned my dream and the bad night I had had to two or three friends, but begged that they would say nothing of it to Mr. Griffith. The next day a letter arrived containing some photos of my son, saying he had had fever, but was better, and hoped immediately to leave for a much more healthy station, and written in good spirits. We heard no more till the 9th of May, when a letter arrived with the news of our son's death from a fresh attack of fever, on the night of the 11th of March, and adding that just before his death he kept calling repeatedly for me. I did not at first connect the date of my son's death with that of my dream until reminded of it by the friends, and also an old servant, to whom I had told it at the time.[3]

The authors of *Phantasms of the Living* asked Mrs. Griffith for more details to see if she frequently dreamed about her son, or of dying. They reasoned that if she worried excessively about her son, it would lessen the remarkable nature of this dream, because she may have had such dreams every night. She answered:

> I have never in all my life, before or since, had any such a distressing dream, nor am I ever discomposed in any way by uncomfortable dreams. I never remember at any time having any dream from which I have had any difficulty in knowing at once, whilst awakening, that I had been dreaming, and never confuse the dream with reality. I also unhesitatingly assure you that I have never had any hallucination of the senses as to sound or sight.

Extraordinary stories like these provide the motivation to study whether such experiences are what they appear to be. To overcome the reasonable doubts that must be maintained for any extraordinary claim, researchers began to study clairvoyance in the laboratory, and from the scientific perspective this is where the story really becomes interesting.

ESP Card Experiments

One of the first researchers to use cards as ESP targets was the French physiologist and Nobel laureate Charles Richet.[4] In 1889, Richet published a report describing his experiments in which a hypnotized person was able to successfully guess the contents of sealed, opaque envelopes at odds far beyond chance. But for the next several decades, most psi researchers focused primarily on thought transference, as described in the preceding chapter, and on mediumship as a means of studying the possibility of postmortem survival.

Eventually, investigators realized that virtually all the interesting evidence for survival-related phenomena could also be explained as telepathy by the medium, so research efforts began to shift to telepathy. In addition, telepathy lent itself to controlled laboratory investigation, whereas survival research did not. It was eventually discovered that psi performance in telepathy tests did not diminish when there was no "sender." It also proved to be nearly impossible to create a test for "pure" telepathy that could not also be explained as clairvoyance. So most researchers began to focus on clairvoyance.

It may seem odd that it took any time at all to go from systematic research on survival phenomena, to telepathy research, and then to clairvoyance, before it was realized that the fundamental issue in all cases was the nature of *psi perception*. But this just illustrates how difficult this topic is to study. Some researchers made these leaps in short order. Others took years. Collectively it took about a half-century to come to what we now see as a "reasonable" approach. Fifty years from now, entirely new "reasonable" ideas may have evolved.

Card Criticisms

The evolution of card tests reflects what researchers were learning about both ordinary and extrasensory perception. For example, some of the earliest tests used cards that were shuffled by hand and then placed face down in a deck. The test participant guessed the identity of the card on the top of the deck, then turned it over. The experimenter recorded both the guess and the actual identity of the card. Then some researchers noticed that it might be possible (and this proved to be the case for some cards) that the act of printing a symbol on the front of the card might leave a slightly raised impression on the back of the card. This impression could allow a participant consciously or unconsciously to guess the symbol with better than chance accuracy.

This objection led to the use of cards enclosed inside opaque envelopes. The participant held the envelope and guessed the symbol on the hidden card. An experimenter recorded this guess, then the test participant opened the envelope and the experimenter recorded the identity of the card. This led to criticisms that because the participant had handled the card, he or she could mark it, say with a thumbnail, to increase the chances of guessing it correctly the next time. Someone intent on cheating might have been able to feel the mark through the envelope.

Over six decades, methods continued to improve to take into account many new objections to the successful results obtained in these experiments. For example, test participants were no longer allowed to handle the cards, and then they were separated from both the cards and the experimenters by opaque screens. Then, participants and experimenters were located in separate rooms, and, later, in separate buildings. This last design feature addressed the criticism of "involuntary whispering," which may have provided a sensory cue about the identity of the card in cases where the experimenter knew the identity of the card and was within earshot of the test participant.

In one of the few cases where skeptics actually tested a criticism, in 1939 psychologists J. Kennedy and W. Uphoff asked twenty-eight observers to record 11,125 mock ESP trials to see if "motivated recording errors" could explain J. B. Rhine's results. They found that 1.13 percent of the data were misrecorded as they expected, with "ESP believers" making recording errors in favor of ESP and "ESP skeptics" making errors in favor of no ESP. Of the errors made by believers, 71.5 percent *increased* the ESP scores, while for skeptics, 100 percent of the errors *decreased* the ESP scores.[5]

Half a century later, in 1978, Harvard University psychologist Robert Rosenthal summarized twenty-seven studies from the behavioral sciences that investigated motivated recording errors. He confirmed that the average overall error rate was about what Kennedy and Uphoff had previously found, about 1 percent.[6] And even though a 1 percent error rate could not explain the results of Rhine's ESP card-guessing experiments, investigators at the time adopted controls against recording errors, such as the use of assistants to double-check the data recorded during the experiment.

By the end of the 1930s, duplicate recording and double-blind data-checking procedures were routinely employed. Later, to overcome charges that experiments could be explained by participant collusion or fraud, assistants were employed to verify that the test participants did not violate the experimental protocols.

Statistical Questions

For a time, a variety of statistical criticisms were leveled at the card experiments.[7] One was the problem of "optional stopping," which may occur when the experimenter decides to end an experiment because the results "look good." Such practices can inflate the apparent success of a study. These issues were settled by prespecifying the number of trials to be collected in each experiment. Another statistical question was whether the procedures used to evaluate the results were appropriate. For example, in a typical card-guessing experiment, people were asked to guess the symbols of shuffled cards containing five each of five geometrical symbols (star, wavy line, square, circle, and cross). Because participants usually did not receive feedback about their results until after they had guessed all twenty-five trials, statistical analysis of the card experiments assumed that the chance of success on any given trial was one in five. This led to a chance expected hit rate of 20 percent over the course of many runs of twenty-five-card decks.

But some critics questioned whether the 20 percent chance assumption was valid. This issue was eventually resolved both by mathematical proof and through empirical "cross-checks." The latter were control tests in which a person's guesses for target cards in run 1 were compared with the actual targets from run 2, the guesses for run 2 with the targets in run 3, and so on. Rhine and his colleagues conducted these cross-checks for all the guesses in twenty-four separate experiments. They found that for the actual experiments the average hit rate was 7.23/25, or 29 percent, a highly significant result, while the control cross-checks averaged 5.04/25, or 20 percent, as expected by chance.[8]

The early statistical criticisms were finally settled by statistician Burton Camp of Wesleyan University. Camp was the president of the Institute of Mathematical Statistics in 1937. In December of that year, he released a statement to the press that read:

> Dr. Rhine's investigations have two aspects: experimental and statistical. On the experimental side mathematicians, of course, have nothing to

say. On the statistical side, however, recent mathematical work has established the fact that, assuming that the experiments have been properly performed, the statistical analysis is essentially valid. If the Rhine investigation is to be fairly attacked, it must be on other than mathematical grounds.[9]

Results

From the 1880s to the 1940s, 142 published articles described 3.6 million individual trials generated by some 4,600 percipients in 185 separate experiments.[10] These figures exclude three studies involving mass ESP tests broadcast over the radio, which added more than a million additional trials and more than 70,000 participants to the sixty-year database of ESP tests.

Figure 6.1 shows the hit rates for a subset of tightly controlled ESP card experiments involving standard, five-symbol card decks. Where chance expected hitting is at 20 percent, the graph shows the point estimates and 95 percent confidence intervals for studies involving cards hidden inside sealed, opaque envelopes (130,000 guesses) and placed behind opaque screens (497,000 guesses), studies where percipients and experimenters were separated by distance (164,000 guesses), and tests where sensory cueing was "shielded" by having percipients guess cards that were selected in the future (115,000 guesses).[11]

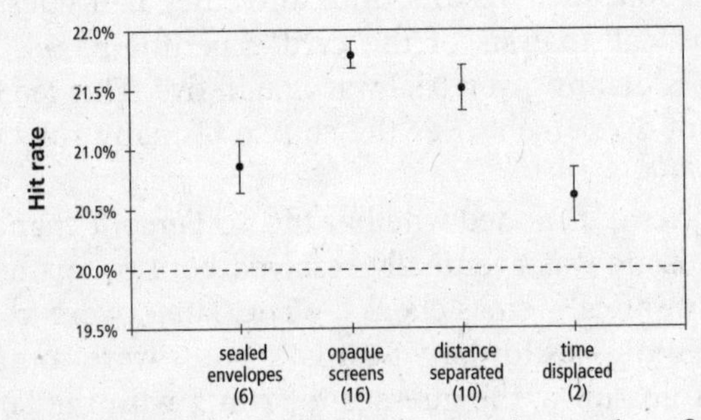

Figure 6.1. Results of high-security ESP card tests. The number of reported experiments is in parentheses.

One might think that the body of evidence summarized in figure 6.1, reflecting only the high-security studies reported by nearly two dozen investigators from 1934 to 1939, and 907,000 trials, would have been sufficient to settle the question about the existence of psi perception. And in fact, these experiments did cause many scientists to take psi phenomena seriously. For example, Professor H. J. Eysenck, chairman of the Psychology Department at the University of London, wrote in 1957:

> Unless there is a gigantic conspiracy involving some thirty University departments all over the world, and several hundred highly respected scien-

> tists in various fields, many of them originally hostile to the claims of the psychical researchers, the only conclusion the unbiased observer can come to must be that there does exist a small number of people who obtain knowledge existing either in other people's minds, or in the outer world, by means as yet unknown to science.[12]

But one reason that the evidence did not establish the reality of clairvoyance as firmly as Rhine had hoped was the suspicion that unsuccessful studies were not being published as often as successful studies. (This "file-drawer" problem has been mentioned before, and it will come up again.) Today we have ways of estimating how many unpublished or unretrieved studies would have been required to reduce the overall ESP card test results to a nonsignificant level. For the thirty-four studies summarized in figure 6.1, which are all the published high-security ESP card studies conducted with five-symbol decks from 1935 to 1939, the file-drawer estimate is 29,000 studies. That is, to reduce this body of evidence to a nonsignificant level we would need to have file drawers crammed with 29,000 unpublished, unsuccessful studies, a ratio of 861 unpublished studies for each published study.

Harvard psychologist Robert Rosenthal has suggested that, for a given body of data, a ratio of five unpublished studies to one published study is sufficient to consider the observed effect "robust." For the ESP card tests, therefore, explanations such as chance, selective reporting, and sensory leakage cannot plausibly explain the results.

If we consider *all* the ESP card tests conducted from 1882 to 1939, reported in 186 publications by dozens of investigators around the world, the combined results of this four-million trial database translate into tremendous odds against chance—more than a billion trillion to one.[13] If we assume that there is a selective-reporting problem in this database and calculate the number of unpublished, unsuccessful studies required to nullify these astronomical odds, we find that the file drawer would need to contain more than 626,000 reports. That's more than 3,300 unpublished, unsuccessful reports for each published report. This again demonstrates that chance results and selective reporting cannot reasonably explain these results.

One of the discoveries made with ESP cards was that psi performance invariably declined with repeated testing. This is not surprising because the ESP card test is tedious to begin with, and being asked to guess the same card symbols over and over, thousands of times, is just plain boring. The mind wanders, motivation diminishes, and after twenty minutes it is difficult to continue focusing on the task. Even the experimenters began to find these studies somewhat monotonous. This led to increasing interest in free-response studies such as the dream-telepathy and ganzfeld-telepathy experiments discussed in the preceding chapter.

Remote-Viewing Experiments

Among the many variations of free-response tests, the "remote-viewing" experiment was reborn in the mid-1970s. These experiments were part of a long genealogy of picture-drawing tests. The history of these experiments is discussed in depth by artist Ingo Swann, who compares many examples of the correspondences between target sketches and responses over the years.[14] Swann shows that the first picture-drawing studies, published by British researchers Fredrick W. H. Myers and Edmund Gurney in 1882, were virtually identical in style, method, and results to the experiments published later by German and French investigators, and by Upton Sinclair in the United States. As we shall see, a hundred years after Myers and Gurney reported their results, researchers working for U.S. government military and intelligence agencies again observed the same results.

Remote Viewing and the Government

Perhaps the best-known remote-viewing research in modern times began in the early 1970s, when various U.S. government agencies initiated a program at Stanford Research Institute (SRI), a scientific think tank affiliated with Stanford University. In the late 1970s, SRI became an independent corporation called SRI International, which is the name it goes by today.

Physicist Harold Puthoff founded the SRI program. He was joined soon afterward by physicist Russell Targ, and a few years later, by another physicist, Edwin May. When Puthoff took another position in 1985, the program came under the leadership of May. In 1990, the entire program moved to a think tank called Science Applications International Corporation (SAIC), a major defense contractor. That program finally wound down in 1994, after twenty-four years of support and about $20 million in funding from U.S. government agencies such as the CIA, the Defense Intelligence Agency, the Army, the Navy, and NASA.

Government agencies saw remote viewing as a possible new source of information. Even if it was only partially correct, it might provide valuable clues to help piece together the information jigsaw puzzles that constitute the typical intelligence operation. Moreover, remote viewing potentially provided a unique intelligence technique in that information could be secretly obtained at a distance and through any known form of shielding. The agencies continued to show interest in remote viewing for more than twenty years because the SRI and SAIC programs occasionally provided useful mission-oriented information at high levels of detail. Given that this information was obtained at virtually no expense, and with no risk of life compared to sending agents into the field, *and* it sometimes provided information otherwise blocked by shielding or hidden structures, it is clear why military and intelligence agencies were interested.

Sometimes the results were so striking that they far exceeded the effects typically observed in formal laboratory tests. In one test conducted at the request of government clients who wished to see how useful remote viewing might be in real intelligence missions, Dr. Edwin May described how a remote viewer was able to successfully describe a target, having no prior information about the target other than that it was "a technical device somewhere in the United States." The actual target was a high-energy microwave generator in the Southwest. Without knowing this, the "viewer" drew and described an object remarkably similar to a microwave generator, including its function, approximate size, and housing, and even correctly noted that it had "a beam divergence angle of 30 degrees."[15]

Most of the classified, mission-oriented remote viewings could not be evaluated as controlled, formal experiments, because that was not their intent. In some cases, however, unexpected information obtained through remote viewing was later confirmed to be correct, and this was important because it demonstrated the pragmatic value of this technique for use in real-world missions.

In one especially interesting test case in the late 1970s, a remote viewer given only latitude and longitude coordinates of a location somewhere in the United States successfully described a secret facility in Virginia whose very existence was highly classified. He was able to describe accurately the facility's interior and was even able to correctly sense the names of secret code words written on folders inside locked file cabinets.[16] A skeptical newspaper reporter later heard this astonishing story and decided to check it out for himself. He drove to the location specified by the map coordinates, some 135 miles west-southwest of Washington, D.C., expecting to find "the base camp of an extraterrestrial scouting party or, at the very least, the command center for World War III."[17] Instead, he found "just a spare hillside, a few flocks of sheep, and lots of droppings." No secret military outpost, no armed personnel, no buildings.

When informed of this, the Navy project officer in charge of the SRI remote-viewing tests was alarmed. He had assumed the test was successful because of reports he had received from the CIA and National Security Agency. A few days later, the project officer abruptly changed his mind, telling the reporter that the test was valid after all and offering excuses such as that the CIA or NSA man tasked with confirming the accuracy of the remote viewing "couldn't read a map," or maybe the psychic had accidentally described a nearby space communications center in West Virginia. What he didn't say was that the newspaper reporter saw exactly what he was supposed to see—flocks of sheep on a hillside. The secret military facility was indeed at that very spot, hidden deep underground.

In this case, as in many similar cases of operational remote viewing, it is not possible to calculate odds against chance. Still, most people would agree that the odds would be extremely small—so small as to justify serious research into whether clairvoyance really does occur under tightly controlled conditions. In addition, "psychic spies" would want to find ways of making it more reliable, and of finding people who are extremely good at it.

Remote-Viewing Procedures

In typical remote-viewing experiments, a "viewer" is asked to sketch or to describe (or both) a "target." The target might be a remote location or individual, or a hidden photograph, object, or video clip. All possible paths for sensory leakage are blocked, typically by separating the target from the viewer by distance, sometimes thousands of miles, or by hiding the target in an opaque envelope, or by selecting a target in the future.

Sometimes the viewer is assisted by an interviewer who asks questions about the viewer's impressions. Of course, in such cases the interviewer is also blind to the target so he or she cannot accidentally provide cues. In some remote-viewing studies, a sender visits the remote site or gazes at a target object during the session; these experiments resemble classic telepathy tests. In other studies there are no senders at the remote site. In most tests, viewers eventually receive feedback about the actual target, raising the possibility that the results could be thought of as precognition rather than real-time clairvoyance. We consider these implications in more detail in the next chapter.

Judging the Results

All but the very earliest studies at SRI (and all of the SAIC remote-viewing experiments) evaluated the results using a method called "rank-order judging." This is similar to the technique employed in the dream-telepathy experiments discussed earlier. After a viewer had remote-viewed a target (a geographic site, a hidden object, a photograph, or a video clip), a judge who was blind to the true target looked at the viewer's response (a sketch and a paragraph or two of verbal description) along with photographs or videos of five possible targets. Four of these targets were decoys and one was the real target.

As we have come to expect, the actual target was always selected at random from this pool of five possibilities to ensure that neither the viewers nor the judges could infer which was the actual target. The judge was asked to assign a *rank* to each of the possible targets, where a rank of 1 meant that the possible target matched the response most closely, and a rank of 5 meant that it matched the least. The final score for each remote-viewing trial was simply the ranking that the judge assigned to the actual target.[18]

Design Evolution

Over the two decades of recent remote-viewing experiments, articles describing these results were published in prominent scientific journals such as *Nature*[19] and the *Proceedings of the IEEE*,[20] and in a few popular books.[21] As expected, the published results prompted dozens of criticisms.[22] The constructive criticisms helped researchers evolve their experimental designs into progressively tighter methods, and eventually a list of design criteria emerged that provided extremely tight conditions for demonstrating proof-of-principle for any form of clairvoyance. They included rules such as: (1) no one who knows the identity of the target should have any contact with the remote viewer until *after* his or her description of the target has been safely secured; (2) no one who knows about the target or whether the session was successful should have any contact with the judge until *after* the judging has been completed; and (3) no one who knows about the target should have access to the remote viewer's responses until *after* the judging has been completed.[23]

SRI Experiments: 1973–1988

In 1988 Edwin May and his colleagues analyzed all psi experiments conducted at SRI from 1973 until that time.[24] The analysis was based on 154 experiments, consisting of more than 26,000 separate trials, conducted over those sixteen years. Of those, just over a thousand trials were laboratory remote-viewing tests. The statistical results of this analysis indicated odds against chance of 10^{20} to one (that is, more than a billion billion to one). As we've seen in the telepathy experiments and ESP card tests, chance is not a viable explanation for such results. In this particular database, clairvoyance may not be the only explanation, especially since some of the early SRI work contained design problems that were identified later. About the same level of psi performance was observed, however, in later remote-viewing experiments, suggesting that design problems couldn't completely explain away the results.

SAIC Experiments: 1989–1993

In 1995, the CIA commissioned a review of the government-sponsored remote-viewing research. The principal authors of the report were Dr. Jessica Utts, a statistics professor at the University of California, Davis, and Dr. Ray Hyman (whom we have met before) from the University of Oregon. The review committee's primary task was to evaluate the remote-viewing experiments conducted at Science Applications International Corporation, although it also reviewed the SRI studies to see if the SAIC experiments replicated the earlier experiments.

The SAIC studies provided a rigorously controlled set of experiments that had been supervised by a distinguished oversight committee of experts from a variety of scientific disciplines. The committee included a Nobel laureate physicist, internationally known experts in statistics, psychology, neuroscience, and astronomy, and a retired U.S. Army major general who was also a physician.

Of ten government-sponsored experiments conducted at SAIC, six involved remote viewing. Because the SRI studies had previously established the existence of remote viewing to the satisfaction of most of the government sponsors,[25] the SAIC experiments were not conducted as "proof-oriented" studies, but rather as a means of learning how psi perception worked.

Results

After studying the SRI and SAIC experiments in detail, the government review committee came to six general conclusions.[26] First, it found that so-called free-response remote viewing, where viewers were allowed to describe whatever came to mind, was more successful than forced-choice remote viewing, where viewers were required to select their responses from a few discrete possibilities (like ESP card symbols). Second, in test after test, psi performance among a small group of selected individuals far exceeded performance among unselected volunteers. This was an important observation, because if design problems accounted for successful experiments—as critics often assumed—then the selected group would not have been able to perform *consistently* better than unselected volunteers.

Third, mass screenings to find talented remote viewers revealed that about 1 percent of those tested were consistently successful. This says that first-class remote-viewing ability is relatively rare, but it probably varies across the general population much like athletic ability and musical talent. Fourth, neither practice nor training consistently improved remote-viewing ability. As with musical talent, some people with natural ability can perform highly effective remote viewing after only a few minutes of instruction, while those without that raw talent find remote viewing difficult or impossible to perform. Fifth, it is not yet clear whether feedback about the remote-viewing target is necessary, but it does provide a psychological boost that increases performance. And sixth, neither the use of electromagnetic shielding nor the distance between the target and the viewer seems to affect the quality of remote viewing.

Jessica Utts ended her review as follows:

> It is clear to this author that anomalous cognition is possible and has been demonstrated. This conclusion is not based on belief, but rather on commonly accepted scientific criteria. The phenomenon has been replicated in a number of forms across laboratories and cultures. . . .
>
> I believe that it would be wasteful of valuable resources to continue to look for proof. No one who has examined all of the data across laboratories, taken as a collective whole, has been able to suggest methodological or statistical problems to explain the ever-increasing and consistent results to date.[27]

And what about the devil's advocate, Ray Hyman? After reviewing the same evidence, he concluded:

> I agree with Jessica Utts that the effect sizes reported in the SAIC experiments and in the recent ganzfeld studies probably cannot be dismissed as due to chance. Nor do they appear to be accounted for by multiple testing, filedrawer distortions, inappropriate statistical testing or other misuse of statistical inference. . . . So, I accept Professor Utts' assertion that the statistical results of the SAIC and other parapsychologists experiments "are far beyond what is expected by chance."
>
> The SAIC experiments are well-designed and the investigators have taken pains to eliminate the known weaknesses in previous parapsychological research. In addition, *I cannot provide suitable candidates for what flaws, if any, might be present.* Just the same, it is impossible in principle to say that any particular experiment or experimental series is completely free from possible flaws.[28]

In other words, as we have seen in the discussion of the ganzfeld-telepathy results, the archskeptic agreed that the results were not due to chance, or selective reporting, or statistical problems, or even to any *plausible* design flaws. He is then left with only one remaining refuge, which is to imply that there *must* be something wrong, presumably because the alternative—genuine psi—is too remarkable to consider. To Hyman's credit, he goes on to suggest, as he did for the ganzfeld-telepathy research, that

> It might be worthwhile to allocate some resources towards seeing whether these findings can be independently replicated. If so, then it will be time to reassess if it is worth pursuing the task of determining if these effects do indeed reflect the operation of anomalous cognition.[29]

So again, the key question is, Have these results been replicated?

PEAR Precognitive Remote Perception

The Princeton Engineering Anomalies Research (PEAR) Laboratory at Princeton University began conducting studies in remote viewing in 1978. The PEAR researchers preferred the term "precognition remote perception" (PRP) to reflect the fact that in many of their trials the targets were randomly selected *after* they were described. In addition, their method of analyzing results differed from the rank-order method used by SRI/SAIC, and their studies involved both a "percipient" (a remote viewer) and an agent (an individual known to the viewer) who visited a site in the field.

While there are differences in some aspects of the PEAR program, qualitatively the results are essentially the same as those reported by the SRI/SAIC program. For example, consider one of the PRP trials where a percipient and an agent were separated by twenty-two hundred miles. Some forty-five minutes before the agent randomly selected a site, the percipient described the following impressions:

> Rather strange yet persistent image of [agent] inside a large bowl—a hemispheric indentation in the ground of some smooth man-made materials like concrete or cement. No color. Possibly covered with a glass dome. Unusual sense of inside/outside simultaneity. That's all. It's a large bowl. (If it was full of soup [the agent] would be the size of a large dumpling!)[30]

The agent turned out to be visiting a large radio telescope at Kitt Peak, Arizona. If we did not know what a radio telescope was, the viewer's word portrait would provide a good description. A radio telescope resembles "a large bowl."

PEAR Method of Evaluation

For a PEAR PRP trial, the percipient was asked to write a short description of the geographic site where an agent *is, was,* or *would be* at a prescribed time. The percipient then filled out a thirty-question "descriptor" form. These descriptors asked whether the perceptive impression was mainly dark or light, mainly indoors or outdoors, whether there were animals in the scene or not, significant sounds or not, and so on.

Meanwhile, the agent typically spent from five to fifteen minutes at the target site, beginning at the prescribed time. He or she wrote down any impressions and filled out the same thirty-item descriptor form. In most cases, the agent also took photographs of the scene to provide a better record of the experience and for future reference.

The target sites were determined in two ways: *instructed* and *volitional*. In the instructed mode, a person not otherwise involved in the experiment randomly selected a site from a large pool of previously selected locations. An electronic random-event generator was typically used to make this selection. This information would be given to the agent sealed in an envelope, with instructions to open the envelope only after leaving the laboratory. With the volitional method, the target site was selected spontaneously by the agent, who was traveling at some distant location unknown to the percipient, and where no preestablished target pool existed.

With both methods, most of the remote perceptions were performed precognitively, before the agent arrived at the site, and even before a site was selected. Obviously, no communication was permitted between the percipient and the agent until both had completed their tasks.

Analytical Method

To analyze the results of a single trial, the researchers matched the percipient's descriptor list against the agent's descriptor list for the actual target.

They then compared the descriptor lists to all other targets in the entire database. This provides an objective, mathematically rigorous way of evaluating the likelihood of each individual trial. Some criticism has been directed at the PEAR PRP methods, primarily because in most cases the percipient and the agent knew each other, and because the percipient knew approximately when and where the agent was going to visit a site.[31] This shared knowledge might have biased one or both of the participants to fill in their descriptor lists in similar ways. The results of such shared knowledge could, in principle, inflate the score obtained in each trial. This same criticism was charged against Upton Sinclair's methods and some of the thought-transference experiments of the late nineteenth century.

In response, the PEAR researchers reanalyzed their data to see whether shared knowledge might have biased the results, especially in the volitional trials. They argued that if this bias were responsible for some extra hits on the descriptors, then it should have resulted in better results for the volitional trials than for the instructed trials. But no statistical differences were found.[32] Thus, while shared biases may have influenced the results in principle, the magnitude of any such bias was too small to be detected. Of greater importance is the finding, shown in figure 6.2, that the PEAR remote-perception studies produced essentially the same results as those seen by many other researchers over the years.

Of the 334 PRP trials that had been published as of 1987, 125 were in the instructed mode and 209 in the volitional mode. The final odds against chance for the PEAR researchers' overall database were 100 billion to 1. For the instructed trials alone, the outcome was a billion to 1, and for the volitional trials, 100,000 to 1. Thus, the results actually ran opposite to the shared-knowledge suggestion, with somewhat greater performance demonstrated in the instructed mode.

Psi Perception So Far

Figure 6.2 summarizes all the telepathy and clairvoyance experiments considered so far. Each experiment produced results far beyond chance, and surprisingly, the results are approximately the same in each case. The only notable differences are in the high-security ESP card tests and the "ordinary-state" ESP studies. Why were these latter studies different?

Ordinary-State Psi

British psychologist Julie Milton, from the University of Edinburgh, Scotland, examined all free-response psi experiments conducted in the "ordinary" state of consciousness, as opposed to studies employing altered states of consciousness such as dreams, the ganzfeld, and hypnosis.[33] Milton's analysis encompassed seventy-eight studies published from 1964 to 1993.

It included some of the early SRI remote-viewing experiments but excluded the SAIC studies, the PEAR PRP studies, and all the dream and ganzfeld experiments. The ESP card studies were not considered because they used a forced-choice design. In sum, the studies Milton considered were reported in fifty-five publications by thirty-five different investigators, and they involved 1,158 participants, most of whom were unselected volunteers.

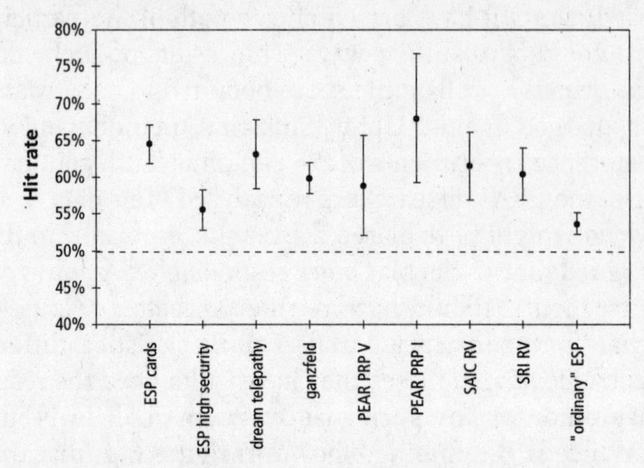

Figure 6.2. Summary results for all clairvoyance tests, with point estimates and 95 percent confidence intervals after the results were transformed into 50-percent-equivalent chance hit rates.[34] These tests include all ESP card experiments reported in 188 studies from 1882 to 1939, involving 4.6 million trials; a subset of high-security ESP card tests reported in 34 studies from 1935 to 1939, involving 907,000 trials; 450 dream-telepathy sessions; 2,549 ganzfeld sessions; 209 PEAR PRP volitional sessions; 125 PEAR instructional sessions; 455 SAIC remote viewings; 770 SRI remote viewings; and 2,682 "ordinary-state" ESP sessions. The label "PEAR PRP V" refers to the PEAR Laboratory's precognitive remote-perception studies in the "volitional" mode, and "PRP I" refers to the same studies in the "instructional" mode. The label "ordinary ESP" at the right end of the graph refers to clairvoyance studies conducted in the ordinary state of consciousness.

She found that the overall effect resulted in odds against chance of ten million to one. The reported effects did not significantly vary among the thirty-five investigators, and analysis of the file-drawer problem showed that 866 unsuccessful, unpublished studies would have to exist to eliminate the overall effect. Thus chance and selective reporting could not plausibly explain these results. Milton did find that two potential design flaws were related to larger psi scores, but the forty-eight studies free of those flaws still resulted in odds of forty thousand to one.

Of particular interest is that the resulting hit rate of 54 percent was quite a bit smaller than the overall effects observed in the other studies. Given the predicted psi-enhancing effects of experiments involving altered states of consciousness, such as dreams and the ganzfeld state, it is not surprising that the "ordinary-state" results were somewhat weaker. Henri Bergson may have been right: perhaps we *do* filter out psi impressions in the ordinary state of mind.

Ordinary and Nonordinary States

If conscious awareness is in fact the end of a long series of perceptual filters, then if we could bypass conscious awareness and gain access to more direct or primitive perceptions, perhaps we would find enhanced psi performance. One way of studying this expectation would be through the use of hypnosis to create a psi-conducive state. That is, what would happen if we used hypnosis to suggest to someone that psi would be experienced easily, safely, and comfortably?[35] Would that enhance clairvoyance?

In 1994, psychologists Rex Stanford and Adam Stein, from St. John's University in New York, published a meta-analysis of ESP studies contrasting the use of hypnosis and an "ordinary-state" condition.[36] They found twenty-nine relevant studies, of which twenty-five provided enough information to calculate the experimental outcomes. These were reported by eleven different investigators in publications appearing from 1945 to 1982. Twenty-three of the studies involved forced-choice methods, and two used free-response designs. Figure 6.3 shows the results. The hypnosis condition resulted in psi effects significantly greater than chance, with odds of twenty-seven hundred to one. By comparison, the "ordinary-state" condition resulted in odds that cannot exclude chance (eight to one). Thus, there is some evidence supporting the idea that by manipulating expectations and bypassing conscious awareness, we can improve psi.

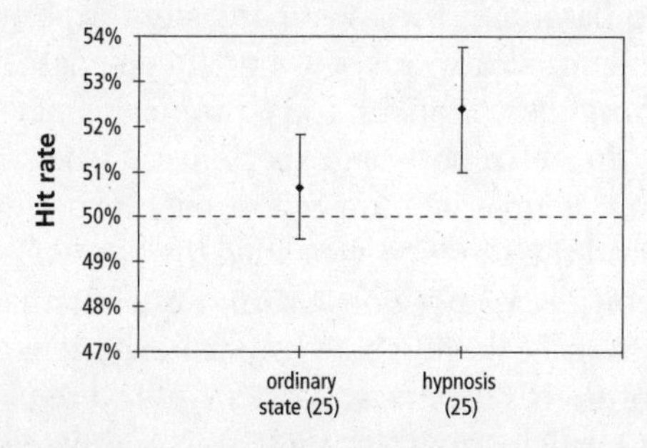

Figure 6.3. Results of ordinary-state experiments versus hypnosis experiments in terms of 50-percent-equivalent hit rates and 95 percent confidence intervals, with the number of experiments in parentheses.

Stanford and Stein then examined whether the psi effects were about the same across the different experimenters, and found that they were not. The eleven experimenters obtained results that were significantly different from each other, which raises two possibilities: these studies were not examining the same effect, or the investigators' methods and skills in inducing hypnotic suggestions differed dramatically. It is known, for example, that successful hypnosis requires strong interpersonal skills and extensive experience. In addition, people vary widely in the degree to which they can be hypnotized, and we do not know in these studies whether the participants were all equally "hypnotizable" (which can be measured by various techniques).

Sheep and Goats

One way of independently checking the results suggested by the hypnosis studies is to examine another form of suggestion, one that is in some ways stronger than conventional hypnotic induction. These are the subtle suggestions induced in us by our culture, our personal experiences, and the beliefs we learned from parents and schools. Together, culture, experience, and beliefs are potent shapers of our sense of reality. They are, in effect, hidden persuaders, powerful reinforcers of our sense of what is real. Our deep beliefs determine what we view as logically reasonable and what we consider to be morally and ethically self-evident.

As we'll explore in more detail in chapter 14, the hidden "hypnosis" of belief actually determines to a greater degree than is commonly known what we can consciously perceive. The hypnosis experiments showed that a slight tweaking of these beliefs resulted in a different performance. Thus, we would expect that people who accept the existence of ESP—for reasons of culture, experience, or belief—will score higher, on average, than people who do not.

This turns out to be one of the most consistent experimental effects in psi research. It was whimsically dubbed the "sheep-goat" effect by psychologist Gertrude Schmeidler, who in 1943 proposed that one reason that confirmed skeptics do not report psi experiences is because they subconsciously avoid them.[37] People who do report such experiences Schmeidler called the "sheep," and the skeptics she called the "goats."

These studies typically had people fill in a questionnaire asking about their degree of belief in ESP and about any psi experiences they may have had. On the basis of their responses, participants were classified as either sheep or goats. All participants then took a standardized psi test, like an ESP card test, after which the results of the sheep and goats were compared. The idea was that the performance of the sheep would be significantly better than that of the goats.

In 1993, psychologist Tony Lawrence from the University of Edinburgh, Scotland, reported a meta-analysis of all sheep-goat forced-choice experiments conducted between 1943 and 1993. Lawrence found seventy-three published reports by thirty-seven different investigators, involving more than 685,000 guesses produced by forty-five hundred participants. The overall results were strongly in favor of the sheep-goat effect, with believers performing better than disbelievers with odds greater than a trillion to one. Analysis of the file-drawer problem showed that it would require some 1,726 unpublished, nonsignificant studies for each published study to eradicate this effect. Thus, the file-drawer problem cannot explain this result. Nor did Lawrence find that the results could be explained by variations in the quality of the studies, or by the presence of a few studies with exceptionally large outcomes. He concluded that "The results of this meta-analysis are quite clear—if you believe in the paranormal you will score higher on average in forced choice ESP tests than someone who does not."[38]

Summary

We can draw three strong conclusions from ESP, remote-viewing, hypnosis, ordinary-state, and sheep-goat clairvoyance tests. First, these experiments exclude chance, selective reporting, and design flaws as alternative explanations. Second, some experiments have been replicated thousands of times by dozens of investigators from the 1880s to the present. And third, the psi effects measured across the various experiments are remarkably similar to one another.

The third conclusion is particularly important. While the methods, hypotheses, and purposes of the studies reviewed here were all somewhat different, each study examined the same underlying phenomenon—the ability to perceive objects and events at a distance, beyond the reach of the ordinary senses. We've seen that essentially the same effects have been repeatedly observed by dozens of investigators using different methods. This is why the late Carl Sagan agreed that some of the scientific evidence provided by psi experiments is persuasive enough to take these phenomena very seriously.

So, the evidence demonstrates that psi perception operates between minds and through space. This is troubling for many scientists, but not unimaginable. After all, it is possible that tomorrow someone will discover some sort of previously overlooked, supersensitive organ that might account for reports of telepathy, and maybe even for clairvoyance. But when we consider psi perception across *time,* all bets are off. The possibility that we can detect or, worse, be *influenced* by events in the future or past is so far beyond current scientific concepts that it staggers the imagination. But before we let our reeling imaginations stop us, let's examine the evidence for psi perception across time.

CHAPTER 7

Perception Through Time

People like us, who believe in physics, know that the distinction between past, present, and future is only a stubbornly persistent illusion.

ALBERT EINSTEIN

In previous chapters, we saw that the perceptual forms of psi are difficult to distinguish clearly in the laboratory. Telepathy in the lab, and in life, can be explained as a form of clairvoyance, and clairvoyance is difficult to localize precisely in time. Concepts like "retrocognition," "real-time clairvoyance," and "precognition" have arisen, blurring the usual concepts of perception and time. It seems that we must think of psi perception as a general ability to gain information from a distance, unbound by the usual limitations of both space *and* time.[1]

As long as we are interested in demonstrating the mere existence of perceptual psi, these conceptual distinctions do not matter. But when we try to understand how these effects are possible, the differences become critical. For example, it's important when theorizing about psi to know if it's actually possible to directly perceive someone's thoughts. Likewise, it's important to know if it's possible to perceive objects at a distance in real time.

Based on the experimental evidence, it is by no means clear that pure telepathy exists per se, nor is it certain that real-time clairvoyance exists. Instead, the vast majority of both anecdotal and empirical evidence for perceptual psi suggests that the evidence can all be accommodated by various forms of precognition. This may be surprising, given the temporal paradoxes presented by the notion of perception through time. But one simple way of thinking about virtually every form of perceptual psi is that we occasionally bump into our own future. That is, the only way that we personally know that something is psychic, as opposed to a pure fantasy, is because sometime in our future we get verification that our mental impressions were based on something that really did happen to us. This means that, in principle, the original psychic impression *could have been* a precognition from ourselves.

The Phenomenon

Nothing puzzles me more than time and space; and yet nothing troubles me less, as I never think about them.
CHARLES LAMB

Abraham Lincoln believed in prophetic omens. He told his friend and biographer, Ward H. Lamon, that shortly after the presidential election in 1860, he looked into a mirror and saw a double image of himself. Lincoln took this as an image of his future and understood that he would be elected to a second term but would die before the end of it.[2] The Cleveland *Plain Dealer* later published an account of Lincoln's beliefs about prophesies, and someone asked him if the newspaper's report was true. He replied, "The only falsehood in the statement is that the half of it has not been told. This article does not begin to tell the wonderful things I have witnessed."

Later, Lincoln reportedly told Lamon of a dream he'd had in which he heard people weeping as if their hearts would break. He could not see the mourners, so he followed the sound through the White House until he arrived at the East Room. "There," according to his account to Lamon, "was a sickening surprise. Before me was a catafalque on which rested a corpse wrapped in funeral vestments. Around it were stationed soldiers who were acting as guards." Lincoln asked the soldiers, "Who is dead in the White House?" They replied, "The president, he was killed by an assassin."

It is less well known that on the night of Lincoln's assassination, General Ulysses S. Grant and his wife, Julia, were supposed to accompany the president to Ford's Theater in Washington. This was an event of great honor, because a few days before, General Grant had accepted the unconditional surrender of the Confederate General Robert E. Lee, and Grant was enjoying a reception in his honor at the capital.

That morning, the day of the assassination, Mrs. Grant felt a great sense of urgency that she, her husband, and their child should leave Washington and return to their home in New Jersey. The general could not leave because he had appointments that day, but Mrs. Grant's sense of urgency grew throughout the day. Even though they were supposed to accompany President and Mrs. Lincoln to Ford's Theater, she insisted that they immediately leave for their home in New Jersey. She repeatedly sent word to her husband throughout the day, begging him to leave. He finally conceded, and when they reached Philadelphia, they heard the news about the assassination. They later learned that not only were they supposed to be sitting in the same box as the president, but they were also on actor John Wilkes Booth's list of intended victims.[3]

Thousands of such anecdotes scratch the surface of a rich lore of historical prophesies, premonitions, and forewarnings. Virtually every culture throughout history has developed techniques of divining the future, and many ancient mythologies are based on the inevitability of foretold destinies. While stories of accurate prophesies leave us in awe, we still need hard evidence that such things are really possible and not just fairy tales. This brings us back to the laboratory.

Forced-Choice Tests

In 1989 Charles Honorton and psychologist Diane Ferrari published a meta-analysis of all "forced-choice" experiments on precognition conducted between 1935 and 1987.[4] In a typical forced-choice precognition study, a person is asked to guess which one of a fixed number of targets will be selected later. The targets could be colored lamps, ESP card symbols, or a die face. Later, one target is randomly selected, and if the person's guess matches the selected symbol, this is counted as a "hit." In many such studies, immediately after the person guesses a symbol the target is randomly generated and presented as feedback.

Note that a test for precognition differs from a test for "psychokinesis"—mind-matter interactions—in only one essential way: Say that you toss a pair of dice, and while the dice are still in the air you wish or *intend* that you get a seven. This would be a psychokinesis test. Now say that you toss a pair of dice, and while the dice are still in the air you guess or *perceive* that you will get a seven. This would be a precognition test. In the first case you tried to *will* a certain result to occur; in the second case you tried to *perceive* what the result would be. The first is relatively active and the second is relatively passive, but the observable *outcomes,* at least with random systems like dice, are identical.

As in all psi experiments where the results depend on a clear definition of "chance expectation," the method of randomly selecting the future symbol is an important feature of these experiments. In early studies, decks of cards were shuffled by hand or machine; in later studies, electronic circuits were used to generate truly random numbers. The basic test is simple and the results are easy to interpret.

Honorton and Ferrari were mainly interested in studying three points: Was there any evidence for precognition? Was the effect related to variations in experimental quality? And did precognition performance vary with moderating variables, such as the type of subject population or the form of feedback employed?

Results

Honorton and Ferrari surveyed the English-language scientific literature to retrieve all experiments reporting forced-choice precognition tests. They found 309 studies, reported in 113 articles published from 1935 to 1987, and contributed by sixty-two different investigators. The database consisted of nearly two million individual trials by more than fifty thousand subjects. The methods used in these studies ranged from the use of ESP cards to fully automated, computer-generated, randomly presented symbols. The most frequently used participants were college students (in about 40 percent of the studies), and the least frequent were the experimenters themselves (in about 5 percent of the studies). People had been tested both individually and in groups.

The future targets were selected in many ways. Some studies used quasi-random methods relying on naturalistic events, like the average daily low temperatures recorded in a large group of cities located throughout the world. Other studies used informal methods such as dice tossing and card shuffling, or more formal techniques such as the use of tables of preprinted random numbers and electronic random-number generators. The time interval between the guesses and the generation of the future target ranged from milliseconds to a year.

The combined result of the 309 studies produced odds against chance of 10^{25} to one—that is, ten million billion billion to one. This eliminated chance as a viable explanation. The possibility of a selective-reporting bias—the file-drawer problem—was also eliminated by determining that the number of unpublished, unsuccessful studies required to eliminate these astronomical odds was 14,268. Further analysis showed that twenty-three of the sixty-two investigators (37 percent) had reported successful studies, so the overall results were not due to one or two wildly successful experiments. In other words, the precognition effect had been successfully replicated across many different experimenters.

Trimmed Analysis

Successfully replicating an effect does not mean that the results observed in different experiments will be *identical,* because there will always be some variations in study designs and participants. Instead, we would expect the results to be *about* the same, known in statistical terms as "homogeneous." In meta-analyses it is not expected that the effects observed in different studies will be homogeneous until the "outliers" are trimmed away. These are studies that for one reason or another produced wildly large or wildly small effects, possibly because of design problems, or the use of dramatically different procedures or personnel, or just by chance. In any case, to be sure that the same results have been replicated, a homogeneous set of effects is commonly created by trimming from the full set 10 percent of the studies producing the largest effects and 10 percent of the studies producing the smallest effects.

After performing this trimming, Honorton and Ferrari were left with 248 studies, and the total number of investigators was reduced from sixty-two to fifty-seven. But the combined effect of the remaining 80 percent of the data still produced odds against chance of a billion to one. This means that these fifty-seven investigators observed precognition effects that were effectively *the same,* and these effects could not be attributed to chance or to selective reporting.

Study Quality

Honorton and Ferrari identified eight elements of good experimental design for precognition studies. These included prespecifying how many samples would be collected, preplanning the method of statistical analysis, using proper randomization methods, and using automated recording. If there was a significant *negative* relationship between study quality and precognition performance, it would tend to support the critical assertion that better-designed studies produce smaller effects. But no such relationship was found for these studies; in fact, the actual relationship was slightly *positive* rather than negative.[5] Another question they asked was whether the extremely poor quality studies showed larger effects than the extremely high quality studies. Again, no differences were found.

A further question was whether study quality improved with time. This would be predicted, as investigators are expected to improve their experimental methods in response to criticisms. In fact, there was a significant positive relationship with odds against chance of ten billion to one.[6] However, even though study quality improved over time, the size of the precognitive effect did not. This indicates that the effects experimenters had been seeing repeatedly for more than half a century were remarkably stable.

Moderating Variables

To explore what factors might have moderated precognition, Honorton and Ferrari studied the performances of several classes of participants: unspecified, mixtures of several different populations, students, children, adult unselected volunteers, experimenters, and people selected on the basis of prior successful performance or some other special abilities. With one exception, the effect size did not appreciably vary among these different populations. Studies using selected participants produced larger effects than studies using unselected participants, with odds against chance of a thousand to one.

Another variable examined the type of feedback provided to the participants. Feedback ranged from none, to time-delayed, to feedback of the results of a series of trials, to trial-by-trial feedback. Trial-by-trial feedback tended to result in better performance, with odds against chance of about one hundred to one. In fact, some 42.6 percent of the studies with trial-by-trial feedback were successful (where only 5 percent would be expected by chance), and none of the studies without feedback were successful.

Honorton and Ferrari then examined the time interval between a participant's response and when the target was generated. Because trial-by-trial feedback had produced larger effects, they expected that shorter time intervals would also produce larger effects. After placing the studies in seven classes of time intervals (milliseconds, seconds, minutes, hours, days, weeks, and months), they did indeed find this relationship, with millisecond feedback being better than the other times, at odds against chance of just under one hundred to one. This relationship may have more to do with psychological factors than with any inherent limitations on how far precognition can "see." That is, in forced-choice tests if feedback is held off for more than a few minutes, the motivational effects of the feedback are lost. The participant is less able to remember what the feedback relates to, and it is expected that motivation to pay close attention to the task would diminish.

Based on their observations, Honorton and Ferrari predicted that selected participants who were tested individually using trial-by-trial feedback would perform better than unselected participants who were tested in groups using no feedback. The first set was called the "Optimal Group" and the second the "Suboptimal Group." As predicted, the Optimal Group performed significantly better than the Suboptimal Group, with odds against chance of a million to one. Seven of the eight Optimal studies were independently successful with odds of at least twenty to one, whereas none of the Suboptimal studies were successful.

The presence of significant moderating variables in these studies was important because it showed that precognition performance was not merely a statistical oddity but varied in ways that "made sense" psychologically. This suggests that there are lawful relationships about precognition performance that will be found through further research. And this in turn may lead to an improved understanding of how precognition works.

Unconscious Precognition

Another way to investigate precognition is by exploring the possibility that the mind is in contact with its own future state, or alternatively that the mind is slightly "spread out" in time. There are many interesting ways to test this idea. One is to see whether future perceptions interfere with present performance on reaction-time tasks, and another is to see whether future emotional states are detectable in present nervous system activity.

Reaction Time

In the early 1980s, Holger Klintman of the Department of Psychology at Lund University, Sweden, was studying a task where a person was shown a patch of color—red, green, blue, or yellow—followed by the name of a color, i.e., the words *red, green, blue,* or *yellow.*[7] Klintman asked the person to speak aloud the name of the patch of color as quickly as possible, and then speak aloud the word that followed as quickly as possible.

If the initial color patch *matches* the subsequent color name—if the color green is followed by the word *green,* for instance—the task can be done quickly and accurately. This is because as soon as the color patch is seen, associations about that color are activated in memory, including associations about the *name* of the color. So when a person is asked to say the name of a color, if the color name *matches* the color patch it is easy because the memory had already been "primed."

But if the initial color patch *mismatches* the subsequent color name—the color green followed by the word *red,* for example—then the task is surprisingly difficult because the mental gears set in motion by the naming of the color patch have to be overcome to correctly speak the mismatching color word. The experience of trying to perform the mismatch task quickly and accurately is exceptionally frustrating. Many people feel uncomfortable or start laughing uncontrollably when they try it. This task is often used to demonstrate the discomfort of cognitive interference.

Psychologists are usually interested in measuring the time it takes to speak aloud the *second* stimulus, the color name, after it is displayed. The prediction is that this reaction time will be faster when the initial color patches match the subsequent names, and slower when they mismatch. Of course, to avoid the powerful priming effects of expectation, the color name following the color patch must be determined randomly from one trial to the next. Neither the test participant nor the experimenter is allowed to know whether any given trial will be a match or a mismatch, so this is a classic double-blind experiment.

Klintman had been conducting this sort of conventional perceptual experiment for some time, measuring the reaction times to the second stimulus, when he decided to measure how much time it took people to speak aloud the color of the *first* stimulus, the color patch. He reasoned that he could use this first reaction time as a baseline or control reaction time, which he would then compare with the second reaction time. This would help him form a more sensitive measure of reaction time. Klintman was surprised to find more variation than he had expected in the first reaction times. He investigated further and was astonished to discover that the initial reaction times were faster when the color patch and the color name *matched,* and slower when the following color name *mismatched.*

After considering and rejecting all conventional explanations for this effect, Klintman decided to test the possibility of what he called "time-reversed interference." By this he meant that the person's precognitive sensing of the future stimulus somehow traveled back in time, causing cognitive interference when the future stimulus was a mismatch. Klintman guessed that interference from the future was causing the first reaction time to slow down.

He devised a double-blind experiment to test this idea, and ran twenty-eight subjects through his procedure. It produced odds against chance of sixty-seven to one in favor of the time-reversed interference hypothesis. So he designed another experiment, and again obtained successful results in favor of time-reversed interference. He ran another experiment, and another. After five successful experiments, each of which used a somewhat different design to provide conceptual replications of the same hypothesized effect, the combined results of the experiments resulted in overall odds against chance of 500,000 to 1. Klintman was satisfied that the observations were not caused by chance fluctuations, nor were they due to experimental artifacts.

He concluded that the effect was dependent upon the *meaning* between the two events; personality was an important variable in predicting who would show the effect; later trials produced poorer results than early trials, probably owing to a combination of fatigue and reduction of novelty; the effect was observed with ordinary student volunteers; the effect was relatively insensitive to the experimenter; and it was repeatable.

In addition, participants were entirely unaware that their performance was being affected by their own future perceptions, suggesting that unconscious nervous system activity may be used to detect precognitive perceptions. Studies relying on unconscious responses may be more effective than those relying on conscious responses by bypassing psychological defense mechanisms that may filter out psi perceptions from ordinary awareness.[8]

Future Feelings

In a recent series of experiments conducted in our laboratory at the University of Nevada, Las Vegas, we've explored unconscious nervous system responses to future events. Strictly speaking, such responses are a subset of precognition known as "presentiment," a vague sense or feeling of something about to occur but without any conscious awareness of a particular event.[9]

The unconscious responses studied in our experiments took advantage of a well-known psychophysical reflex known as the "orienting response," first described by Pavlov in the 1920s. The orienting response is a set of physiological changes experienced by an organism when it faces a "fight or flight" situation. For human beings, the response also appears in less dangerous contexts, such as when confronting a novel or unexpected stimulus. The classical orienting response is a series of simultaneous bodily changes that include dilation of the pupil, altered brain waves, a rise in sweat gland activity, a rise/fall pattern in heart rate, and blanching of the extremities.[10] These bodily changes momentarily sharpen our perceptions, improve our decision-making abilities, increase our strength, and reduce the danger of bleeding. This makes sense from an evolutionary perspective because when our ancestors were challenged by a tiger, the ones who survived were suddenly able to see and hear exceptionally well, make very fast decisions, become unusually strong, and not bleed as easily as usual.

It's relatively easy to produce an orienting response on demand by showing a person an emotionally provocative photograph. Stimuli like noxious odors, meaningful words, electrical shocks, and sudden tactile stimuli are also effective. Because a person's general level of arousal is affected cumulatively by successive stimuli, the strength of the orienting response tends to diminish after three to five emotional pictures in a row. In our study, to prevent participants from "habituating," we randomly interspersed the photos used to produce the orienting responses within a pool of twice as many calm photos.

Design

We had a participant, say "Pattie," sit in a comfortable chair approximately two feet from a color computer monitor.[11] On the first and second fingers of her left hand, we attached electrodes to record fluctuations in skin conductance, known as "electrodermal activity." On the pad of the third finger of her left hand, a device was attached to record both heart rate and the amount of blood in her fingertip. Signals from these electrodes were monitored by a computer.

After the electrodes were attached, Pattie rested her wired-up left hand comfortably in her lap. In her right hand, she held a computer mouse, and when ready to begin she pressed the mouse button. As illustrated in figure 7.1, this caused the computer to select one target photo at random out of a large pool of possibilities, but it showed only a blank screen. After five seconds of the blank screen, the selected photo was displayed for three seconds; this was followed by a blank screen for five seconds, and this was followed by another five-second rest period. After the rest period, a message informed Pattie that she could press the mouse button again whenever she felt ready for the next trial.

Meanwhile, during this eighteen-second recording period, the three physiological responses were continuously monitored. In this experiment, participants viewed forty pictures in a single session, one picture at a time. On each successive trial, the computer randomly selected one target photo from a pool of 120 high-quality digitized color photographs. The target photos were divided into two subjective categories, *calm* and *emotional*. Calm targets consisted of pleasant pictures of landscapes, nature scenes, and cheerful people. Emotional targets consisted of arousing, disturbing, or shocking pictures, including erotic photos and autopsies.[12]

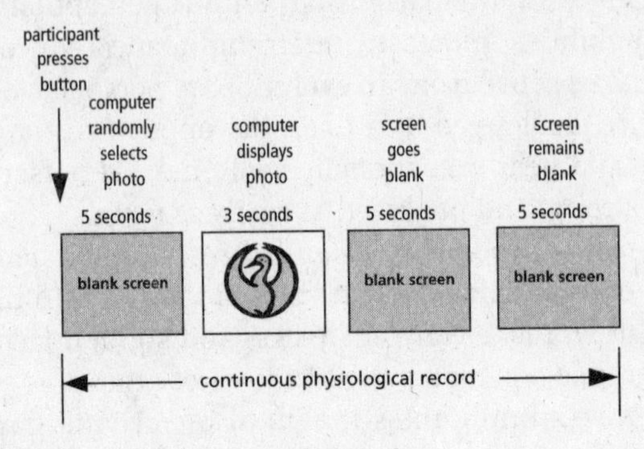

Figure 7.1. Illustration of experimental procedure.

The analytic technique applied to the data was called a "superposed epoch analysis." This is an averaging procedure in which the eighteen seconds of continuous physiological data from each trial (called a recording *epoch*) was averaged for all trials in which calm pictures were presented, then separately averaged for all trials in which emotional pictures were presented.

Results

Bypassing the mathematical and technical details, figure 7.2 shows the basic result for electrodermal activity for one female participant, "SD." As expected by the classical orienting response, shortly after seeing emotional targets, SD's electrodermal (i.e., sweat-gland) activity increased. After seeing calm targets, her electrodermal activity remained calm.

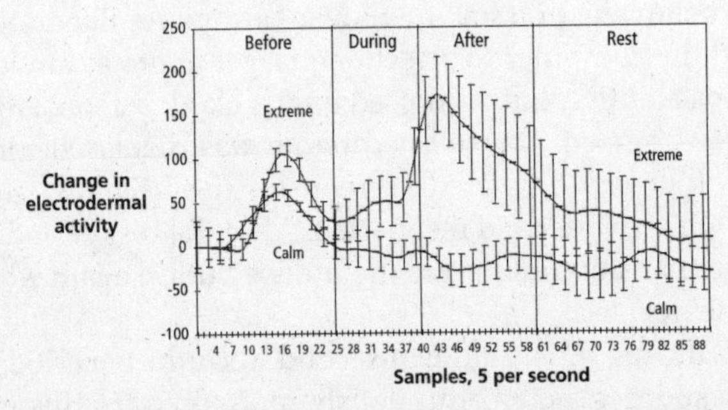

Figure 7.2. Superposed epoch analysis for one participant's electrodermal activity, with 65 percent confidence intervals. The four sections correspond to *before, during,* and *after* the target was displayed and a *rest* period. Presentiment is seen as higher electrodermal activity before the participant viewed the emotional targets.

Now we come to the interesting part. Before SD saw both types of pictures, her electrodermal activity began to rise, revealing that she was anticipating the subsequent target. What's surprising is that her electrodermal activity increased *more* if the future picture was going to be emotional. This difference is what we called a "presentiment effect," and it's a close analog in the autonomic nervous system to what Klintman found in reaction time: it appears that a person's "future" experience can affect his or her nervous system in the present. This is clear in figure 7.2, because there we can see the past, present, and future on a single graph.

Figures 7.3, 7.4, and 7.5 show the combined results of changes in electrodermal activity, heart rate, and finger blood volume for twenty-four participants who viewed a total of 900 pictures, 317 of which were emotional and 583 of which were calm.[13] Notice that this experimental design has a built-in control: the physiological results observed in the during-display and after-display conditions *must* reflect what is expected according to the orienting response; otherwise, something would be wrong with the analysis technique or with the measurements. Some people have idiosyncratic responses that do not follow the expected orienting response, and we can check this by seeing how they actually responded to the calm and emotional target pictures.

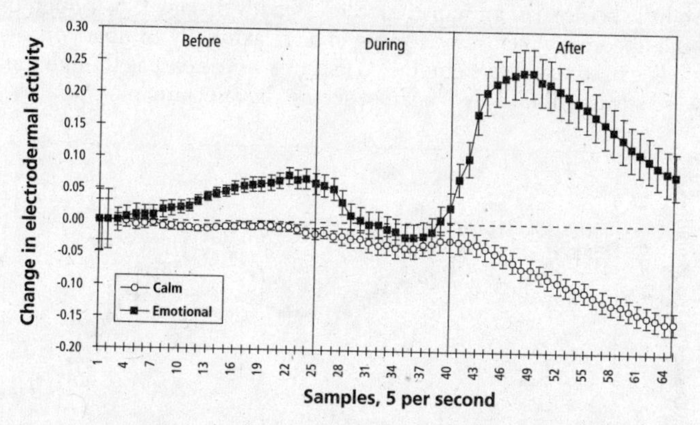

Figure 7.3. As expected by the classical orienting response, electrodermal activity after display of the target pictures was much higher for emotional pictures than for calm pictures. Electrodermal activity *before* display of emotional pictures was also higher than before display of calm pictures. This graph and figures 7.4 and 7.5 show the combined results for twenty-four participants who contributed a total of nine hundred trials in two experiments. Confidence intervals are 65 percent.

As expected by the classical orienting response, *after* the participants viewed emotional pictures, their autonomic nervous systems reflected the expected (average) reaction: heart rate dropped, blood volume in the finger dropped, and electrodermal activity increased. By comparison, responses to calm pictures just showed that they remained relaxed. These results confirmed that the experimental method was working as planned. The important observation was that *before* the emotional pictures were seen, the participants "pre-acted" to their own future emotional states. When asked after the experiment if they were consciously aware of the upcoming pictures, nearly all participants said no, supporting the idea that presentiment is largely an unconscious process.

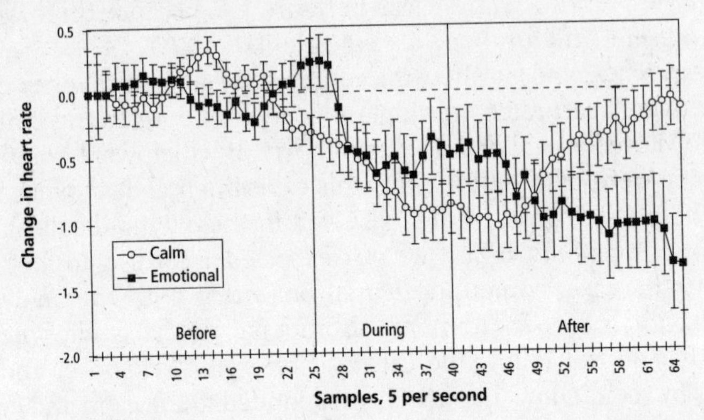

Figure 7.4. Before seeing the calm pictures, participants' average heart rate increased a little because of anticipation, then steadily dropped, as though they "knew" the upcoming picture was going to be relaxing. By comparison, average heart rate began to rise *before* emotional pictures were seen, as though the participants were steeling themselves against seeing the shocking picture.

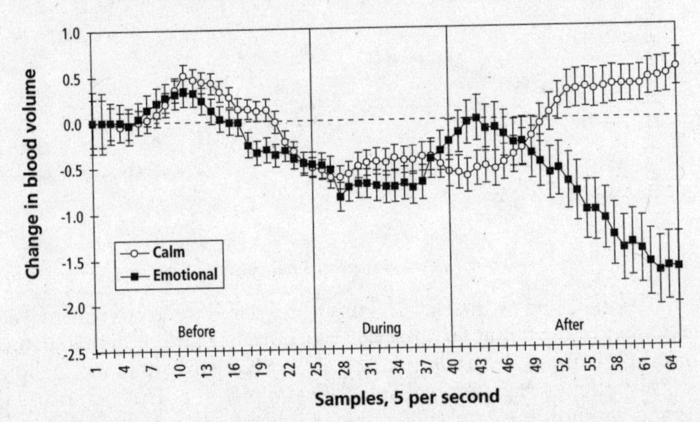

Figure 7.5. About two seconds after display of the target pictures, blood volume in the finger dropped for emotional pictures, but not for calm pictures. Notice also that blood volume dropped about one second *before* display of emotional pictures.

INDEPENDENT REPLICATIONS

As we've emphasized, in science the proof of the pudding is independent replications. After we reported these presentiment studies at the annual conference of the Parapsychological Association in August 1996, Professor Dick Bierman, a psychologist at the University of Amsterdam, attempted to replicate the experiment. We gave him copies of the target photos we had used, and he worked with his own electrodermal hardware and software.[14]

Bierman presented the target pictures to sixteen participants, each of whom saw forty pictures for either 3.0 seconds, which he called the "long" condition, or 0.2 seconds, which he called the "short" condition. The pooled results, shown in figure 7.6, successfully replicated the results that we had observed. Electrodermal activity was significantly higher before presentation of emotional pictures, as compared to the same activity before presentation of calm pictures.

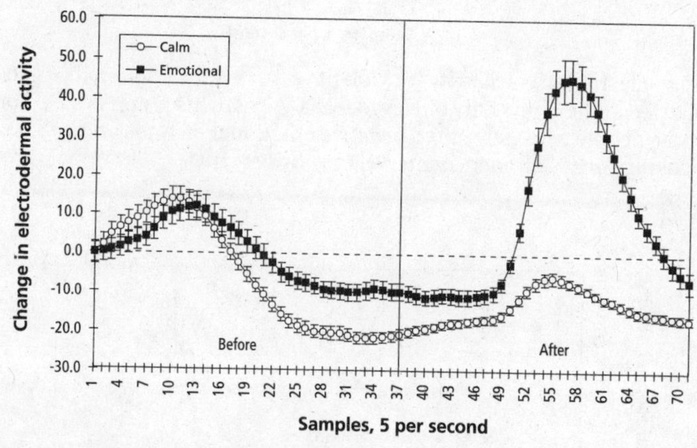

Figure 7.6. Results of Professor Bierman's replication experiment. Confidence intervals are 65 percent.[15]

To examine the presentiment results in more detail, we separated the emotional pictures into two types: positive (generally erotic themes) and negative (violent and injury themes). Figure 7.7 shows the results for participants run in our laboratory. We saw a clear difference in electrodermal activity before presentation of positive versus negative emotional pictures, and essentially no difference after presentation. Figure 7.8 shows the same analysis for the participants run by Professor Bierman, again indicating similar results. These differences would be difficult to account for by any form of "normal" explanation, such as some sort of ordinary anticipation effect. The findings also suggest—as would be expected if presentiment truly does reflect foreknowledge of future events—that the autonomic nervous system is not just "pre-acting" to a future *shock* to the nervous system, but is pre-acting to the emotional *meaning* of the future event.

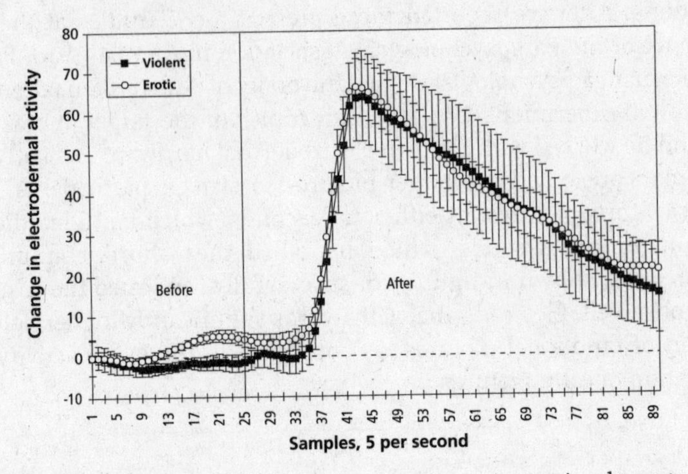

Figure 7.7. Electrodermal activity for violent versus erotic emotional targets in data collected at the University of Nevada, Las Vegas. This graph is based on thirty-three people who viewed 158 negative emotional pictures and 278 positive emotional pictures. Confidence intervals are 65 percent.[16]

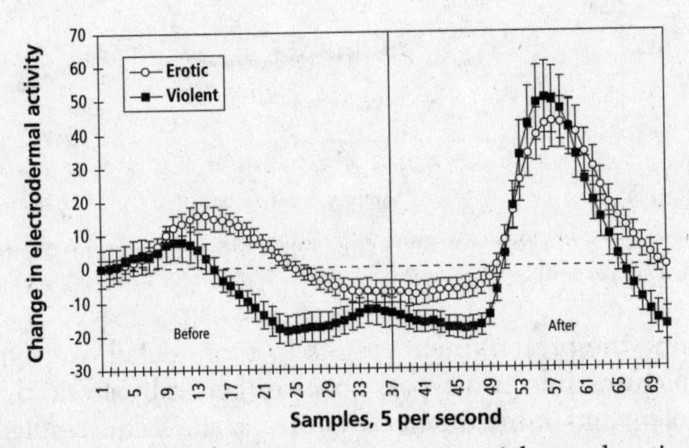

Figure 7.8. Electrodermal activity for people viewing violent and erotic emotional pictures in data collected by Professor Bierman, University of Amsterdam. Confidence intervals are 65 percent.

When Is the Present?

Let anyone try, I will not say to arrest, but to notice or attend to, the present moment of time. One of the most baffling experiences occurs. Where is it, this present? It has melted in our grasp, fled ere we could touch it, gone in the instant of becoming.
WILLIAM JAMES

William James may have been onto something. The present may not be where—or when—we think it is. Klintman's reaction-time studies and the physiological presentiment experiments confirm the results of the forced-choice precognition meta-analysis. They suggest that under certain circumstances we can consciously or unconsciously respond to events in our future, events that we have no normal way of knowing.

As this chapter was being written, neuroscientists from the University of Iowa College of Medicine reported an experiment on intuitive hunches in the prominent journal *Science*. They measured electrodermal activity in ten normal people and in six brain-damaged patients while they individually played a card game. The game involved four card decks and instructions to simply select a deck and turn over a card, one at a time. Some cards would result in winning money, and others in losing money. Two of the decks were "bad" in that they had a higher percentage of losing cards, and the other two were "good," having a higher percentage of winning cards.

Remarkably, without knowing that there were good or bad decks, or anything about the distribution of losing cards within each deck, both the normal people and brain-damaged patients "began to generate anticipatory [electrodermal activity] whenever they pondered a choice that turned out to be risky, *before they knew* explicitly that it was a risky choice."[17] In discussing this surprising effect, the researchers speculated that "The bias mechanism identified here is distinct from other neural mechanisms whose integrity is crucial for decision-making. . . . In other words, we propose an addition to mechanisms already recognized as necessary for proper reasoning. . . ."[18] Given the evidence for precognition described in this chapter, it is entirely possible that the additional mechanism—something beyond what conventional neuroscience has recognized so far—is psi.

Shortly after psi experiments began in the 1880s, it became apparent that differences between telepathy, real-time clairvoyance, and precognition were actually differences in *semantics*—our descriptions of the phenomena—rather than in any fundamental properties of psi perception itself. Then experiments began to be conducted on mind-matter interaction effects, and these seemed to be something completely different.

CHAPTER 8

Mind-Matter Interaction

Life is infinitely stranger than anything which the mind of man could invent. We would not dare to conceive the things which are really merely commonplaces of existence.

SHERLOCK HOLMES TO DR. WATSON

Does mental intention affect the physical world? In a trivial sense, the answer is obviously yes. An automotive engineer imagines a new way to build a car, and several months or years later it appears. This transformation from mental into physical is not considered remarkable because the sequence of events is well understood.

But a similar question can be asked that is no longer self-evident: does mental intention directly affect the physical world, without an intermediary? This question concerning the ultimate role of the human mind in the physical world has intrigued philosophers for millennia. Indeed, the concept that mind is primary over matter is deeply rooted in Eastern philosophies and ancient beliefs about magic. For the past few hundred years, such beliefs have been firmly rejected by Western science as mere superstition. And yet, the fundamental issues remain as mysterious today as they did five thousand years ago. What is mind, and what is its relationship to matter? Is the mind caused, or is it causal?

Consciousness

Speculations about the nature of consciousness have substantially increased in the last few years. Each discipline has its own views of what consciousness may be, and many articles have been contributed by neuroscientists, cognitive scientists, computer scientists, and biologists. In physics, the inescapable fact that the simple act of observation changes the nature of a physical system caused virtually all the founders of modern physics, including Werner Heisenberg, Erwin Schrödinger, and Albert Einstein, to think deeply about the strangely privileged role of human consciousness.[1]

A growing number of contemporary physicists have continued the tradition of speculations about consciousness, mind, and matter,[2] for some of the implications of modern physics are perplexing to say the least. As physicist Bernard d'Espagnat wrote in an article in *Scientific American*, "The doctrine that the world is made up of objects whose existence is independent of human consciousness turns out to be in conflict with quantum mechanics and with facts established by experiment."[3] Many other articles on this topic have been published in scientific journals, including the *American Journal of Physics, Physics Letters, Scientific American, Foundations of Physics,* and *Physical Review.*[4] A recent expression of the problem, which is directly relevant to psi research, can be found in a speculation about quantum theory by physicist Euan Squires, published in 1987 in the *European Journal of Physics:*

> If conscious choice can decide what particular observation I measure, and therefore into what states my consciousness splits, might not conscious choice also be able to influence the outcome of the measurement? One possible place where mind may influence matter is in quantum effects. Experiments on whether it is possible to affect the decay rates of nuclei by thinking suitable thoughts would presumably be easy to perform, and might be worth doing.[5]

Given the distinguished history of speculations about the role of consciousness in quantum mechanics, one might think that the physics literature would report a substantial number of original experiments on this topic. Surprisingly, a search revealed only three studies.

Experiments

The first was an article by MIT physicists Hall, Kim, McElroy, and Shimony, who in 1977 reported an experiment based upon "taking seriously the proposal that the reduction of the wave packet is due to a mind-body interaction, in which both of the interacting systems are changed."[6] Their experiment examined whether one person could detect if another person had previously observed a quantum mechanical event (in this case, gamma emission from sodium-22 atoms).

Their idea was based on the concept that if one person's observation actually changed the physical state of a system, then when another person observed the same system later, the second person's experience may differ according to whether the first has or has not looked at the system. Their results, based on a total of 554 trials, did not support the hypothesis. The observed number of "hits" obtained in their experiment was exactly the number expected by chance.

The second study was referred to by the MIT physicists. They mentioned that a previous experiment at MIT using radioactive cobalt-57 was successful, with forty hits out of sixty-seven trials, a 60 percent hit rate where chance expectation was 50 percent.[7]

The third study was a long-term investigation reported by Princeton University engineer Robert Jahn, psychologists Brenda Dunne and Roger Nelson, and their colleagues.[8] In 1986, they reported the results of millions of trials collected from thirty-three people during seven years of experimentation. They used electronic random-number generators, a type of electronic coin-flipper, as the physical target. These experiments, involving long-term data collection with unselected individuals, provided persuasive evidence of a relationship between mental intention and the output of these random physical devices.

Thus, of three relevant experiments reported in mainstream physics journals, one described results exactly at chance and two described positive effects. Given the fantastic theoretical implications of such an effect, it seems rather strange that no further experiments of this type can be found in the physics literature. But this is not to say that no such experiments have been performed. In fact, hundreds of conceptually identical experiments have been conducted by psi researchers.

Actually, it's not so strange that the mainstream physics literature mentions only three reports. Even though theoretical physicists have seriously discussed the possibility of mind-matter interaction, a scientific taboo about *empirically* studying such topics—referred to by Einstein as "spooky" effects at a distance—prevails and reflects a host of underlying assumptions about the way nature *ought* to work. We discuss the origins of these assumptions in chapter 15, but in any case, because of the insular nature of scientific disciplines and the general uneasiness about parapsychology, the vast majority of psi experiments are unknown to most scientists. In the past, a few skeptics conducted superficial reviews of this literature and alleged that they found flaws in one or two experiments, but no one bothered to examine the entire body of evidence.

Motivations

There are pragmatic as well as scientific and philosophical reasons for studying mind-matter interaction effects. In the next chapter, we'll explore some of the evidence for and implications of these effects in living organisms. Here, we're more interested in mind-matter interactions in *nonliving* systems.

One practical reason for wondering whether mind influences matter is the possibility that this form of "active" psi may be related to why computers and other complex machines sometimes fail. Why are some individuals extremely adept at handling machines, while others gain reputations for mysteriously causing them to break? Why are "meaningful failures" so commonplace in engineering circles that such instances are half-seriously called examples of "Murphy's Law"? Is it possible that, under certain circumstances, computer-system failures may be "psi-mediated"?[9]

This is not merely an academic question. It is becoming increasingly important to determine why complex systems sometimes fail. A growing number of critical applications—nuclear power plants, air traffic control, intensive care units, among others—depends completely on the proper functioning of computers. Great strides have been taken in the design of fault-tolerant computers, and today the causes of the great majority of computer-system failures can be traced to one of two broad categories: human factors and machine factors. Human factors include poor user-interface design, stressful work environments, logical or functional design errors, and software bugs.[10] Machine factors include circuit-board failures, power surges, and electromagnetic interference.[11]

Unfortunately, it's not possible to assign every failure to a known category.[12] While some unexplained cases can undoubtedly be solved with sufficient detective work, as computer systems become more complex, distributed, and interdependent, determining the ultimate cause of failure becomes much more difficult. In fact, recent work on nonlinear dynamic systems theory (part of the broader realm of complexity and chaos theory) indicates that there are severe limits on our ability to predict the future of supposedly deterministic systems, including computers.[13] Even specially designed, highly redundant, fault-tolerant computer systems sometimes fail in completely mysterious ways.[14] So, besides examining the known human and machine factors for possible sources of system failures, it is a good idea to explore a less well understood intermediary factor: direct human-machine interaction.

Gremlins and Angels

Some people are renowned for their ability to fix machines quickly. Others are prohibited from coming near electronic equipment during important demonstrations, for fear that the equipment will fail. Some psychologists have referred to the latter phenomenon as the "gremlin effect."[15] Indeed, the apparent tendency of things to go wrong at the worst possible time is so prevalent that in engineering circles Murphy's Law is regarded as a "first principle."

Many gremlin legends are undoubtedly a result of selective memory and superstition, but after we have sifted through the odd coincidences, a residue of anecdotes and a small body of research suggest that the "lab lore" may have some basis in fact. Among the hundreds of anecdotes about unusual human-machine interactions is an amusing story told by physicist George Gamow, who described the "Pauli Effect" as follows:

> It is well known that theoretical physicists are quite inept in handling experimental apparatus; in fact, the standing of a theoretical physicist is said to be measurable in terms of his ability to break delicate devices merely by touching them. By this standard Wolfgang Pauli was a very good theoretical physicist; apparatus would fall, break, shatter or burn when he merely walked into a laboratory.[16]

Other experimenters, such as Thomas Edison, were legendary for their ability to get complex laboratory apparatus to work inexplicably fast.[17] Of course, not all "computer gremlins" and "computer angels" are due to unexplained causes. For example, during a session of the Supreme Soviet in the 1980s, President Mikhail Gorbachev suggested that the 470 Deputies try out the new, automated voting system in the Kremlin, which came complete with giant projection screens suspended at both ends of the hall. Gorbachev gave the signal for the Deputies to vote, all eyes turned to the giant screens, and they saw . . . nothing. According to an article in *Time* magazine, Gorbachev reportedly said, "The machine doesn't work. Get out your old weapons,"[18] referring to the credentials cards that were traditionally used for voting in the Supreme Soviet. Later in the same meeting, the system was successfully demonstrated. This time the technicians remembered to turn on the power.

Stories about fickle machines run the gamut from the sublime to the ridiculous. One perfectly outrageous story concerns a Soviet supercomputer that supposedly electrocuted a man who beat it in a chess game. According to the story, the supercomputer was ordered to stand trial for the murder of a chess champion who was electrocuted when he touched the metal board that he and the machine were playing on. Soviet police inspector Alexei Shainev reportedly told reporters in Moscow, "This was no accident—it was cold-blooded murder."[19]

The decision to put the computer on trial "stunned legal experts" around the world, but the Soviets were convinced that the computer had "the pride and intelligence to develop a hatred for Gudkov [the chess champion], and the motive and means to kill him." The police investigator supposedly explained, "The computer was programmed to win at chess, and when it couldn't do that legitimately, it killed its opponent." He continued, "It might sound ridiculous to bring a machine to trial for murder, but a machine that can solve problems and think faster than any human must be held accountable for its actions."[20]

This story is ridiculous, yet in real life many people personify their automobiles and personal computers and even harbor secret feelings of affection or suspicion toward them. If one accepts the evidence for psi in mind-machine interactions, then a real Frankenstein or a real HAL from *2001: A Space Odyssey* begins to sound less like pure fantasy and more like a possibility.[21] Underlying these technological speculations remains the original question: is there any evidence that mind can *directly* influence matter? One of the first long-term, systematic investigations of this question involved the tossing of dice.

Active and Passive Wishing

Forty thousand years ago, our ancestors believed that destiny could be revealed by casting bones or influenced by sacrifice and prayer. The practice of astragalomancy, or divination by dice, was universally employed in ancient times, with evidence for "casting lots" ranging from African tribes, to the Inuit, to the Maya. The interrelated concepts of chance and destiny figured significantly in the beliefs of early peoples, as reflected, for example, in Shiva, the Hindu god who is often portrayed in statues throwing dice to determine humanity's fate.

Today, sophisticated men and women still "roll the bones" in casinos, and still fervently wish for favorable destinies. What does it mean to "wish"? One type of wish is *passive* in the sense of hoping that the fates will shine favorably upon us. The second type of wish is *active* in the sense that we mentally *intend* to force or entreat the fates to behave in a certain way. This distinction between active and passive wishes also extends to the concepts of good and bad luck. Some people view luck passively, as fate, or the action of favorable or unfavorable forces outside their control. Others view luck actively, as exercising their will upon the world. The same distinction is found in interpretations of psi experiments testing mind-matter interaction versus precognition. Some researchers see the results of mind-matter interaction experiments as being due to the participant simply selecting favorable moments to interact with the system through (passive) psi perception, while others see the results as evidence that the system is actually forced to behave in ways that conform to the participant's will.

While the differences between these two possibilities may seem obvious, in certain kinds of psi experiments a great deal of subtlety is required to distinguish between them. Just as no one (so far) has been able to design an experiment that will cleanly separate pure telepathy from clairvoyance, when the target of mental influence is a random system, no one has been able to design an experiment that will cleanly separate "pure" precognition from "pure" mind-matter interaction.

In fact, the closest anyone has come to testing pure mind-matter interaction is in the investigation of claims where something unusual happened that would *not* have happened by itself. The targets in such experiments are typically stable objects, which by definition do not spontaneously fluctuate.

To date, the main investigations on stable systems have involved mentally bending strips of metal, like spoons, or moving small objects. While claims of such spectacular events have captured the public's attention, unlike for most of the other experiments discussed here there haven't been any systematic, controlled replications of these claims by multiple investigators. So while some of the evidence for "macropsychokinesis" is interesting and probably warrants further study, in the critical court of science the jury is still out when it comes to the reality of metal bending and moving small objects.

Tossing Dice

Iacta alea est. (The die is cast.)

JULIUS CAESAR

Is it possible to control mentally how "the die is cast"? Many skeptics believe that the answer is no, for gambling casinos generally enjoy huge profits. Casinos are profitable, however, primarily because the odds are strongly stacked in favor of the house. With gambling wagers running in the billions of dollars each year, and gamblers always losing more than winning in the long run, the gambling industry is guaranteed healthy profits. Still, while gamblers who tend to win or lose consistently undoubtedly differ in innate mathematical abilities and memory skills, one wonders whether some of the consistent winners occasionally violate chance expectation (assuming a fair game) by taking advantage of some sort of direct mind-matter interaction.

Starting in 1935, researchers stimulated by the work of J. B. Rhine, Louisa Rhine, and their colleagues at Duke University began to test the idea that the fall of dice may be influenced by mental intention. Over the next half-century, some fifty-two investigators published the results of 148 such studies (in English-language publications).[22] The basic dice-tossing experiment is simple: A die face is prespecified, then a die (or group of dice) is tossed while a person "wills" that face to turn up. If the person's mental intention matches the resulting die face, a "hit" is scored. If more hits are obtained than expected by chance, this is taken as evidence for mind-matter interaction.

By 1989 dice experiments had been reviewed and criticized numerous times over the years, but in spite of all the experiments and reviews, no clear consensus had emerged.[23] Sustaining the controversy was a combination of beliefs that the mind-matter interaction effect was exceptionally difficult to replicate, and therefore the "effect" was highly suspect, and that behind the apparently straightforward dice-tossing task lies a bewildering array of pitfalls, any one of which could legitimately cast doubt on the experimental results. Another factor was that nearly all the reviews had focused on the presence of design flaws in a few experiments, rather than examining the entire empirical database.

Meta-analysis

In 1989 psychologist Diane Ferrari and I, then at Princeton University, used meta-analysis to assess the evidence for mind-matter interaction effects in dice experiments.[24] In our analysis of the evidence, we considered specifically whether mental intention had caused a prespecified die face to land face up after being tossed.

As in any meta-analysis, it was necessary to include every study that could be retrieved; otherwise, it is far too easy to allow personal biases to influence the selection of the "good" studies and leave behind the "bad." We examined all the relevant English-language journals for dice experiments up to 1989,[25] and for each study recorded the number of participants in the test, the die face they were aiming for, the total number of dice tossed, and so on. From this information, for each study we calculated a 50-percent-equivalent chance hit rate. In addition, for each study we marked the presence or absence of a series of thirteen quality criteria, such as whether the study employed automatic recording, whether witnesses were present, and whether control tests were performed.[26]

Our literature search located seventy-three relevant publications, representing the efforts of fifty-two investigators from 1935 to 1987. Over this half-century, a total of 2,569 people had attempted to mentally influence 2.6 million dice throws in 148 different experiments, and just over 150,000 dice throws in 31 control studies where no mental influence was applied to the dice. The total number of dice tossed per study ranged from 60 to 240,000; the number of participants per study ranged from 1 to 393.

Figure 8.1 summarizes the results of these studies by the year in which the experiment was conducted. The overall hit rate for all *control* studies (i.e., studies in which no one tried to influence the tossed dice) was 50.02 percent, and the confidence interval was well within chance expectation, resulting in overall odds against chance of two to one. But for all experimental studies, the overall hit rate was 51.2 percent. This does not look like much, but statistically it results in odds against chance of more than a billion to one.

Addressing the Criticisms

As we've seen, an often-cited criticism of combined experimental results is that the overall effects might have been due to only a few investigators who reported the bulk of the studies. For the dice experiments, the number of studies conducted per investigator ranged from one to twenty-one, with the majority of the investigators (64 percent) reporting one, two, or three studies. To test the criticism, we calculated the overall odds against chance only for those twenty-five investigators reporting three or fewer studies (totaling forty-two studies). The result remained highly significant, with odds against chance greater than a billion to one. So overall success was not due to a few exceptional investigators.

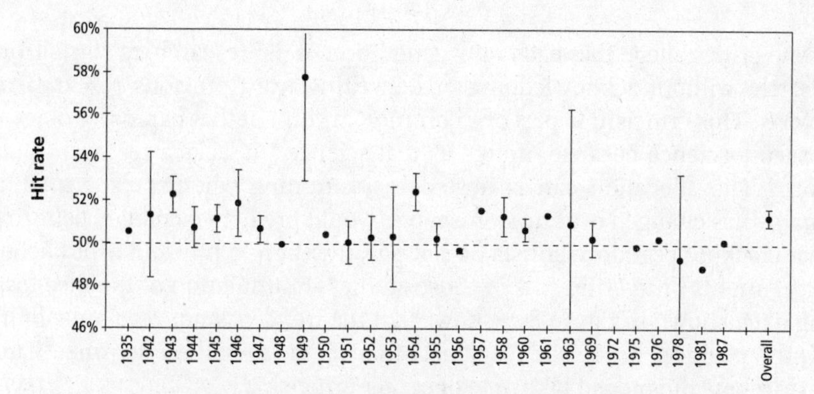

Figure 8.1. Fifty-percent-equivalent hit rates for all experimental dice-tossing studies, listed by year, with 95 percent confidence intervals. For years with a single study, the hit rate is indicated as a single point with no confidence interval. The overall combined hit rate is shown at the right.

But maybe the overall hit rate was enlarged by the results of a few extreme *studies,* perhaps because those studies were flawed in some way. To check this, we deleted the "outlier" studies by applying a standard trimming procedure, just as Honorton and Ferrari did for their precognition meta-analysis.[27] In our analysis, it was necessary to delete 52 studies (or 35 percent of the total 148 studies) to produce a homogeneous set of effects. Compare this 35 percent with exemplary studies in the physical sciences, where it is sometimes necessary to discard as much as 45 percent of the data to achieve a homogeneous distribution (as discussed in chapter 4). The overall effect observed in the remaining 96 studies still resulted in odds against chance of more than three million to one. Therefore, the experimental effect was independently replicable even when outliers were discarded, meaning that essentially the *same* effect had been *repeatedly observed* in 96 studies.

But maybe the successful experiments were published more often than nonsignificant studies, and that's why we saw such large results. To assess the effect of the "file drawer" of nonsignificant studies, we calculated the number of unpublished, unsuccessful studies that would be needed to reduce the observed odds down to odds of less than twenty to one. For these experiments, the file-drawer number was 17,974. That is, for *each* study we found, 121 additional, unretrieved, and unsuccessful studies would have been required to nullify the observed effect. That many studies would have required each of the fifty-two investigators involved in these experiments to have conducted one unpublished, nonsignificant study per month, every month, for twenty-eight years. This isn't a reasonable assumption; thus, selective reporting cannot explain these results.

Quality

Some critics allege that each new generation of psi researchers starts from scratch, without acknowledging or benefiting from previous researchers' efforts. This criticism is part of a common argument that parapsychology is a pseudoscience because, unlike a "real science," it lacks a research tradition.[28] The allegation can be tested by examining whether experimental quality has changed over time. A skeptic would predict no change, believing that crackpot pseudoscientists do not pay attention to previously published experiments or to criticisms of those studies. Examination of the dice-tossing data, shown in figure 8.2, shows that the trend was not zero but significantly positive, with odds against chance of a million to one. Later researchers did indeed take note of earlier criticisms.

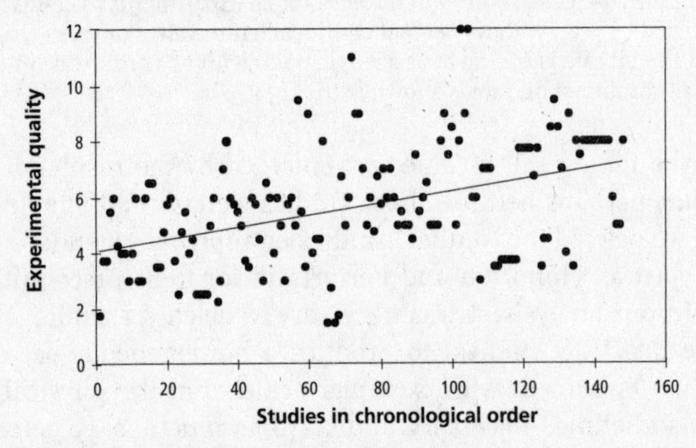

Figure 8.2. Quality of dice-tossing experiments significantly improved over time, with odds against chance of greater than a million to one.

Skeptics have also suggested that experimental effects will plummet as experimental quality improves. This critique reflects the assumption that if the "perfect" experiment is conducted, the results will reflect the true state of affairs, namely, that psi does not exist. We tested this argument by looking at the relationship between hit rates (in this case, averaged by year) and the study quality averaged per year. We found that the relationship was essentially flat, so the critique is not valid.

Die-Face Analysis

In the 1930s, J. B. Rhine and his colleagues recognized and took into account the possibility that some dice studies may have been flawed because the probabilities of die faces are not equal. With some dice, it is slightly more likely that one will roll a 6 face than a 1 face because the die faces are marked by scooping out bits of material. The 6 face, for example, has six scoops removed from the surface of that side of the die, so it has slightly less mass than the other die faces. On any random toss, that tiny difference in mass will make the 6 slightly more likely to land face up, followed in decreasing probability by the 5, 4, 3, 2, and 1 faces. Thus, an experiment that relied exclusively upon the 6 face as the target may have been flawed because, unless there were also control tosses with no mental intention applied, we could not tell whether above-chance results were due to a mind-matter interaction or to the slightly higher probability of rolling a 6.

To see whether this bias was present in these dice studies, we sifted out all reports for which the published data allowed us to calculate the effective hit rate separately for each of the six die faces used under experimental and control conditions. In fact, the suspected biases were found, as shown in figure 8.3. The hit rates for both experimental and control tosses tended to increase from die faces 1 to 6. However, most of the experimental hit rates were also larger than the corresponding control hit rates, suggested something interesting beyond the artifacts caused by die-face biases. For example, for die face 6 the experimental condition was significantly larger than the control with odds against chance of five thousand to one.

Because of the evidence that the die faces were slightly biased, we examined a subset of studies that controlled for these dice biases—studies using design protocols where die faces were equally distributed among the six targets. We referred to such studies as the "balanced-protocol subset."

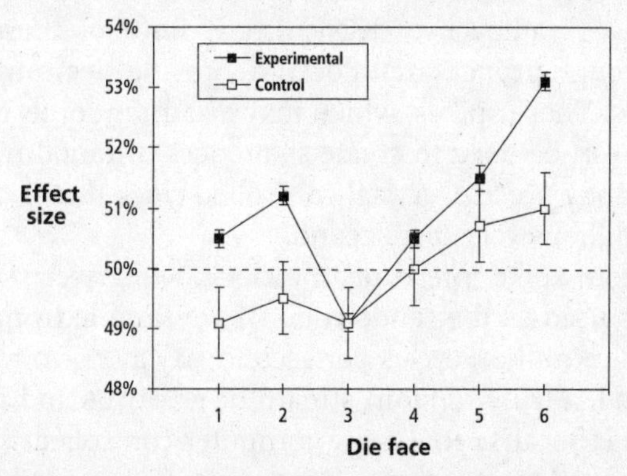

Figure 8.3. Relationship between die face and hit rates for experimental and control conditions. The error bars are 65 percent confidence intervals.

Sixty-nine experiments met the balanced-protocol criteria. Our examination of those experiments resulted in three notable points: there was still highly significant evidence for mind-matter interaction, with odds against chance of greater than a trillion to one; the effects were constant across different measures of experimental quality; and the selective-reporting "file drawer" required a twenty-to-one ratio of unretrieved, nonsignificant studies for each observed study. Thus chance, quality, and selective reporting could not explain away the results.

Dice Conclusions

Our meta-analysis findings led us to conclude that a genuine mind-matter interaction did exist with experiments testing tossed dice. The effect had been successfully replicated in more than a hundred experiments by more than fifty investigators for more than a half-century. If all this was so, then we might reasonably expect that there ought to be corroborating evidence from other experiments, using other types of physical targets. And there is.

Tossing Bits

Experiments involving random-number generators (RNGs) are the modern equivalents of dice studies. An RNG is an electronic circuit that creates sequences of "heads" and "tails" by repeatedly flipping an electronic "coin" and recording the results. A participant in a typical experiment is asked to mentally influence the RNG's output so that in a sequence of predefined length, it produces, say, more "heads" than "tails." Actually, most RNGs produce sequences of *bits* (the numbers 1 and 0); thus a person's task usually involves wishing for an RNG to produce more 1's or more 0's, depending on the instructions.

Modern RNG circuits usually rely upon one of two random sources: electronic noise or radioactive decay times. Both of these are physical sources that, through proper circuit design, provide electronic spikes at unpredictable times. These spikes, which may occur randomly a few thousand times a second, can be used to create sequences of random bits by having the spike interrupt a precise, crystal-controlled clock that is counting at the rate of, say, 10 million cycles per second.

When a random spike interrupts the clock, whichever state the clock is in ("1" or "0") is used as the random bit. If we sample from the clock at a slower rate than 10 million cycles per second, say at 1,000 randomly timed spikes per second, a truly random stream of 1,000 1's and 0's can be produced per second. Because RNGs are computer-controlled, even at a rate of 1,000 bits per second the random sequence can be recorded perfectly.

Participants in these tests often get feedback about the distribution of random events in the form of a digital display, audio feedback, computer graphics, or the movement of a robot's arms.[29] Most modern RNGs are technically highly sophisticated, employing features such as electromagnetic shielding, environmental fail-safe alarms, and fully automatic data recording.[30]

On Randomness

The RNG experiment is sometimes called a test of "micropsychokinesis," meaning mind-matter interaction on a very small scale. Why are microscopic random systems used as experimental targets instead of stable macroscopic objects, such as bars of metal? Surely, if the mind is able to influence matter, we could just observe the effect directly—someone mentally bending a spoon, for example—and avoid the use of statistics altogether. There are four parts to the answer. First, substantial laboratory experience shows that reliable detection of large-scale mind-matter interactions in nonliving systems is extremely rare. There are a few cases, but the vast majority of the evidence is anecdotal, or was collected under uncontrolled conditions. Second, even stable macroscopic systems like heavy weights and pendulums randomly fluctuate in microscopic ways. Thus any experiment involving extremely precise measurements ultimately has to rely on statistical methods. Third, the RNG experiment was historically designed to refine experiments involving the use of dice, not to look for large-scale effects. And fourth, random systems are psychologically "easier" to influence mentally than are massive objects, because such influence does not violate any physical conservation laws. That is, the behavior of a random physical system is defined not by the outcome of a single event but by the collective behavior of the entire system.

This last point is important. It leaves room in the behavior of single events for unusual things to happen without violating the overall behavior of the system as a whole. Since most RNG experiments consist of relatively short periods of data collection, unusual events may occur during those periods without violating the long-term stability of the RNG itself. Ever since the advent of probabilistic physical theories (such as quantum mechanics, stochastic electrodynamics, statistical mechanics, and thermodynamics), it has been recognized that physical laws are fundamentally statistical; they are based on *tendencies* of events, not on certainties. So there are no longer any absolutes that can "violate the law." Uncommon events, like the spontaneous "unmixing" of cream from a cup of coffee, may be unusual, but they are not physically impossible.

The main advantage of RNGs over dice is that the RNG is more amenable to fully automated, fast data collection and analysis. In addition, the source of randomness in some RNGs can be traced directly to quantum mechanical uncertainties. This has allowed physicists to explore quantum interpretations of observer effects in physical systems that would theoretically look like what we call micropsychokinesis. Modern RNG studies were in fact pioneered by physicist Helmut Schmidt, who began these studies in the 1960s when he was with Boeing Laboratories.[31] Today, most RNG experiments are based on Schmidt's original ideas and are completely automated, including the presentation of instructions, the provision of feedback on a trial-by-trial basis, and data storage and analysis. This automation eliminates the possibility that experimenters may inadvertently contaminate the data set; it also allows experimenters to act as participants in their own studies without fear of introducing recording biases into the data.

Meta-analysis

Assuming that an RNG is designed to generate random sequences of 0's and 1's, then the null hypothesis (that is, the idea that mind-matter interaction does *not* exist) is equivalent to observing an average chance hit rate of 50 percent. If a collection of all similar RNG studies produces an average effect that is greater than 50 percent with odds greater than chance, then we know that something interesting is going on.

In 1987, Princeton University psychologist Roger Nelson and I conducted a comprehensive meta-analysis of the RNG experiments.[32] All experiments in the meta-analysis asked the same question: is the output of an electronic RNG related to an observer's mental intention in accordance with prespecified instructions? In each experiment, if the outcome was in accordance with the mental "aim," this was assigned an appropriately measured hit rate of greater than 50 percent. If the outcome was opposite to the mental aim, this resulted in less than 50 percent. The chance expected outcome was, of course, exactly 50 percent.

Quality Assessment

In our meta-analysis, we assigned each experiment a single quality score derived from a set of sixteen criteria. These criteria were developed from many published criticisms about RNG experiments. The quality criteria assessed the integrity of the experiment in four general categories—procedures, statistics, the data, and the RNG device—and they covered virtually all design criticisms ever raised about RNG experiments. The criteria were similar to those used for the dice tests, including factors such as whether control tests were conducted, whether data were automatically recorded and double-checked, and whether a tamper-resistant RNG was used.

Results

From a wide range of sources, we found 152 references dating from 1959 to 1987. These reports described a total of 832 studies conducted by sixty-eight different investigators, including 597 experimental studies and 235 control studies. Of the 597 experimental studies, 258 were reported in a long-term investigation generated by the Princeton University PEAR laboratory, which also reported 127 of the control studies.

The overall experimental results produced odds against chance beyond a trillion to one. Control results were well within chance levels with odds of two to one. In terms of a 50 percent hit rate, the overall experimental effect, calculated per study, was about 51 percent, where 50 percent would be expected by chance. Point estimates for these results (excluding the single, long-term PEAR experiment) are shown averaged by year in figure 8.4. The right-most line in the figure estimates the overall effect of just under 51 percent for all studies combined.

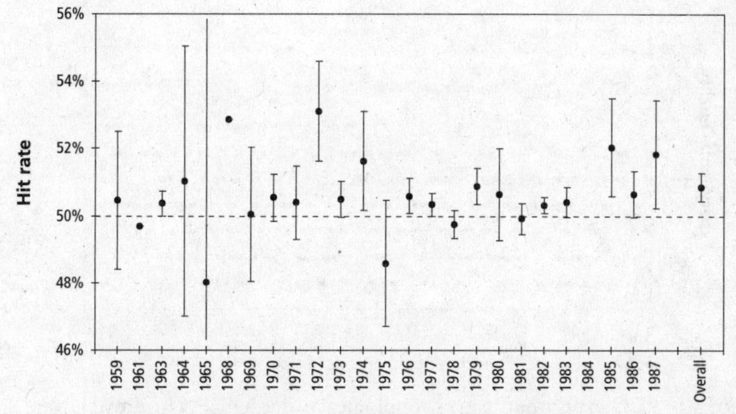

Figure 8.4. Yearly hit-rate point estimates and 95 percent confidence intervals for RNG studies of mind-matter interaction. In some cases the confidence intervals are so small that they are obscured by the point-estimate dots.

Now, we can directly compare the overall results of the dice experiments and the RNG experiments, as shown in figure 8.5.[33] Both experimental (E) and control (C) results are shown. We see that the dice study results and the RNG study results are remarkably similar, suggesting that the same mind-matter interaction effects have been repeatedly observed in nearly five hundred dice and RNG experiments for more than five decades.

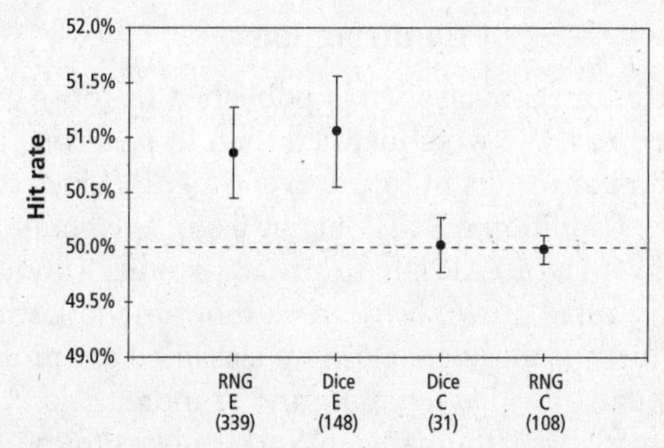

Figure 8.5. Hit-rate point estimates and 95 percent confidence intervals for dice and RNG studies of mind-matter interaction. The number of experiments is shown in parentheses.

Addressing the Criticisms

As observed in the dice studies, and shown here again in figure 8.6, the experimental quality of the RNG studies improved over time. These quality improvements demonstrated that researchers were paying close attention to one another, and to previous skeptical criticisms.[34]

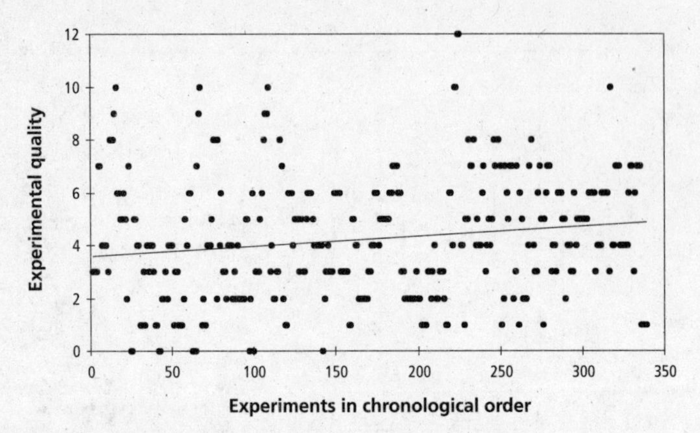

Figure 8.6. RNG experiments in chronological order. Quality improved over time with odds against chance of five hundred to one.[35]

We also tested the assertion that as experimental quality improved, effect sizes would decrease, ultimately declining to the "true" value of zero (meaning, no psi). We found that the observed hit rates were *not* related to experimental quality.[36] Finally, we found that the number of unreported or unretrieved RNG studies required to reduce the RNG psi effect to a nonsignificant level was 54,000—about ninety times the number of studies actually reported.

Confirmations

The preceding RNG meta-analysis was published in 1989, and the results at that time were clear. We were justified then in *predicting* that we would continue to see similar results in future experiments. Since then, the greatest accumulation of additional RNG data by a single laboratory has been at the Princeton University PEAR lab. From 284 studies conducted in 1989, its database in 1996 had grown by nearly a thousand to 1,262, contributed by 108 people. Thirty of these people were classified as "prolific" contributors because they had provided ten thousand or more trials.

Princeton University mathematician York Dobyns found that the seven years of new PEAR RNG results closely replicated the preceding three decades of RNG studies reviewed in the meta-analysis.[37] That is, our 1989 prediction had been validated. Because the massive PEAR database provides an exceptionally strong confirmation that mind-matter interactions really do exist, we can confidently use it to study some of the factors influencing these effects.

Psychologist Roger Nelson and his colleagues found that the main RNG effect for the full PEAR database of 1,262 independent experiments, generated by 108 people, was associated with odds against chance of four thousand to one.[38] He also found that there were no "star" performers—this means that the overall effect reflected an accumulation of small effects from each person rather than a few outstanding results from "special people." This finding confirms the expectation that mind-matter interaction effects observed in the hundreds of studies collected in the 1989 RNG meta-analysis were part of a widespread ability distributed throughout the population, and were not due to a few psychic "superstars" or a few odd experiments. Further analysis of the PEAR data showed that the results in individual trials were best interpreted as small changes in the probabilities of individual random events rather than as a few instances of wildly large effects. This means that the results cannot be explained by unexpected glitches in the RNG devices, or by strange circumstances in the lab (like a circuit breakdown). Rather, the effects were small but consistent across individual trials, and across different people.[39]

If we accept that one person can affect the behavior of an RNG, another question naturally arises: would two people together produce a larger effect? The PEAR database included some experiments where cooperating pairs used the same mental intention on the same RNG. Analysis of these data found that, on average, the effects were indeed larger for pairs than for individuals working alone. However, two people didn't automatically get results that were twice as large as one person's results. Instead, the composition of the pairs was important in determining the outcome. Same-sex pairs, whether men or women, tended to achieve null or slightly negative outcomes, whereas opposite-sex pairs produced an effect that was approximately twice that of individuals. Moreover, when the pair was a "bonded" couple, such as spouses or close family members, the effect size was more than four times that of individuals.

There were also some gender differences. PEAR lab psychologist Brenda Dunne found that women tended to volunteer more time to the experiments, and thus they accumulated about two-thirds of the full database, compared with one-third for men. On the other hand, their effects were smaller on average than those of men, with odds of the difference being due to chance at eight hundred to one.[40]

The PEAR database also allowed an examination of what happened when the same individuals tried the same experiment many times. An analysis demonstrated a clear relationship between the size of the effect and the sequence in which the experiments were conducted.[41] The first experiment typically resulted in a large effect. This declined in the second and third replications, followed by a recovery of a significant effect that seemed to plateau at a stable performance level for that person. This pattern of high initial result, followed by a decline, followed by a rise back to a stable measure, is reminiscent of the so-called decline effect often observed in the ESP card test results. This same pattern also resembles how changes in motivation modulate performance in many other types of human skill.[42]

Another factor investigated in the PEAR studies, and replicated in a few studies in the rest of the literature, was the effect of distance between the participant and the RNG itself. In RNG experiments, the person is usually located near the RNG. But in about one-fourth of the PEAR database, and in a dozen or so experiments from other laboratories, the participants were in "remote" locations ranging from an adjacent room to thousands of miles away from the RNG. Analysis of the effect of distance found no systematic differences; the results of local and remote experiments were indistinguishable—essentially no decline in effects as a function of distance.[43] A smaller subset of the PEAR studies was also conducted with time delays, where the participant made his or her efforts *before* the RNG data was generated. Again, there was no falloff in scoring related to the timing of the mental effort. Other investigators, notably physicist Helmut Schmidt, have reported similar evidence for time- and space-independence in RNG experiments.[44]

Mind-Matter Summary

After sixty years of experiments using tossed dice and their modern progeny, electronic RNGs, researchers have produced persuasive, consistent, replicated evidence that mental intention is associated with the behavior of these physical systems.[45] We know that the experimental results are not due to chance, selective reporting, poor experimental design, only a few individuals, or only a few experimenters. We are now beginning to see how the magnitude of the effects varies under differing conditions.

It's interesting to note how skeptical responses to these experiments have evolved over the six decades they have been reported. From the 1930s to the 1950s, the standard skeptical view was that the results were plainly impossible. This opinion was based on the shortsighted assumption that the scientific models of the day were unconditionally correct. And this meant that any apparently successful psi studies could be explained only by chance, fraud, or design flaws.[46] Skeptics directed basically the same criticisms at the results of psi perception experiments.

As decades passed, and as improvements in experimental design addressed testable criticisms, replications continued to show small but persistently successful outcomes. More sophisticated criticisms arose, and were resolved in subsequent studies, until today virtually no serious criticisms remain for the best RNG experiments. Informed skeptics agree that *something* interesting is going on.

Of course, disagreements still arise over interpretations of the data, but after hundreds of replications, the existence of an interesting effect has been demonstrated beyond a reasonable doubt. We have reached the point where, as Sir Arthur Conan Doyle said via the formidable Sherlock Holmes, "When you have eliminated the impossible, whatever remains, however improbable, must be the truth."

Independent Observers

After all the skeptical criticisms have been addressed, one always remains, like a smirk on a fading Cheshire cat: the results were due to fraud, or to a massive collusion among the hundred or so investigators reporting these results. This degree of paranoia hardly seems worthy of a response, but a method of addressing even this last criticism was developed by physicist Helmut Schmidt and psychologists Robert Morris and Luther Rudolph from Syracuse University.[47] Their clever RNG experiment involved the participation of skeptical third-party observers.

In these studies, allegations of collusion, fraud, and motivated inattention were no longer possible because the independent observers played critical roles in the experiments: they selected which trials would be subjected to mental effort and which would be the controls. If such studies were successful, then fraud or collusion would have to extend beyond the experimenters and include the independent observers! At some point, when the line is crossed from rational discourse to paranoia, it is time to end the debate.

In 1993, Helmut Schmidt, the "father" of the modern RNG studies, wrote of a successful study conducted using the third-party experimental design: "The present study confirms the existence of the [psi] effect under particularly well-controlled conditions where the participation of independent observers precludes experimenter error, or even fraud."[48] These third-party-observer RNG studies have now been replicated five times, resulting in overall odds against chance of twelve thousand to one.[49] Because of such research, RNG studies have begun to attract the attention of mainstream physicists. For example, theoretical physicist Henry Stapp of the University of California at Berkeley published an article in 1994 attempting to show how mind-matter interaction effects might be consistent with a generalization of quantum theory.[50]

Stapp's paper was published in the prominent journal *Physical Review,* and it attracted the attention of the British science magazine *New Scientist.* A reporter asked Stapp why, given the heresy of psi and the double heresy of meddling with quantum theory, he wrote the paper. The article reports that "Stapp, while acknowledging the inherent heresy of the idea, thought it worth further consideration. One reason for his interest was that he acted as an independent monitor of Schmidt's experiments."[51]

The study Stapp referred to resulted in odds against chance of one thousand to one. As we have seen, many RNG experiments have provided similar outcomes, and the overall results for all RNG studies exceed odds of trillions to one. Perhaps other theorists are not attempting to explain these results, even if they accept them as real, because they do not see why anyone would be interested in microscopic statistical changes in the behavior of electronic circuits.

Perhaps interest will increase when enough people realize that these RNG experiments merely scratch the surface of what appears to be a more fundamental phenomenon. Imagine, for example, that the tiny effects observed with RNGs also influenced living organisms. What if instead of asking someone in the lab to influence an inanimate electronic circuit, we asked him or her to mentally affect a distant person's nervous system? What then?

CHAPTER 9

Mental Interactions with Living Organisms

Prayer is not an old woman's idle amusement. Properly understood and applied, it is the most potent instrument of action.

MOHANDAS K. GANDHI (1869–1948)

Mental healing techniques, including prayer and spiritual healing, are as popular today as they have been throughout history and across all cultures. A January 1992 cover story on prayer in *Newsweek* reported that more than three-quarters of all Americans pray at least once a week. Even among the 13 percent who are atheists or agnostics, nearly one in five prays daily, "siding, it seems, with Pascal, and wagering that there is a God who hears them."[1] A March 1994 cover story on prayer in *Life* magazine reported that "Nine out of 10 Americans, ignoring speculation that God is dead, pray frequently and earnestly—and almost all say God has answered their prayers."[2] A similar poll conducted in June 1996 for a *Time* cover story on "faith and healing" found that 82 percent believed that prayers could heal.[3]

From the conventional medical perspective, this widespread behavior is a benign but presumably irrational diversion, at least when it comes to praying for a distant person. It simply reflects the innate human tendency to regress toward primitive, magical thinking in the face of modern medicine's failure to cure many chronic health problems. Some observers point to the $30 billion a year spent on alternative therapists, faith healers, and dubious remedies sold in health-food stores as further evidence of this primitive predisposition.

Of course, this enormous financial outpouring also reflects a deep, unsatisfied need: people are desperately seeking cures that modern medicine cannot provide. And beyond the motivations of religious faith, perhaps people ask others to pray for them because it costs nothing and provides the feeling that at least *something* is being done for a condition that medicine cannot treat. The question is, does it work?

The orthodox assumption that praying for a *distant* person cannot heal, or that any form of distant mental healing is impossible, is based on the conventional belief that the mind is simply an emergent property of the physical brain. This being the case, the mind is localized within the brain and is entirely dependent on the workings of the physical brain.[4] If this is true, then a healer in location A cannot affect the physiology of a patient in a distant location B, because without some sort of physical or psychological intervention there is no mechanism by which the patient can be affected.[5] Hence, distant mental healing is impossible, and praying for others to be healed is no more than wishful thinking. This logic is unassailable if the standard scientific assumptions are correct. But are they?

Belief Becomes Biology

William James said near the end of the nineteenth century, "No mental modification ever occurs which is not accompanied or followed by a bodily change." A hundred years later, Norman Cousins summarized the modern view of mind-body interactions with the succinct phrase "Belief becomes biology."[6] That is, an external suggestion can become an internal expectation, and that internal expectation can manifest in the physical body.

While the general idea of mind-body connections is now widely accepted, forty years ago it was considered dangerously heretical nonsense. The change in opinion came about largely because of hundreds of studies of the placebo effect, psychosomatic illness, psychoneuroimmunology, and the spontaneous remission of serious disease.[7] In studies of drug tests and disease treatments, the placebo response has been estimated to account for between 20 to 40 percent of positive responses. The implication is that the body's hard, physical reality can be significantly modified by the more evanescent reality of the mind.[8]

Evidence supporting this implication can be found in many domains. For example:

- Hypnotherapy has been used successfully to treat intractable cases of breast cancer pain, migraine headache, arthritis, hypertension, warts, epilepsy, neurodermatitis, and many other physical conditions.[9]

- People's expectations about drinking can be more potent predictors of behavior than the pharmacological impact of alcohol.[10] If they think they are drinking alcohol and expect to get drunk, they will in fact get drunk even if they drink a placebo.

- Fighter pilots are treated specially to give them the sense that they truly have the "right stuff." They receive the best training, the best weapons systems, the best perquisites, and the best aircraft. One consequence is that, unlike other soldiers, they rarely suffer from nervous breakdowns or post-traumatic stress syndrome even after many episodes of deadly combat.[11]

- Studies of how doctors and nurses interact with patients in hospitals indicate that health-care teams may speed death in a patient by simply diagnosing a terminal illness and then letting the patient know.[12]

- People who believe that they are engaged in biofeedback training are more likely to report peak experiences than people who are not led to believe this.[13]

- Different personalities within a given individual can display distinctly different physiological states, including measurable differences in autonomic-nervous-system functioning, visual acuity, spontaneous brainwaves, and brainware-evoked potentials.[14]

While the idea that the mind can affect the physical body is becoming more acceptable, it is also true that the mechanisms underlying this link are still a complete mystery. Besides not understanding the biochemical and neural correlates of "mental intention," we have almost no idea about the limits of mental influence. In particular, if the mind interacts not only with its own body but also with distant physical systems, as we've seen in the previous chapter, then there should be evidence for what we will call "distant mental interactions" with living organisms. And there is.[15]

When the mental intention is to beneficially affect a distant organism's physiological (or psychological) condition, we will call this "distant mental *healing*." Some of the many variants of distant mental healing are known as "spiritual healing," "prayer," "faith healing," "divine healing," and "bioenergy therapy." In contrast to local healing techniques such as "laying-on-of-hands," where a healer lightly touches or passes his or her hands near a patient's body, with distant mental healing the practitioner simply directs his or her healing thoughts or intentions to a patient at a distance.

Spontaneous Case Studies

At the extreme end of credulity, distant mental healing has been credited with producing instantaneous cures of even the most advanced cases of malignant illnesses. More credible are claims that distant mental healing assists in alleviating pain, promoting spontaneous remission, and accelerating the normal process of recuperation from disease or injury. Unfortunately, mental healers are rarely trained in conventional medical diagnosis, and their records—if indeed they keep any—are rarely useful for judging the merits of their claims.

In addition, it seems that many (but not all) of the truly astounding cases of mental healings occurred long ago and far away, so we cannot tell how many rounds of embellishment the stories have gone through. As a result, most of the amazing cases of distant mental healing proclaimed by healers, or by disciples of healers, cannot be rigorously assessed. Still, reviews of the literature provide intriguing evidence for the continuing reality of direct mental healing.

For example, a report in 1983 in the *British Medical Journal* described reliable witnesses to modern cases of unusual cures attributed to distant mental healing.[16] Similarly, a thorough search of the literature in 1993 for cases of spontaneous remission of disease found thirty-five hundred references in more than eight hundred medical and scientific journals in twenty languages.[17] The spontaneous-remission literature suggests that extraordinary forms of healing are widespread, and are probably more common than is generally believed.

Clinical Tests

To overcome the difficulties in assessing spontaneous cases of distant mental healing, researchers began to conduct clinical trials. This shift in focus resembles the trend we've seen in the other realms of psi research. From 1984 to 1993, psychologist Jerry Solfvin, psychologist Sybo Schouten, psychiatrist Daniel Benor, and physician Larry Dossey each independently reviewed clinical studies of distant mental healing in detail.[18] They all agreed that the evidence was intriguing but that the typical quality of the clinical studies was poor. This made it difficult to reach more than tentative conclusions.

Sybo Schouten, for example, reviewed psychic-healing studies primarily conducted in his native country of the Netherlands, where an estimated one thousand professional mental healers annually provide two million patient contacts. After considering about a dozen of the best available clinical studies, Schouten concluded cautiously:

> It looks as if psychic healing does have an effect on the health of the patients. The effects seem much stronger for subjectively experienced states of health than for objectively measured health criteria. It appears very important that the patient knows that treatment is applied. The effect due to the method itself is weak or non-existent, whereas psychological variables associated with the patient and with the healer-patient interaction contribute most to the healing effects.[19]

In an often-cited clinical study, in 1988 physician Randolph Byrd reported a double-blind study of intercessory prayer in coronary-care-unit patients at San Francisco General Hospital.[20] Byrd sent the names, diagnoses, and conditions of 193 randomly selected patients to people of various religious denominations who were asked to pray for them. A similar group matched for age and symptoms was not prayed for. The primary findings of this study were remarkable: The prayed-for patients were five times less likely to require antibiotics and three times less likely to develop pulmonary edema. None of the prayed-for group required endotracheal intubation, and fewer patients in the prayed-for group died.

Byrd's study was praised for providing solid clinical evidence that prayer works, but (of course) it was also criticized. The criticisms included suspicions of bias owing to Byrd's religious motives (which is not exactly an experimental design issue, but motivations are always subject to suspicion); the fact that no one verified whether the praying groups actually prayed as they were supposed to; the lack of information about the type of prayer strategies employed; and the experimenters' failure to assess whether people outside the experiment prayed for members of the control group.[21]

These design issues aside, no significant differences were found between how long the two groups spent either in the hospital or on the coronary care unit, and the few measures that did favor the prayed-for group showed only a 5 to 7 percent improvement over the controls. Thus, while this experiment was a laudable attempt to study distant mental healing for patients in need under real-life conditions, it did not indicate that distant mental healing was a particularly robust method. Physician Larry Dossey, in reviewing this and other studies involving prayer, came to the conclusion that "The Byrd experiment is suggestive but inconclusive and inherently ambiguous. It simply contains too many problems that prevent us from drawing firm conclusions about the possible power of prayer. In fact all the human prayer studies we have examined so far fall into this category."[22]

Because of difficulties in assessing and interpreting the effects of distant healing in clinical studies, many other researchers have concentrated on controlled laboratory studies. The problem in examining these experiments is not the lack of studies but rather their profusion. Dozens of studies can be found where the mentally influenced organisms have included everything from bacteria to human beings.[23] But unlike the simple hit or miss results used to judge success in telepathy, clairvoyance, precognition, dice, and RNG studies, the outcome measures in studies involving living organisms ranged from mortality rates, to the speed of wound healing, to unconscious physiological responses, to biochemical changes, and beyond.

Laboratory Experiments

A few efforts have been made to review this massive experimental literature.[24] At least 130 publications describe controlled experiments on living systems ranging from enzymes to cell cultures, bacteria, plants, mice, hamsters, dogs, and human beings. Of these studies, 56 reported results with odds against chance of one hundred to one or better, where only one or two such studies would be expected by chance. Because the odds of obtaining 56 successful studies out of 131 are well beyond a trillion to one, either this database reflects an extremely robust effect or it reflects a file-drawer problem. The latter is probably the case, but given the huge odds against chance in these studies, selective reporting alone probably cannot explain the overall success rate.

In late 1996, the Office of Alternative Medicine at the National Institutes for Health convened a study group to conduct a formal meta-analysis of all such studies published in all languages. The process will take a few years to complete. In the meantime, because there have been more studies of distant mental interactions with human beings than with any other living system, preliminary analysis of these studies provides some hints on what we can expect from the NIH review.

Experiments with Human Beings

Distant-mental-interaction experiments with human beings have usually involved measurement of various unconscious physiological factors. Incidentally, it is a little-known curiosity that in 1929 a German psychiatrist named Hans Berger invented the electroencephalograph (EEG) as a way of studying whether telepathy might be explained by brain waves.

Some of the earliest experiments studied "the feeling of being stared at." Laboratory investigations of this age-old phenomenon, discussed in more detail later, began around the turn of the century.[25] Modern experimental studies began to be published about forty years ago, and this approach to studying psi effects gained popularity in the 1960s.[26]

Two types of experiments involving measurements of psi influences on the human body have been conducted: investigations of physiological (bodily) *correlates* of conscious psi perception, and the use of physiological measures as unconscious *detectors* of psi. The majority of these studies used physiological measures in an agent–percipient (or sender–receiver) paradigm, similar to that used in remote-viewing and telepathy experiments. They examined the autonomic or central nervous system of a receiver while a remote sender attempted to influence the receiver with emotional or other meaningful information.

In 1963, consciousness research pioneer Charles Tart measured skin conductance, blood volume, heart rate, and verbal reports in a sender–receiver study. He as the sender received random electrical shocks to see if remote receivers could detect those events. As is typical in these studies, the receivers' physiology reacted significantly to the remote shocks, but there was no evidence that they were *consciously* aware of the events.[27] In independent experiments later, engineer Douglas Dean at the Newark

College of Engineering in New Jersey, psychologist Jean Barry in France, and Icelandic psychologist Erlendur Haraldsson at the University of Utrecht in the Netherlands all observed significant changes in receivers' finger blood volume when a sender, located sometimes thousands of miles away, directed emotional thoughts toward them.[28] At about the same time, the journal *Science* published a study by two physiologists who reported finding significant correlations in brain waves between isolated identical twins.[29]

Braud's Experiments

The largest systematic body of experiments has been reported by psychologist William Braud and his colleagues, primarily anthropologist Marilyn Schlitz.[30] Braud's experiments, conducted over seventeen years, mostly at the Mind Science Foundation in San Antonio, Texas, focused on people attempting to influence the nervous system of remote percipients. In 1991 Braud and Schlitz summarized all their studies to date: thirty-seven experiments employing seven different physiological response systems, such as blood pressure and muscle tremor. Altogether, these studies comprised 655 sessions, with 449 people or animals acting as receivers, 153 people acting as senders, and 13 principal experimenters.

The thirty-seven experiments combined resulted in odds against chance of more than a hundred trillion to one. Fifty-seven percent of the experiments were independently significant (with odds better than twenty to one), where 5 percent would be expected by chance. These and the other laboratory studies on living systems provide strong support for the idea that people can unconsciously respond to distant mental influences.

Figure 9.1 summarizes a subset of fifteen studies reported by Braud and Schlitz (listed as studies 1–13b),[31] along with the results of four replications by other investigators (listed as studies 14–17), and the overall combined results. In all these studies, the receiver was wired up to a monitor that continuously measured the conductivity of his or her skin, called "electrodermal activity." Skin conductivity is effective in detecting unconscious fluctuations in emotion, and is a central component in lie-detector hardware. In these studies, the sender is instructed at random times, usually by a computer, to attempt to arouse or to calm the distant person solely by thinking about that person. At other randomly selected times, the sender is instructed to direct his or her attention elsewhere to provide "no-mental-influence" control periods. The sender and receiver are always isolated by distance, and sometimes also by special soundproof and electromagnetically proofed rooms.

The outcome measure in most of these studies is the proportion of the receiver's total electrodermal activity occurring in the instructed direction (calm or aroused) divided by the total electrodermal activity in that session.

If there was no mental interaction effects, this would result in a chance proportion of 50 percent. Individual sessions typically consist of 10 to 20 one-minute periods of randomly alternating mental-influence and control periods, for a total of about 15 to 20 minutes.

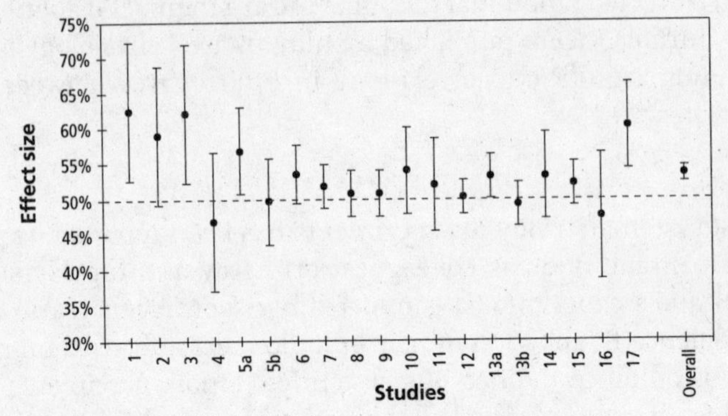

Figure 9.1. Point estimates and 95 percent confidence intervals for results of all known experiments studying distant mental influence on human electrodermal activity, as of early 1997.[32]

As figure 9.1 shows, the distant-mental-interaction effects on electrodermal activity were fairly consistent across these studies, including the replications. The average effect size over a total of four hundred individual sessions was about 53 percent, compared to a 50-percent-equivalent figure for chance. This provides evidence for successful remote influence of electrodermal activity with odds of 1.4 million to 1. Remember, these "effect sizes" are actually measurements comparing electrodermal changes during mental-influence periods versus control periods.

The Effect Size

There is an increasing emphasis in meta-analysis on reporting experimental effects in terms of "normalized effect sizes." The procedure for calculating these effect sizes will not be detailed here. It turns out, though, that the average effect size for the electrodermal studies is about 0.25, where the possible effect sizes range from −1 (absolute success, but opposite to the predicted direction) to +1 (absolute success).[33]

To understand the meaning of a 0.25 effect size, it may be useful to consider the example of a medical treatment claimed to change the outcome of an illness. A medical treatment that ordinarily produced a 37.5 percent survival rate would be enhanced by a treatment with an effect size of 0.25 to such an extent that the new survival rate would become 62.5 percent. Thus, the small effect seen in figure 9.1 actually has important practical consequences. When we add a distant mental "treatment" with an effect size of

0.25 to a conventional treatment, the majority of a population that would have *died* would instead become a majority that would *live.*

The 0.25 effect size for distant mental interactions may also be compared with the effect sizes obtained in recent placebo-controlled studies of the drugs propranolol and aspirin, which were based on testing 2,108 and 22,071 people, respectively. Both of these studies were stopped before they reached their planned end-points because in both cases the drugs were found to have beneficial results, and it was considered unethical to continue the clinical trials (which withheld the drugs from the control groups). The equivalent effect size for the propranolol study was a mere 0.04, and for the aspirin study (mentioned in chapter 4), 0.03. When we compare these effect sizes to the 0.25 effect size in distant-mental-interaction studies, we see that some psi effects recorded in the laboratory are much larger than many people realize.

The Feeling of Being Stared At

The "feeling of being stared at" is the focus of a subset of distant-mental-interaction studies. This is a particularly interesting belief to investigate because it is related to one of the oldest known superstitions in the Western world, the "evil eye," and to one of the oldest known blessings in the Eastern world, the *darshan,* or gaze of an enlightened master. Most ancient peoples feared the evil eye and took measures to deflect the attraction of the eye, often by wearing shiny or attractive amulets around the neck.

Today, most fears about the evil eye have subsided, at least among educated peoples. And yet many people still report the "feeling of being stared at" from a distance. Is this visceral feeling what it appears to be—a distant mental influence of the nervous system—or can it be better understood in more prosaic ways? In the laboratory today, the question is studied by separating two people and monitoring the first person's nervous system (usually electrodermal activity) while the second person stares at the first at random times over a one-way closed-circuit video system. The stared-at person has no idea when the starer is looking at him or her.

Figure 9.2 shows the results for staring studies conducted over eight decades.[34] Similar to William Braud's electrodermal studies but conducted in a context that more closely matched common descriptions of "feeling stared at," these studies resulted in an overall effect of 63 percent where chance expectation is 50 percent. This is remarkably robust for a phenomenon that—according to conventional scientific models—is not supposed to exist. The combined studies result in odds against chance of 3.8 million to 1.

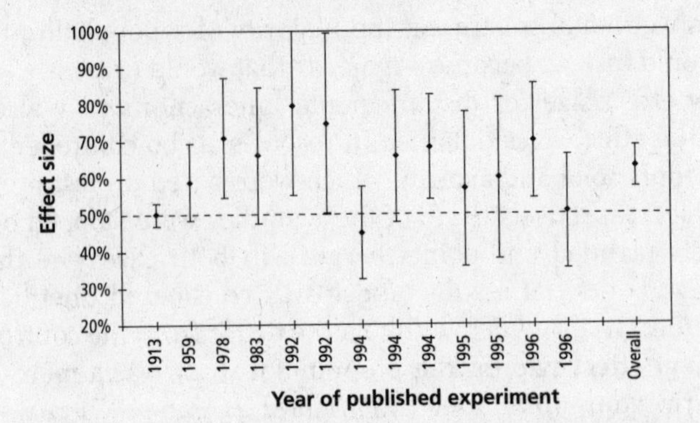

Figure 9.2. Effect sizes for studies testing the "feeling of being stared at," where 50 percent is chance expectation. Confidence intervals are 95 percent.

Summary

Given the evidence for psi perception and mind-matter interaction effects discussed so far, we could have expected that experiments involving living systems would also be successful. The studies discussed here show that our expectations are confirmed. The implications for distant healing are clear.

All the experiments discussed so far have been replicated in the laboratory dozens to hundreds of times. They demonstrate that some of the "psychic" experiences people report probably do involve genuine psi. Now we move outside the laboratory to examine a new type of experiment, one that explores mind-matter interaction effects apparently associated with the collective attention of groups.

CHAPTER 10

Field Consciousness

We allow our ignorance to prevail upon us and make us think we can survive alone, alone in patches, alone in groups, alone in races, even alone in genders.

MAYA ANGELOU

In preceding chapters, we have seen that when a person directs his or her attention toward a remote, physical object, the behavior of that object changes in interesting ways: tossed dice no longer fall at random, electronic circuits behave strangely, and the human nervous system responds to unseen influences. Here we consider what happens when groups of people, ranging from dozens to billions, focus their minds on the same thing. We'll see that some aspects of the mind-matter interaction effects witnessed in the laboratory appear to "scale up" to influence the world at large. As outlandish as it may seem, this may even include the weather.

Neuroscientists believe that consciousness is spawned by the rich interconnections and complex information exchange among billions of neurons in the brain. If this is true, then considering the rich interconnections and information exchange among billions of intelligent minds on Earth, might this imply that the world itself has something like a "global" mind?[1] Recent experiments suggest that the answer may be yes.

In our laboratory, we've been studying a phenomenon dubbed "field-consciousness" effects. These experiments were pioneered by Dr. Roger Nelson at Princeton University and have been replicated by Professor Dick Bierman at the University of Amsterdam. The experiments suggest that groups of people, ranging from a dozen in a small workshop to billions watching a live television broadcast, may affect the physical world in unexpected ways. The experiments also imply—a theme that will be explored in more detail later—that there is a fundamental interconnectedness among all things, including individual and "mass minds." To explore the rationale for these experiments, we begin with a brief overview of the concepts of fields and field consciousness.

Physical Fields

The classical notion of a physical field developed in the seventeenth century as a way of understanding phenomena such as gravity and electromagnetism. These mysterious forces appeared to affect objects at a distance by some invisible means. Unfortunately, the idea of invisible forces acting at a distance closely resembled religious ideas about spirits, which were anathema to the mechanistic theories developing in the seventeenth century, so physicists of that era looked for other, more "rational" explanations. Because it was thought to be impossible to have action-at-a-distance with no intermediaries to convey the energies, they imagined forces as acting on objects through the exchange of energy packets, much like colliding billiard balls.

According to Newtonian (classical) physics, the only way the sun and earth could interact was by assuming that the earth was actually responding to something invisible in its local vicinity, and likewise for the sun. This invisible "something" was called a field, which was interpreted as a modification of the space surrounding the sun and the earth, a modification that was inferred by how particles—for example "gravitons" in the case of gravity—behaved. In other words, the earth was not really attracted to the sun through some invisible force at a distance, but through the mutual exchange of graviton particles.

The development of quantum mechanics in the late 1920s expanded the classical notion of fields in a way that would have shocked Newtonian physicists. Quantum fields do not exist physically in space-time like the classically inferred gravitational and electromagnetic fields. Instead, quantum fields specify only *probabilities* for strange, ghostlike particles as they manifest in space-time. Although quantum fields are mathematically similar to classical fields, they are more difficult to understand because, unlike classical fields, they exist outside the usual boundaries of space-time.

This gives the quantum field a peculiar *nonlocal* character, meaning the field is not located in a given region of space and time. With a nonlocal phenomenon, what happens in region A instantaneously influences what occurs in region B, and vice versa, without any energy being exchanged between the two regions. Such a phenomenon would be impossible according to classical physics, and yet nonlocality has been dramatically and convincingly revealed in modern physics experiments. In fact, those experiments are independent of the present formulation of quantum mechanics, which means that any future theory of nature must also embody the principle of nonlocality. We'll return to nonlocality again in chapter 16.

Consciousness Fields

Just as the individual is not alone in the group, nor any one in society alone among the others, so man is not alone in the universe.

CLAUDE LÉVI-STRAUSS

The idea that consciousness may be fieldlike is not new.[2] William James wrote about this idea in 1898, and more recently the British biologist Rupert Sheldrake proposed a similar idea with his concept of morphogenetic fields.[3] The conceptual roots of field consciousness can be traced back to Eastern philosophy, especially the *Upanishads,* the mystical scriptures of Hinduism, which express the idea of a single underlying reality embodied in "Brahman," the absolute Self. The idea of field consciousness suggests a continuum of nonlocal intelligence, permeating space and time. This is in contrast with the neuroscience-inspired, Newtonian view of a perceptive tissue locked inside the skull.

One of the more controversial modern claims about the effects of field consciousness was proposed by the founder of transcendental meditation, Maharishi Mahesh Yogi. As David Orme-Johnson, dean of research at Maharishi International University, put it:

> Stressed individuals create an atmosphere of stress in collective consciousness that reciprocally affects the thinking and actions of every individual in that system. . . . Crime, drug abuse, armed conflict, and other problems of society are more than just the problem of individual criminals, drug users, and conflicting factions in society. Such problems are more fundamentally symptoms of stress in collective consciousness.[4]

Transcendental meditation researchers have reported that the so-called Maharishi effect has been replicated in forty-two studies, some published in mainstream sociology journals.[5] As expected, sociologists have criticized the designs of these studies. One of the main criticisms is that in many of these studies the variables of interest were indices of social order, such as crime, war hostilities, traffic accidents, and quality of life. These indices are influenced by dozens of external factors, and even when obvious influences such as day-of-the-week, holidays, and seasonal effects are accounted for, they are still notoriously difficult to take into account.

To further complicate things, the Maharishi effect predicts that the social-ordering effect, say reduction in crime, is proportional to the number of meditators who are "generating" coherent consciousness through their meditations. But because the number of meditators on a day-to-day basis in many of the transcendental meditation studies was not constant, there were unavoidable interactions between the number of meditators on a given day and fluctuating values of the various social indices.

One way to avoid the design problems encountered by the transcendental meditation researchers would be to keep one of the variables fixed. This could be either the number of meditators or the "target" of consciousness-induced order. Beyond this, as philosopher Evan Fales and sociologist Barry Markovsky of the University of Iowa suggested after reviewing the Maharishi effect, "Presumably, if the material world can be influenced in purposive ways by collective meditation, *inanimate* detectors could be constructed and placed at varying distances from the collective meditators."[6] This is essentially the approach that we took, although our motivations were based upon a logical extension of laboratory research on mind-matter interactions using random-number generators, and not by the claims of the transcendental meditators.

Properties of Consciousness

Whatever else consciousness may be, let us suppose that it also has the following properties, derived from a combination of Western and Eastern philosophies.[7] The first property is that consciousness extends beyond the individual and has quantum field–like properties, in that it affects the probabilities of events. Second, consciousness injects *order* into systems in proportion to the "strength" of consciousness present. This is a refinement of quantum physicist Erwin Schrödinger's observation about one of the most remarkable properties of life, namely, an "organism's astonishing gift . . . of 'drinking orderliness' from a suitable environment."[8]

Third, the strength of consciousness in an individual fluctuates from moment to moment, and is regulated by focus of attention. Some states of consciousness have higher focus than others. We propose that ordinary awareness has a fairly low focus of attention compared to peak states, mystical states, and other nonordinary states.[9]

Fourth, a group of individuals can be said to have "group consciousness." Group consciousness strengthens when the group's attention is focused on a common object or event, and this creates coherence among the group. If the group's attention is scattered, then the group's mental coherence is also scattered.

Fifth, when individuals in a group are all attending to different things, then the group consciousness and group mental coherence is effectively zero, producing what amounts to background noise. We assume that the maximum degree of group coherence is related in some complicated way to the total number of individuals present in the group, the strength of their common focus of attention, and other psychological, physiological, and environmental factors.

Sixth, physical systems of all kinds respond to a consciousness field by becoming more ordered. The stronger or more coherent a consciousness field, the more the order will be evident. Inanimate objects (like rocks) will respond to order induced by consciousness as well as animate ones (like people, or tossed dice), but it is only in the more labile systems that we have the tools to readily detect these changes in order. In sum, when a group is actively focused on a common object, the "group mind" momentarily has the "power to organize," as Carl Jung put it.[10]

This leads us to a very simple idea: *as the mind moves, so moves matter.* For our measure of matter, we looked for changes in order, or coherence, in physical systems. This is easiest if we monitor physical systems that are by nature truly random. While a rock should experience fluctuations in order and disorder because of the fluctuations of many minds, it is difficult to measure changes in a rock within the timescale of the experiment, so we must rely upon quickly changing physical systems such as the electronic random-number generators (RNGs) with which we are already familiar.

Measurements

In the basic field-consciousness experiment, we measure fluctuations in a group's attention while simultaneously measuring fluctuations in the behavior of one or more physical systems. Note that the experimental protocol does not require a group specifically to focus its *intention,* or directional attention, toward a specified physical target. In fact, attempting to maintain such a focus may arouse powerful defense mechanisms, doubts, and fears that block the very effects we wish to observe.

Changes in order are easily detectable in random physical systems because under ordinary conditions, and by definition, a random system on average has zero order. If order *does* appear, it can be detected immediately using fairly simple statistical methods. Fluctuations in order are expected to occur by chance, of course, but in this case we are not as interested in any particular fluctuation as *when those changes take place* in relationship to changes in the group's attention.

Experiments

From March 1995 through July 1996, we examined field-consciousness effects in a series of eight experiments conducted during (1) a personal growth workshop in March 1995 involving a dozen participants; (2) the live broadcast of the Sixty-seventh Annual Academy Awards in March 1995, with an estimated one billion people watching worldwide; (3) a comedy show at a Las Vegas casino in September 1995, with about forty people present; (4) the announcement of the O. J. Simpson verdict in October 1995, with about 500 million people listening or watching; (5) the Superbowl football game in January 1996, with about 200 million people listening or watching; (6) prime-time television shows broadcast on the four major television networks in the United States one Monday night in February 1996, with an estimated ninety million people watching; (7) the Sixty-eighth Annual Academy Awards in March 1996, with about one billion people watching worldwide; and (8) the Opening Ceremonies of the Centennial Olympic Games in July 1996, with about three billion people watching worldwide.

Exploring Mind and Matter

A variety of technical approaches and designs were employed in these experiments to study various aspects of field-consciousness effects, so rather than specifying the exact methods used in each experiment, I will describe the general approach. For the "matter" part of the experiment, we programmed one or more electronic random-number generators (RNGs) to generate 400 random bits (0's and 1's) every six seconds. We called each group of 400 bits a "sample," which was roughly equivalent to flipping a fair coin four hundred times, then recording the number of heads and tails that resulted.

The RNGs were instructed to continuously collect samples about an hour before the event (say, an hour before the live TV broadcast of the Academy Awards), during the event itself, and for an hour after the event. Data collected before and after the broadcast were used as controls and were examined to ensure that the RNG was working correctly. This approach typically yielded a few thousand samples for each RNG for each experiment. For each collected sample, we examined the number of 1's produced in 400 random bits. This number was transformed by a standard statistical formula into a measure of the amount of order in the random sequence. This, in effect, measured the degree of statistical equilibrium, or balance, in the electronic circuit every six seconds.

In some experiments, when there were periods of time that were clearly of either high or low interest, judges were asked to log in a notebook when, according to their subjective impression, these events occurred, along with the content of each event. For example, during the Academy Awards broadcast, the few minutes before, during, and after the presenter announced, ". . . and the Oscar for Best Picture goes to . . . ," would probably have been judged as a high-interest period. For the same broadcast, commercial breaks would probably have been judged as low-interest periods. In other experiments, if there were fewer periods that were clearly of low interest, the entire event was considered to be of high interest, and control data generated after (or before) the event acted as a low-interest comparison.

Workshop

Our first study took place in a personal growth workshop using a technique called Holotropic Breathwork™. This is a powerful therapeutic method developed by psychiatrist Stanislav Grof. It involves a combination of deep-breathing techniques, listening to rhythmic music, and focused massage.[11] The session took place on March 4, 1995, for nine hours, in Las Vegas, Nevada, and twelve people were present.[12] A single electronic RNG was used to continuously measure fluctuations in physical order. Of several ways of examining the resulting data, the simplest method was to consider the entire nine-hour session as a high-interest period.[13] After the workshop, the same RNG was run for a nine-hour period starting at 11:00 P.M., alone in a room, as a low-interest control.

Figure 10.1 shows the results of this experiment for the last seven hours of both data streams. The two curves indicate the degree of order induced into the random data in terms of odds against chance. The more order, the greater the odds against chance. For binary random data, where over the long term the number of 1's and 0's should be the same, we would expect the odds to fluctuate around one in two. This is what we see for the control data. By contrast, if randomness was affected during the workshop due to the intense, coherent attention of the group, then the odds against chance would progressively get higher. And this is what we see for the data collected during the workshop. The odds against chance increased to about one thousand to one by the end of the workshop. This means that if the same seven-hour data sequence were run a thousand times, we would see a result this extreme, or more extreme, only once by chance. The chance results observed with the control data suggest that there was nothing unusual about the RNG or about the method we used to analyze the results.

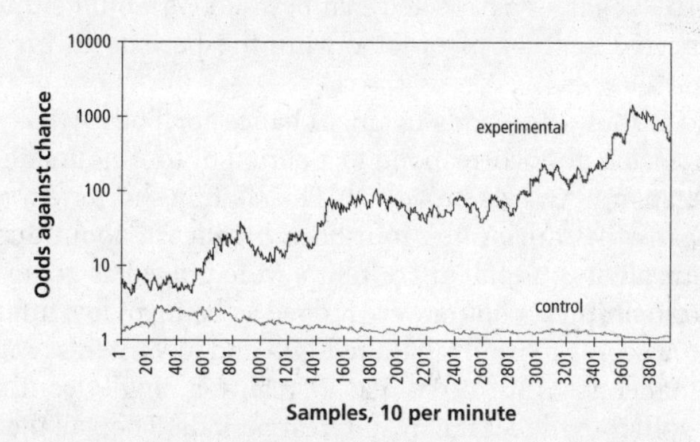

Figure 10.1. Cumulative odds against chance for random data collected during a personal growth workshop (labeled experimental) and control data collected for the same length of time after the workshop ended. The graph shows that order was impressed into the random-number generator during the workshop, as predicted by a field-consciousness effect.

This experiment indicated that during the workshop small but consistent degrees of order were somehow impressed into the random data generated by the RNG. By analogy, imagine that someone in the workshop room was flipping a coin four hundred times every six seconds. Say that she was instructed to keep track of how many times heads appeared in the sequence. Before and after the workshop, she would be counting numbers like 48 heads, 51, 50, 46, 53, and so on, averaging around the expected 50. During the workshop, however, she would begin to record numbers like 54, 50, 52, 58, 53, and so on, averaging around 55. Such an unexpected series of events is one way that coherence can express itself in a random physical system—too many binary events of the same type, i.e., too many heads in a row.

Academy Awards 1995

Participants in the next experiment were the estimated one billion people in 120 countries who viewed the live television broadcast of the Sixty-seventh Annual Academy Awards on March 27, 1995. To assess the fluctuations in group coherence during this broadcast, my assistant and I independently kept minute-by-minute logs of events shown in the program, and we judged whether we thought each noted event was interesting, and likely to attract the attention of the viewing audience, or uninteresting, and likely to bore the audience. We called the interesting segments "high coherence" and the uninteresting segments "low coherence."

We ran two independent RNGs in this experiment to test the prediction that mass-consciousness effects are truly nonlocal. One RNG was about twenty meters from me, as I watched the broadcast in my home, and the other was twelve miles away, running by itself, in my lab at the University of Nevada, Las Vegas. We expected that both RNGs would simultaneously show unexpected degrees of order during the periods of high audience interest.

Figure 10.2 shows the odds against chance for both RNGs combined. The two curves together correspond to a period of four hours, the length of the actual Academy Awards broadcast. The slightly shorter curve for "high interest" means that during the four-hour broadcast, about one thousand samples, equivalent to a total of 1.7 hours, were judged as being of high interest. The remaining 2.3 hours were judged as being of low interest.

Figure 10.2 seems to imply that the high- and low-interest data were collected simultaneously, but of course in reality a single, continuous data stream was split into high- and low-interest periods. Then all the high interest-segments were analyzed together *as though* they took place continuously, and all the low-interest segments were analyzed in the same manner. Thus, the continuous curves in figure 10.2 are really composed of discontinuous periods in time, pasted together into continuous curves for ease in visualizing and interpreting the outcomes.

Figure 10.3 shows the same analysis for the same two RNGs run after the Academy Awards broadcast ended, again for four hours. This graph, analyzed exactly the same way as the data shown in figure 10.2, indicates that the RNGs operated according to chance expectation when no mass events were happening.

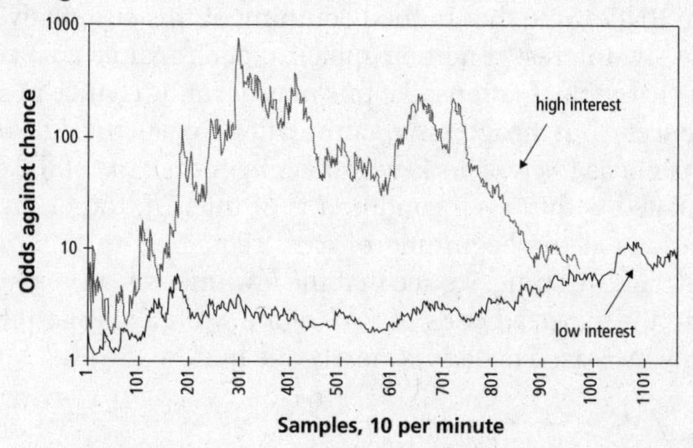

Figure 10.2. Odds against chance for both RNGs combined, for periods of high and low audience interest during the 1995 Academy Awards broadcast.

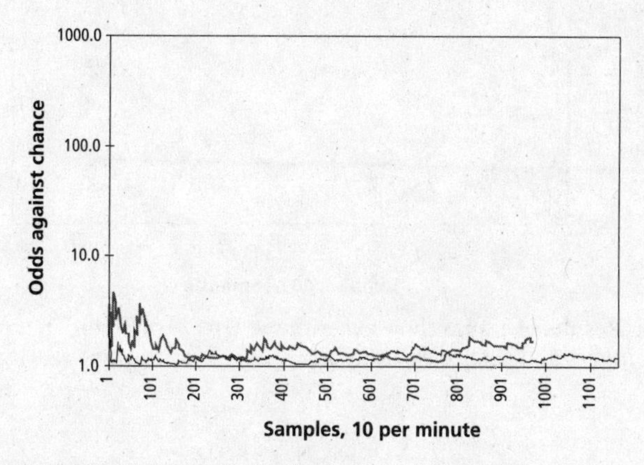

Figure 10.3. Odds against chance for both RNGs combined, run as a control test after the Academy Awards broadcast.

Comedy Show

Our third study took place at a comedy and stage hypnotism show in a Las Vegas casino on the night of September 8, 1995. About forty people were in the audience, including two research staff members from our laboratory. They took an RNG and a notebook computer with them to the show. The RNG was started up just before the show began and stopped a few minutes after the show ended, running for a total of eighty minutes. One researcher rated her subjective impressions about the show about once a minute, recording the clock time and a rating of either high attention or low attention for each event.

Figure 10.4 shows the results, which again confirm the idea that the periods of high group interest were associated with unexpected amounts of order in the RNG. Note that in the beginning of this cumulative graph the odds for the low-interest condition spiked at odds against chance of a thousand to one. Large fluctuations like this may occur by chance in shorter random sequences, but progressive cumulation of such odds over longer sequences is not as likely. This is why the long-term *trend* of the data as they are accumulated within each condition is of interest, rather than momentary fluctuations at the beginning of each sequence. In this case, as data were cumulated over time, we see that the low-interest curve eventually settled down to the expected odds of two to one while the high-interest curve progressively increased to odds of nearly one hundred to one.

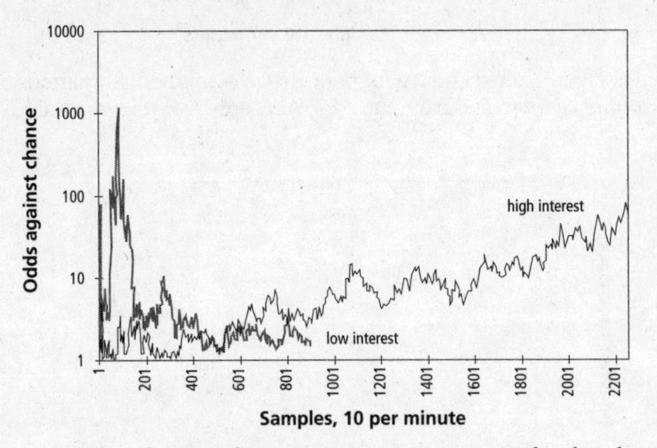

Figure 10.4. Results of casino show experiment. Up is toward order; down is toward randomness and chance expectation.

O. J. Simpson Verdict

On October 3, 1995, something like a half-billion people worldwide watched or listened to the live broadcast of the verdict in one of the most celebrated murder trials in U.S. history—the O. J. Simpson case. According to the Reuters news service, the viewing audience for this single event exceeded the ratings of three of the five Superbowl telecasts between 1991 and 1995. This is impressive since the Superbowl is traditionally one of the year's highest-rated television events in the United States.

We took advantage of this unusual event to test the proposed nonlocal property of field consciousness. We predicted that around the time of the verdict, we would see unexpected behavior in RNGs located anywhere in the world. To test this idea, we asked Dr. Roger Nelson at Princeton University and Professor Dick Bierman at the University of Amsterdam to run RNGs in their labs during the event. We then combined their data with the outputs of three additional RNGs run in our laboratory.

We expected that the unusual degree of mass attention focused on this event would cause the combined output of five independent RNGs simultaneously to show unexpected order when the verdict was announced. The results, shown in figure 10.5, suggest that something unusual did occur in all five RNGs precisely when the verdict was announced. The graph indicates that around the time that the TV preshows began, at 9:00 A.M. Pacific Time, an unexpected degree of order appeared in all the RNGs. This soon declined back to random behavior until about 10:00 A.M., which is when the verdict was supposed to be announced. A few minutes later, the order in all five RNGs suddenly peaked to its highest point in the two hours of recorded data precisely when the court clerk read the verdict.

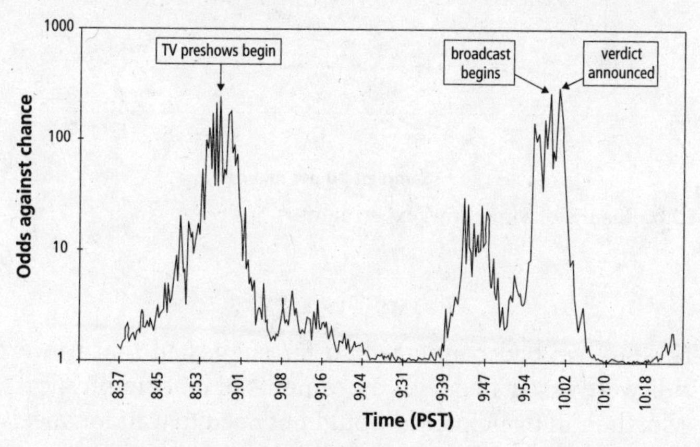

Figure 10.5. Results of O. J. Simpson experiment.

Superbowl XXX

In our next experiment, we were interested in seeing whether fluctuations in mass attention would simultaneously affect *different forms* of matter. To test this, we ran three independent RNGs and three new random generators based on detection of background radioactive particles. All six devices were run during the live broadcast of Superbowl XXX in January 1996.

The results, shown in figure 10.6, indicated that, once again, the combined odds against chance for the amount of order observed in three RNGs and three radiation counters during the *low* attention periods hovered around chance, but the odds against chance during *high* attention periods moved toward greater order. We speculated that the results observed in this experiment were not as dramatic as those observed in previous experiments because it was difficult to distinguish high from low interest periods during the Superbowl. In most TV broadcasts, the commercials usually correspond to low interest periods and the programs to high interest periods. But in the Superbowl, the commercials traditionally rival the actual football game for interest. In addition, the Superbowl broadcast is a fast-moving show, and the fast scene changes themselves tend to attract attention. These factors may have blurred the clear distinctions between high and low interest, which reduced the field-consciousness effects observed in the previous studies.

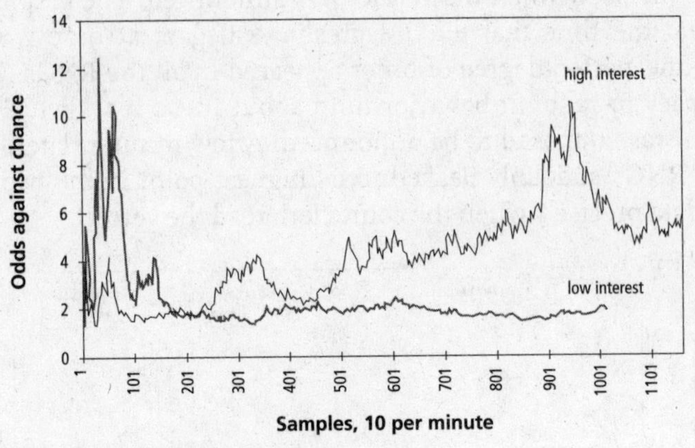

Figure 10.6. Results of Superbowl experiment.

Prime-Time TV

The experiments we had conducted so far suggested that movements of "mass mind" were associated with movements of *order* in physical systems. If this was so, then in principle we would not need to wait for special, mass-broadcast events, because there are predictable movements of many minds every day. During the "prime-time" hours of 8:00–11:00 P.M., about fifty to ninety million minds in the United States follow a predictable ebb-and-flow of attention corresponding to periods of TV programs versus commercials.

To estimate fluctuations in mass interest, we examined all shows broadcast by the four major TV networks over prime time on the night of February 5, 1996. We tracked minute-by-minute when the programs were broadcast and when commercials occurred, then took an average of these periods for the six half-hours in prime time. Figure 10.7 shows this estimate as a number ranging from 0 to 1, where 0 means commercials only, and 1 means programs only. As we might expect, commercials tended to cluster, on average, around the beginning, middle, and end of prime-time half-hours. Our estimate of fluctuations in matter were formed by combining the outputs of three independent RNGs run over the three hours of prime time, using methods similar to those used in previous experiments, then forming half-hour averages as we did for the commercials.

Because the field-consciousness idea predicts that higher audience attention is related to higher order, we predicted that we would see a positive relationship between our measures of mind and matter. In fact, the resulting relationship was substantial, corresponding to odds against chance of about one hundred to one.[14] Notice that the odds against chance for the behavior of the RNGs in this experiment were not particularly unusual, rising to odds of only nine to one at eighteen minutes and then again at twenty-two minutes into the average half-hour. But recall that the point of this experiment was not that the RNGs would show unexpected behavior by themselves, but that fluctuations in audience attention and fluctuations in the RNGs would be significantly *related to each other*. And this is what we observed.

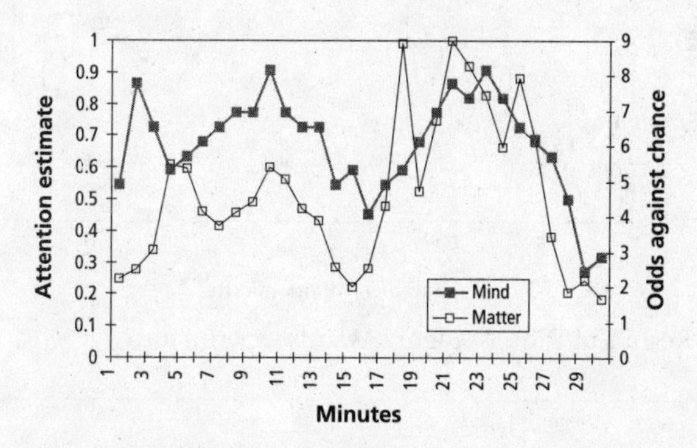

Figure 10.7. Results of prime-time TV experiment. The left axis is the measure of mass-audience attention; the right axis is a measure of fluctuations in matter, displayed in terms of odds against chance.

Academy Awards 1996

Our next study was conducted during the live broadcast of the Sixty-eighth Annual Academy Awards, on March 25, 1996, in hopes of replicating the results of our first Academy Awards test. This time we ran three independent RNGs before, during, and after the broadcast. The results (figure 10.8) indicated, once again, that the outputs of the RNGs during the broadcast cumulated to greater odds against chance during the high-interest periods than during the low-interest periods. The data recorded before and after the live broadcast did not show a similar separation.

Olympics Ceremonies

Our final experiment was conducted before, during, and after the live broadcast of the Opening Ceremonies of the Centennial Olympic Games in July 1996, witnessed by an estimated three billion people worldwide. Because the opening ceremony proved to be of fairly high interest throughout the broadcast, we compared the five-hour data sequence from two independent RNGs to a similar five-hour sequence recorded from the same equipment immediately after the broadcast ended.

The two curves in figure 10.9 display those data. As we've come to expect, data collecting during the Olympics broadcast showed progressively more order while control data recorded after the broadcast showed chance-expected behavior.

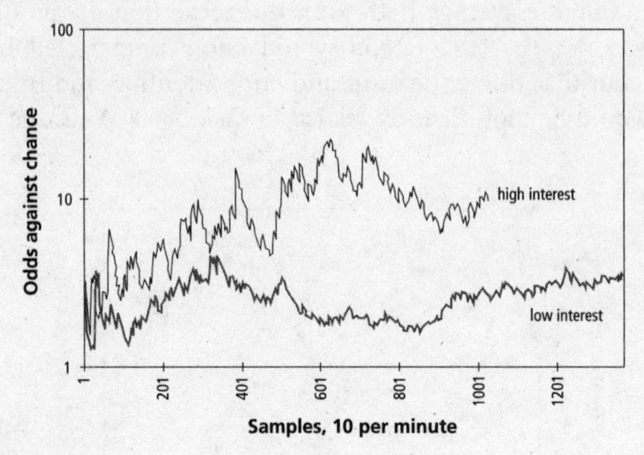

Figure 10.8. Results of 1996 Academy Awards experiment.

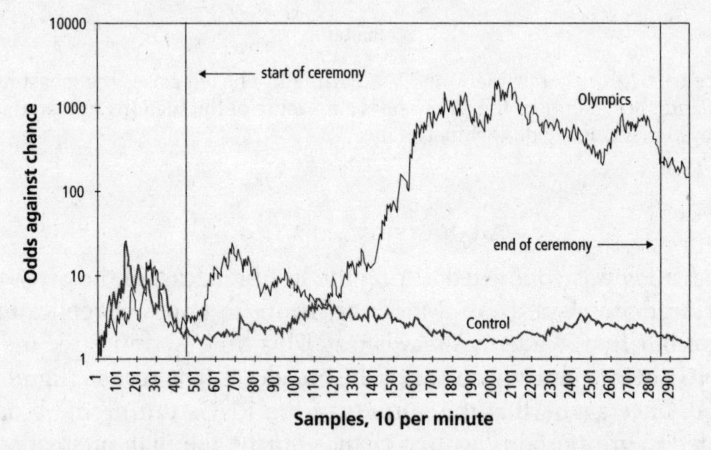

Figure 10.9. Results of Olympics experiment. The curve labeled "Olympics" refers to the combined results of two independent random data streams. "Control" refers to a data stream recorded on one of the RNGs immediately after the broadcast ended.

Replications

Twelve other field-consciousness studies similar to the ones I've described here have been independently reported by psychologist Roger Nelson at Princeton University, psychiatrist Richard Blasband in California, and psychologist Dick Bierman at the University of Amsterdam.[15] Blasband's and Nelson's methods differed from ours primarily in their selection of group settings likely to have high group coherence, like workshops and therapy sessions. They then measured the number of times the RNG exhibited unexpected order during those meetings. Overall, they reported highly significant results comparable to those we found in our experiments (odds against chance of ten thousand to one). Professor Bierman conducted experiments at a major European soccer match and at a home experiencing poltergeist disturbances using methods similar to ours. He observed results similar to those reported by Nelson, Blasband, and us.

In one of his tests of the field-consciousness idea, Roger Nelson predicted that his RNG system would show unexpected behavior during the 1995 performances of a yearly musical review by the San Francisco Bay Revels. He selected five segments of the celebration as being most likely to have a profound and engaging effect on the audience.[16] Seven performances in San Francisco were recorded with Nelson's RNG system without him being present, and the designated pieces (which were repeated seven times) showed the predicted effect with odds against chance of one hundred to one. Three new performances of the Revels in Cambridge, Massachusetts, again replicated the trend.

Wishing for Good Weather

One of the assumptions with which we began our experiments was that field-consciousness effects would introduce order into any form of physical system. Nelson imagined the same thing, and he thought of a very clever "natural" test of this assumption.[17] After having attended many of the graduations at Princeton University, he noticed that around the time of the graduation—when thousands of parents and alumni were in town and many activities were planned to take place outdoors, including the graduation ceremony itself—the weather seemed to be "too good."

Given the effects observed in the field-consciousness studies, he wondered whether "wishing for good weather" might actually make the rain stay away. In fact, in the summer of 1996 when President Clinton was invited to give a commencement address at Princeton University, the local newspaper commented on the legendary good weather in a report on contingency plans in case it rained on graduation day:

> . . . the third scenario is the Monsoon scenario, where it rains hard and commencement has to be moved to Jadwin Gym. Traditionally, this never happens at a Princeton University commencement. Those few times in recent years when precipitation is not only forecast but seems imminent, the rain has miraculously held off.

To test whether collective wishing made a difference, Nelson examined the historical weather data for the days before, during, and after graduation at Princeton University for a period of thirty years. He paid most attention to the daily precipitation data recorded in the Princeton, New Jersey, area, and in six surrounding towns, which acted as "control" locations. He predicted that on the day of graduation there would be more sunshine and less rain in Princeton than on the days before or after.

Nelson's analysis revealed that on average, over thirty years, there was indeed less rain around graduation days than a few days before and after graduation, with odds of nearly twenty to one against chance. An identical analysis for the average rainfall in six surrounding towns showed no such effect. Over thirty years, about 72 percent of the days around graduation had no rain at all in Princeton, whereas only 67 percent of the days in the surrounding towns were dry.

Curiously, on graduation day itself, the average rainfall was slightly higher in Princeton than in the surrounding towns, owing to a massive downpour of 2.6 inches on June 12, 1962. The average rain in the surrounding towns on that same stormy day was only 0.95 inches. What makes this even stranger is that the members of the Princeton Class of '62 reported that the massive rain that day held off until after the ceremony had ended![18] As Nelson then pointed out, this study prompts us to reconsider the old witticism, "Everyone talks about the weather, but nobody does anything about it." Perhaps we *can* do something about it.

Questions About Field Consciousness

When a fashion, a dance, a song, a slogan or a joke sweeps like wildfire from one end of the continent to the other, and a hundred million people roar with laughter, . . . there is the overpowering feeling that in this country we have come nearer the brotherhood of man than ever before.

ERIC HOFFER

The studies described here support ideas about deep interconnectedness espoused by physicists, theologians, and mystics.[19] Mind and matter may be part of what physicist Victor Mansfield describes as "a radically interconnected and interdependent world, one so essentially connected at a deep level that the interconnections are more fundamental, more real than the independent existence of the parts."[20]

The common link between mind and matter, as observed in these experiments, is *order*. Order expressed in the mind is related to focused attention, and order in matter is related to decreases in randomness. We found that the *object* of the focused attention does not seem to be particularly important, only that *something* is sufficiently interesting to hold the attention of a group. Similarly, the exact nature of the physical system used to detect the mass-consciousness effect does not seem to be particularly important, provided that it is a system that naturally fluctuates in some way and can be measured.

In these experiments the "mass mind" we looked at represented only a small fraction of the world's population. One might ask, Why didn't the mental noise produced by all the other minds in the world overwhelm the effects we observed? Or put another way, "How do you know that your results are caused by the group you are measuring, and not by one billion Chinese celebrating their New Year on the other side of the planet?" Good question.

The answer is that we assumed that the mental noise generated by everyone else in the world was *random in time* with respect to the events of interest in our experiments. Therefore, all the other minds did not systematically affect the specific results that we were monitoring. While other groups were undoubtedly involved in many interesting high-focus activities, we assumed that those activities did not occur at the same times as the events in our experiments. In other words, we speculated that we detected field-consciousness effects because we knew precisely when, where, and how to look for them.

We also assumed that the field-consciousness effect was *nonlocal,* meaning that it would not drop off with distance. This is a testable assumption that we haven't fully investigated yet. It may turn out that the Chinese New Year didn't affect these experiments because the mind-matter interaction effect does decline with distance from the group of interest. Of course, if the group is scattered all over the globe during a worldwide broadcast event, then we may not be able to detect any distance effects.

The results of these experiments also bear some resemblance to Jung's concept of "synchronicity," or meaningful coincidences in time.[21] As with synchronicity, we seem to be witnessing meaningful relationships between mind and matter at certain times. But synchronicity, according to Jung, involves *acausal* relationships, and here we were able to *predict* synchronistic-like events. Jung believed that people could experience but not understand in causal terms how synchronicities occurred:

> We delude ourselves with the thought that we know much more about matter than about a "metaphysical" mind or spirit and so we overestimate material causation and believe that it alone affords us a true explanation of life. But matter is just as inscrutable as mind. As to the ultimate things we can know nothing, and only when we admit this do we return to a state of equilibrium.[22]

We are more confident than Jung about what may be possible because it appears that with clever experimental designs, some aspects of Jung's *unus mundus* (one world) are in fact responsive to experimental probes, and some forms of synchronistic events can be—paradoxically—planned. We expect that Nature will reveal to us anything we are clever enough to ask for, but we also know that the revealed information is usually shrouded in unstated (and often unexamined) assumptions. At a minimum, we're beginning to glimpse that past assumptions about rigid separations between mind and matter were probably wrong.

Is Gaia Dreaming?

The idea of the world as an organism has been called the Gaia hypothesis, named after the mythical Greek goddess of the earth.[23] Do field-consciousness effects suggest that there may be a mind of Gaia? Just as the individual neurons in a brain would find it hard to believe that they are participating in the complex dance we call the "conscious mind," perhaps the individuals of earth are participating in the dance of Gaia's mind, and our experiments detected this dance.

Perhaps Gaia's mind is usually scattered, her attention distributed over innumerable objects of interest as the individual elements of her mind (i.e., all of us) conduct our daily business. As each of our minds twinkles and glitters over the course of a day, collectively Gaia's mind is dreaming, or musing aimlessly. But under exceptional circumstances—during worldwide, live television broadcasts, for instance—when many minds are focused on the same object, unbeknownst to us a grand alignment occurs. During these brief, shining moments, the billions of individually glittering minds reassemble into a whole, and the unity of Gaia's mind becomes brilliantly manifest. At such uncommon times (but becoming more common every day), Gaia in effect awakens, and we see this reflected in our random systems because they suddenly start behaving in statistically unexpected ways.

These studies also have profound implications for the understanding of social order and disorder. They suggest that a previously unsuspected cause of global violence and aggression may literally be the chaotic, malevolent thoughts of large numbers of people around the world. For example, the idea of a jihad, a holy war against infidels, which is fervently maintained by millions throughout the world, may not only *directly* (e.g., through terrorist acts) but also *indirectly* disrupt the social order around the world. By contrast, peaceful protests such as those embodied by Gandhi and Martin Luther King, which fostered noble intentions among groups, may have been successful not only for psychological reasons, but also for physical reasons that we are only now beginning to glimpse.

In sum, we've speculated that field-consciousness effects are pervasive but normally invisible unless we know where, when, and how to look for them. Another place where psi may be silently expressing itself is in an immensely popular enterprise built upon wishes—casino gambling.

CHAPTER 11

Psi in the Casino

If ever proof were needed against the existence of telepathy, psychokinesis, precognition or any other form of psychic power, the gambling halls of Las Vegas seem to provide the perfect place to find it. . . . Judging by the faces masked in concentration, it can hardly be said that the gamblers are not exerting every psychic effort to win. And yet still the cash flows into the pockets of the casino owners in an even, predictable stream.[1]

So opens an article entitled "Psychic Powers, What Are the Odds?" in the popular British science magazine *New Scientist*. At face value, such an argument certainly seems to provide a simple coup de grâce to the claims of psi abilities. But when examined more closely, the relationship between psi and the casino is much more interesting.

When parapsychologists dream, they dream about thousands of highly motivated people participating in psi experiments, twenty-four hours a day, in dozens of laboratories, worldwide. They dream that these laboratories will be exquisitely sensitive to human needs and desires, yet maintain stringent controls, fraud-proof testing conditions, and obsessive attention to data collection and verification. Some parapsychologists fantasize that people might even pay for the opportunity to participate in these experiments.

This dream is a reality today in gambling casinos. Except for being profit-oriented, many gambling games are essentially identical to psi experiments conducted in the laboratory.[2] If one accepts the evidence for precognition and psychokinesis, as discussed in chapters 7 and 8, then it is entirely reasonable to expect that those abilities ought to manifest to some degree in the casino as well. That is, at least some percentage of gambling winnings ought to be psi-mediated.

In 1968, Walter Tyminski, president of Rouge et Noir, a corporation specializing in offering advice on the mathematics of casino gambling, and Robert Brier, a philosopher at J. B. Rhine's Institute for Parapsychology, explored the idea that psi might be useful in the casino.[3] They tested a statistical technique to enhance psi-based predictions of the outcomes of roulette games, craps, and baccarat, and they achieved consistent, above-chance results. Tyminski and Brier's methods, however, required long sequences of repeated guessing, combined with an error-correcting scheme, to make a single prediction. The results, while promising, were not sufficiently outstanding to cause people to rush to the casino and attempt to use precognition. So the question remains, Does psi manifest in the casino under normal play by highly motivated gamblers?

For decades this question remained an untestable speculation because casinos, like most businesses, tightly control financial information. It is exceptionally difficult to gain access to casinos' daily profit-and-loss records, even for research purposes. This problem is compounded for research on possible psi-mediated effects, because most casinos are dubious about promoting study of anything that may affect their profits. Fortunately, an executive at the Continental Casino in Las Vegas was personally intrigued by parapsychology and was kind enough to provide us with daily gaming data that allowed us to test the "psi in the casino" hypothesis.[4]

Seeking Psi

To conduct a detailed search for psi in the casino, we would ideally like the daily win/loss data per player, per individual play, per game. We would like these data tracked for thousands of players over thousands of days, and we would like the empirically determined chance payout percentages of the various casino games to compare against the observed payouts. While this level of detail is of interest to casinos, the technology to track gaming behavior to this degree has only recently become available.[5]

The data made available to us were daily figures called "drop" and "result" for slot machines, roulette, keno, craps, and blackjack. "Drop" is casino jargon for the amount of money dropped on the table, i.e., the amount of money bet on the game. "Result" is the amount of money kept by the casino after the game is over. These figures allowed us to calculate the daily *payout percentage,* i.e., the win/loss percentage from the gambler's point of view, per game.[6]

For example, let's say that on Monday the *drop* (players' bets) for all slot machines was $50,000 and the *result* (what the casino kept) was $20,000.

That means the players won $30,000, and thus the payout percentage from the gamblers' perspective was $30,000/$50,000, or 60 percent. In other words, for each dollar played in a slot machine, on average the gambler's return was sixty cents.

While *drop* follows predictable weekly and seasonal cycles, daily payout percentages should not—according to chance—depend on the amount of money dropped; thus these daily values should be completely independent of any known periodicities. If, however, psi *were* widely distributed in the population, then daily fluctuations might not be entirely random. The proposed nonrandomness might manifest in the form of correlations between psi and external factors.[7] But what cycles or factors should we examine?

Behavior, Psi, and Geomagnetism

One factor is related to the fact that the earth, like a bar magnet, is surrounded by a magnetic field. But instead of being static and unchanging, the earth's geomagnetic field (GMF) is in constant flux. It is buffeted by highly charged solar particles, by interactions with the magnetic fields of other planets, and by movement of the earth's molten core.

For decades, the conventional wisdom about GMF and human behavior was that "biomagnetic effects on man are very small and are negligible as compared with other physical environmental stimuli."[8] This conclusion was based on the reasonable assumption that the energy absorbed by the human body due to geomagnetic fluctuations was below the "thermal limit." This means that the effects were so minuscule that cellular functioning was not influenced or disrupted in any way, and so no physiological and certainly no behavioral effects were thought to be possible.

More recent research suggests, however, that electromagnetic and magnetic flux well below the thermal limit, but shaped with certain patterns and complex frequencies, do indeed affect biology ranging from single cells to human physiology and behavior.[9] A small but growing body of literature suggests that variations in very weak, "extremely low frequency" (ELF) electromagnetic and geomagnetic fields affect some forms of human behavior.[10]

Even though we do not know why, the fact that these fields affect human behavior is demonstrable by examining historical data. Take, for example, analyses of accidents. An examination of when 362,000 industrial accidents and 21,000 traffic accidents took place showed significant correlations with ELF variations.[11] In addition, numerous studies have found correlations between changes in planetary GMF and some forms of unusual and abnormal human behavior.[12]

Of particular interest here is the growing literature in parapsychology suggesting that perceptual psi, both in the lab and spontaneously in life, *im-*

proves as GMF fluctuations *decrease*.[13] There are already more than a dozen studies showing that psi performance is better on days when the GMF is quiet. These studies range from spontaneous cases of "crisis telepathy" to performance in laboratory ganzfeld experiments. Some researchers have speculated that this relationship occurs because when the GMF is quiet, the brain is also quiet. The link between magnetic fields and the brain may be ferromagnetic elements located in the brain (possibly as vestigial navigational abilities, similar to tiny deposits of magnetic materials known to be in the brains of homing pigeons).[14] If so, the GMF–psi link would be reminiscent of the reasoning behind the ganzfeld technique: if external influences are quiet, then detection of psi will improve.[15]

If psi *were* operating in the casino, and fluctuating in relationship to environmental factors, we would expect to find a correlation between daily fluctuations in casino payouts and daily fluctuations in GMF. Following what we had already seen in previous studies, we predicted a negative correlation: casino payouts should *increase* when the GMF *decreased*.

Human Behavior and the Moon

There is something haunting in the light of the moon; it has all the dispassionateness of a disembodied soul, and something of its inconceivable mystery.

Joseph Conrad (1857–1924)

Another factor that we explored was the synodic lunar cycle, or the cycle from full moon to new moon. Researchers investigating moon–behavior relationships have most often compared lunar-cycle data against indices of abnormal and extreme behavior such as homicide, criminal activity, disturbances in psychiatric settings, and telephone calls to 911 crisis centers. Researchers have less often explored relationships with fire alarms, ambulance runs, children's unruly behavior, and drug intoxication.[16]

Some of these studies found significant relationships,[17] while others reported only small, inconsistent correlations.[18] For example, in 1979 skeptical psychologists Frey, Rotton, and Barry studied fourteen types of calls to police and fire departments over two years.[19] They found significant "but very small" lunar effects in six out of fifty-six tests but concluded that those few effects were essentially due to chance.

Reviews of the scientific literature on moon–behavior relationships have been generally negative. A meta-analysis published in 1985 by two psychologists concluded that lunar-phase influences were "much ado about nothing," and the authors hoped that their report would be "much *adieu* about the full moon."[20] In a later report, the same authors stated that after dividing the lunar cycle into four equal sections, they found that activities usually termed "lunacy" accounted for 25.7 percent instead of the chance expected 25 percent. They were "not impressed by a difference that would require 74,477 cases to attain significance in a conventional . . . analysis."[21] Although a 0.7 percent increase may not sound like much, in a city with a population of, say, one million, it could translate into hundreds of additional 911 crisis calls per day during "adverse" lunar phases.

A review published a few years later, in 1992, examined the relationship between suicides and lunar cycles and concluded that

> A consideration of the 20 studies examined here indicates that a knowledge of lunar phase does not offer the clinician any increase in ability to predict suicide and does not contribute to the theoretical understanding of suicide.[22]

In sum, lunar myths and lore have endured for millennia while modern science has remained skeptical. Contemporary popular articles on lunar–behavioral effects range from the uncritically dismissive[23] to the uncritically credulous.[24]

Magic and the Moon

> It is the very error of the moon;
> She comes more near the earth than she was wont,
> And makes men mad.
> SHAKESPEARE, *OTHELLO*

In spite of the lack of scientific consensus, surveys continue to show that many people believe in lunar–behavioral relationships. Thus, as with psi, it seems that human experience on this issue is at odds with conventional scientific wisdom. Parapsychologists have certainly learned the folly of ignoring human experience just because current scientific theories cannot adequately explain those experiences; thus it is worthwhile considering the historical links between the lunar cycle and magic. By magic, I mean the primeval origins of what we now call psi.[25]

The relevance of the moon for our study is the observation that religious ceremonies and magical rituals throughout history were often precisely timed to match certain phases of the lunar month. The moon figured prominently in medieval talismans, good-luck charms, and magic. The "witching hour" was midnight under a full moon, because that was when magical forces were supposed to be most powerful. Using secrets from the Cabala, lunar charms were designed to enhance fertility, favorably start new ventures, and heighten psychic powers.

During the centuries that religious and ceremonial practices were being timed to coincide with propitious lunar phases, it was also common knowledge that human and animal behavior was affected by the moon. Pliny the Elder, a Roman naturalist of the first century, wrote that "we may certainly conjecture that the moon is not unjustly regarded as the star of our life. . . . The blood of man is increased or diminished in proportion to the quantity of her light."[26] Nearly two thousand years later, modern medical researchers have reported that postoperative bleeding peaks around the time of the full moon.[27]

From medieval times, it was considered dangerous to sleep in the moonlight or even to gaze at the moon. Sir William Hale, a chief justice of England, wrote in the seventeenth century that "the moon hath a great influence in all diseases of the brain . . . especially dementia."[28] Two hundred years later, in writing England's Lunacy Act of 1882, Sir William Blackstone, the great English lawyer, defined "a lunatic, or *non compos mentis*," as "one who hath . . . lost the use of his reason and who hath lucid intervals, sometimes enjoying his senses and sometimes not, and that frequently depending upon the changes of the moon."[29]

Dipsomania, or periodical alcoholism, was associated with lunar cycles in some of the early psychiatric literature. In light of the legal treatments of lunacy, it is interesting to note that the nefarious "Son of Sam," a serial killer in New York City in the 1970s, murdered five of his eight victims on nights when the moon was either full or new.[30] Public fascination with "creatures of the night," including vampires and werewolves, continues to the present day, suggesting that this age-old folklore will remain in the forefront of our imagination for generations to come. Contemporary surveys confirm that many people still believe that strange behavior peaks around the time of the full moon.[31]

Experimental Psi and the Moon

We located one published experiment suggesting that psi performance in the laboratory varies with the lunar cycle. In 1965 neurologist Andrija Puharich proposed that psi performance might be related to gravity. To test his prediction, he needed to carry out an experiment under changing gravitational conditions. One way to do this was to conduct an experiment every day over the lunar cycle, because the sun-moon system predictably changes the gravitational forces (i.e., tidal forces) felt on earth.

Puharich's predictions for a telepathy test were coincidentally in accordance with expectations from magical folklore. He proposed that perceptual psi would increase around the full moon, decrease at the half-moons, then rise again around the new moon. The experimental results confirmed Puharich's prediction, as illustrated in figure 11.1.[32]

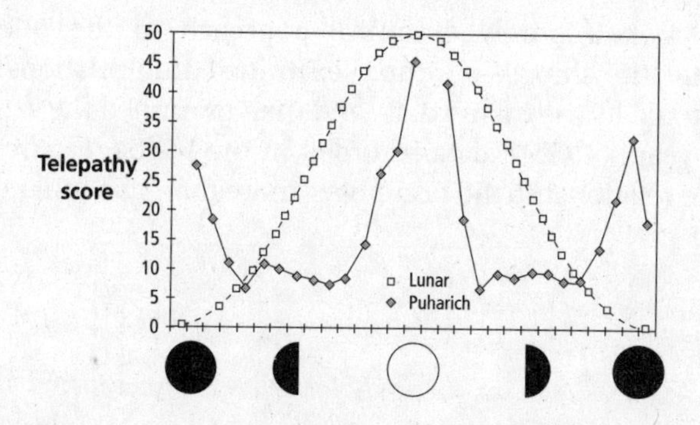

Figure 11.1. Outcome of Puharich's (1965) telepathy test. The left scale shows the score in a telepathy test. The smooth curve is the lunar cycle, labeled on the bottom of the graph with symbols indicating the new, half, and full moons.

Lunar–Solar–GMF Relationships

An alternative explanation for Puharich's result is that the observed effect, rather than being a gravitational effect per se, might reflect a complex relationship between the lunar cycle and the GMF. The suggestion is feasible because fluctuations in GMF have been linked to numerous periodic factors, including long-term "secular variations" related to structures in the earth's core and shorter-term "external variations" such as solar activity, a daily GMF cycle, and, of primary interest here, the synodic lunar cycle.[33]

A flurry of studies in the 1960s, published mainly in the geophysical literature, suggested the existence of a lunar–GMF correlation.[34] Later analyses demonstrated that these apparent correlations were probably due to fluctuations in the solar "wind."[35] It turns out that there is a close coincidence between the length of the lunar synodic month (29.53 days) and the rotational period of the sun, so what originally appeared to be a lunar–GMF association might have been confounded by solar effects. Then, later analyses showed that the moon passes through the earth's magnetosphere around the time of the full moon. This led to new speculations about lunar–GMF relationships. For example, Stanford University geophysicist Anthony Fraser-Smith published evidence of a clear lunar–GMF relationship during total lunar eclipses in data recorded after 1932.[36]

To further confuse the situation, while the geophysical relationships are not yet clarified, biological systems seem to be exquisitely sensitive to tiny energetic effects that might otherwise seem negligible. For example, the marine mollusk responds differently to geomagnetic fields according to the phase of the moon,[37] and both human beings and rats display different thresholds for convulsions according to changes in magnetic fields and the position of the moon during solar eclipses.[38]

Thus, we took a purely empirical approach to the question of a lunar–GMF relationship. We simply examined this relationship for the four years covered by the casino data, and then examined the same relationships for ten years of GMF data recorded in the 1980s. Figure 11.2 shows that a negative relationship did occur over more than 120 lunar cycles.

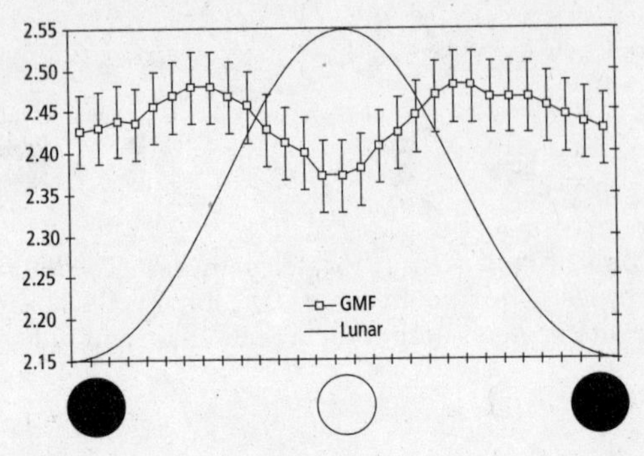

Figure 11.2. Graph of lunar cycle and GMF for 1980 to 1989. The error bars are 65 percent confidence intervals. The left ordinate is the natural log of the daily average GMF "Ap" index.

The Casino Study

Figure 11.3 lists the variables that we used in our analysis of the casino data. The data cover daily values for the four years 1991 through 1994.

VARIABLE	DAILY DATA
	CASINO GAMES
Roulette	drop, result, payout percentage
Keno	drop, result, payout percentage
Blackjack	drop, result, payout percentage
Craps	drop, result, payout percentage
Slots	drop, result, payout percentage
COMBO	average payout percentage for the above five games
	GEOPHYSICS
GMF	natural log of the mean geomagnetic planetary Ap index

Figure 11.3. Variables used in the casino study.

In figure 11.3, the phrase "payout percentage" refers to the payout percentage from the gambler's point of view, i.e., *p% = (drop − result)/drop*. The value of *drop* must be positive, reflecting the fact that real money is dropped on the table or in the slot machine, but *result* can be positive or negative because the casino can win or lose money. The primary term of interest is "COMBO," the combined average of the daily payout percentages for the five casino games.

If *result* is negative, it means the casino lost money to the gamblers, and *p*% will be greater than 100 percent. If *result* is positive, it means that money remained on the table after the day was done, the casino earned a profit, and *p*% will range between 0 percent and 100 percent. In this data set, daily payout percentages on the various games ranged from about 5 percent to more than 400 percent. The large payout percentages represented times when one or more gamblers hit jackpots or numerous gamblers had an unusual run of luck.

Figure 11.4 shows the overall payout averages for each of the casinos games and for COMBO. The very small confidence intervals indicate that gambling payouts are very stable over many years. From this graph we see that for each dollar played in roulette, on average a predictable seventy-seven cents was returned to the gambler.

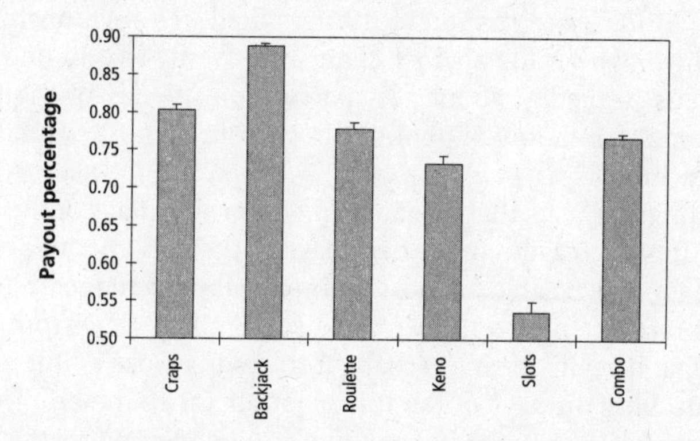

Figure 11.4. Payout-percentage means for the casino games, with 65 percent confidence intervals.

ANALYSES

Our study was primarily interested in two relationships. We predicted, first, that the relationship between lunar cycle and average payout percentage (called "COMBO") would be *positive,* and second, that the relationship between geomagnetic field (GMF) strength and COMBO would be *negative*. We based the first prediction on magical lore and Puharich's experimental results and the second on previous literature suggesting that perceptual psi improves on days of lower GMF.

It is important to point out that these correlations were conducted *with respect to the lunar cycle* by using an analysis centered on the day of the full moon. That is, for the variables COMBO and GMF, we first determined the average of all daily GMF values that fell *on* the day of the full moon. Then we formed a new average for the values that fell *one day after* the full moon, and then one day before, two days after, and so on, until we had determined daily averages for each of the twenty-nine days. Because the database contained four years of daily data, each of these daily averages was based on data covering forty-nine to fifty lunar cycles, which provided good average estimates for each day of the synodic cycle.

Results

I always find that statistics are hard to swallow and impossible to digest. The only one I can ever remember is that if all the people who go to sleep in church were laid end to end they would be a lot more comfortable.

MRS. ROBERT A. TAFT

The lunar cycle–payout percentage relationship was predicted to be positive. As shown in figure 11.5, these data resulted in a relationship that was indeed positive, with odds against chance of twenty-five to one.[39] Average casino payouts peaked at about 78.5 percent on the day of the full moon, and they dropped to a low of about 76.5 percent about a week before and after the new moon.

This finding suggests that by gambling on or near days of the full moon, and by avoiding the casino on or near days of the new moon, over the long term gamblers may be able to boost their payout percentage by about 2 percent. If this relationship continues to be seen in new data from other casinos, then what it really means is that gamblers may lose a little slower than usual by gambling on days of the full moon, because the empirical payout percentage is always going to be less than 100 percent no matter when they play. Casino managers have nothing to worry about.

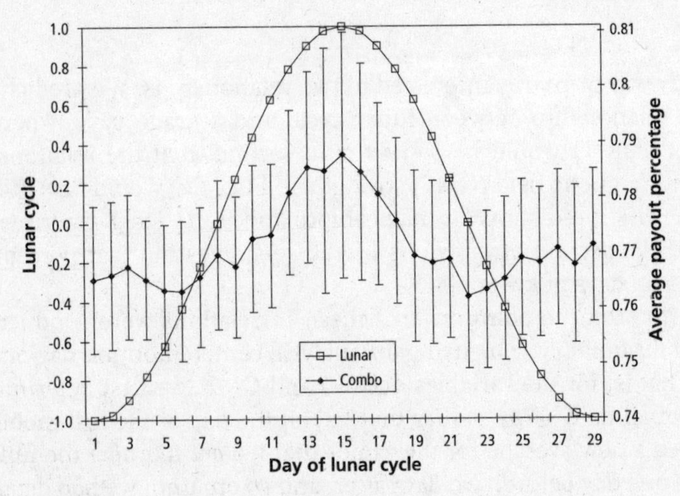

Figure 11.5. Lunar cycle–payout percentage relationship, with 65 percent confidence intervals.

The GMF–payout percentage relationship was predicted to be negative. As shown in figure 11.6, these data resulted in a nearly significant negative relationship, with odds against chance of fourteen to one.[40]

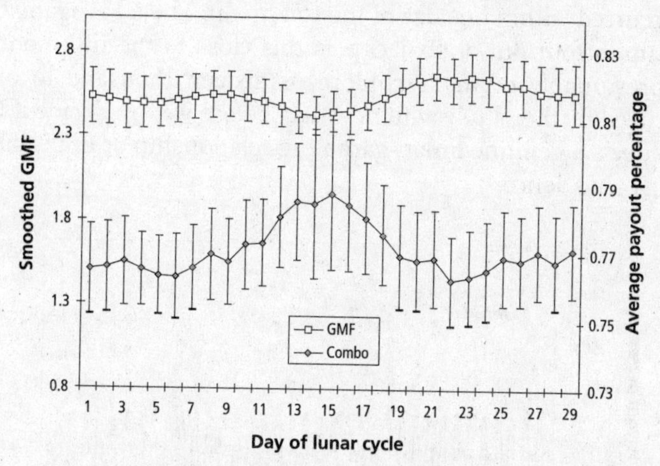

Figure 11.6. GMF–payout percentage relationship, with 65 percent confidence intervals.

Slot-Machine Analysis

To investigate the "lunar effect" in more detail, we examined the distribution of slot-machine payouts by themselves. Figure 11.7 shows that the largest payouts occurred around the time of the full moon. The larger confidence intervals around the time of the full moon indicate that the larger values there were probably due to only a few jackpots rather than to systematically higher payout rates.

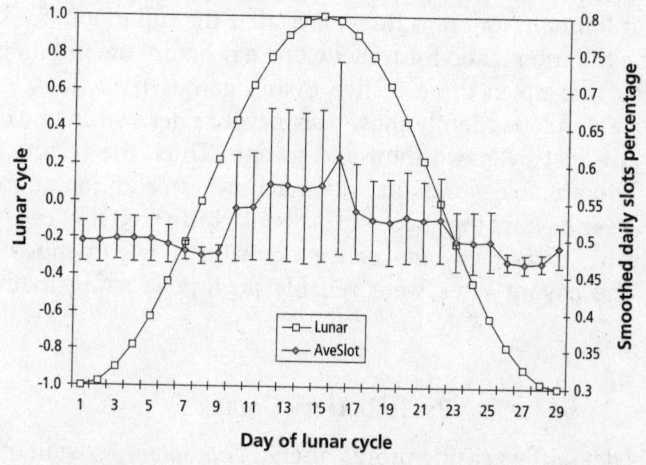

Figure 11.7. Smoothed slot-machine payout percentages by lunar cycle, with 65 percent confidence intervals.

This supposition was confirmed by a closer examination of the daily time-course of slot-machine payouts (figure 11.8), which showed that four of the six major jackpots recorded over the course of the four-year database[41] actually occurred within one day of the full moon. The odds against chance of seeing up to four out of six jackpots this close to the full moon, when jackpots presumably occur at random, is sixteen thousand to one.[42] As usual, we need further study with new data before we can decide if this correlation reflects a genuine lunar–gambling relationship or if it is simply an interesting coincidence.

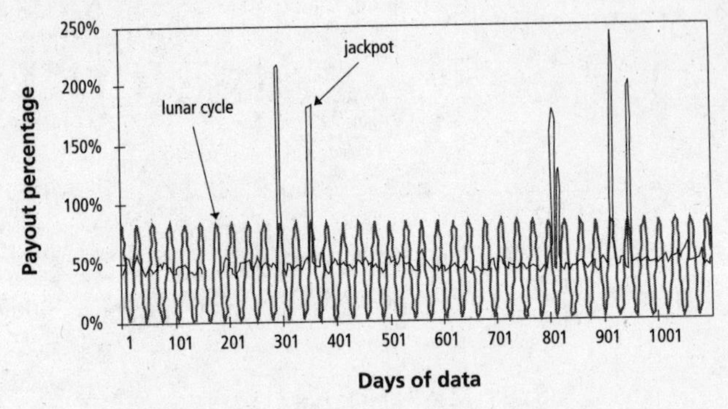

Figure 11.8. Lunar cycle and slot-machine payout percentages. The ordinate is the payout percentage; the abscissa is the number of days. The sine wave is the lunar cycle.

Peak Payout Days in Other Games

Next we analyzed the results of each of the other games independently. We found that the peak average payout rate for blackjack occurred three days before the full moon, for craps three days after the full moon, for keno one day after the full moon, and for roulette one day before the full moon.

The odds that up to three of five casino games (i.e., slots, keno, and roulette) would independently show peak payout rates within one day of the full moon are just over two thousand to one. Thus, the results observed here held for both table games and slot machines. In addition, more sophisticated analyses beyond the scope of this book confirmed that not only were fluctuations in casino payouts consistent with a twenty-nine-day lunar cycle, but the payout rates were reliably *predictable* using mathematical models.[43]

Psi in Lottery Games

If psi really does affect casino profits, then it should also exist in other types of mass games, like lotteries. To test this idea, we looked at daily payouts from the "Pick 3" lottery game, popular in many U.S. states. In this game, the player guesses three digits, and if the guess matches the winning number randomly selected the next day, the player win. We requested lottery information for the year 1993 from fifteen states, but only six states (California, Illinois, Kentucky, Michigan, Missouri, and Virginia) provided data that allowed us to form the daily payout percentage—the ratio of the money won each day to the total money collected that day.

As with casino payout percentages, the "Pick 3" lottery payouts do not have day-of-week biases. Payout percentages are based only on the number of winning lottery tickets per day, and the number of winners is, according to conventional assumptions, a pure chance event. Thus, although there are typically more lottery *players* on Friday and Saturday, the number of *winners,* and therefore the payout percentages, are not affected by the day of the week. Of course, on days when the winning lottery number matched numbers that people commonly select, like 711, or 123, then the payoff percentages tended to be quite large.

We observed that for the year 1993 there was a *positive* relationship between the planetary GMF and the lunar cycle, with odds of one hundred to one against chance.[44] This being the case, we predicted based on previous observations that there should be a *negative* relationship between the lunar cycle and average lottery payouts.

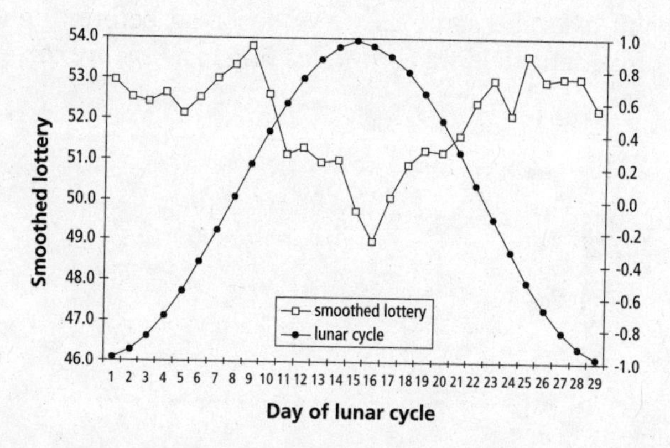

Figure 11.9. Relationship between the days of the synodic lunar cycle and lottery payout rates.

Figure 11.9 shows the result, which confirmed the predicted negative relationship with odds against chance of 130 to 1.[45] In addition, lottery payouts for five of the six states (Illinois, Kentucky, Michigan, Missouri, and Virginia) independently resulted in negative relationships with the lunar cycle. Of those relationships, the result for Michigan was significantly negative with odds of twenty-five to one. This finding supports the idea that it may not be the moon per se that seems to affects psi in gambling, since in this case the lottery payouts *decreased* around the time of the full moon. Instead, there may be one or more "hidden" geophysical relationships that are associated with geomagnetic fluctuations.

French and Russian National Lotteries

If our investigations were correct, and psi does indeed manifest in the pragmatic world of casino and lottery profits, then one might expect to find corroborating evidence from other sources. Such evidence does exist.

Russian computer scientist Mark Zilberman studied the national lotteries in the former Soviet Union and in France for the decade of the 1980s. Specifically, he examined 509 daily draws of the "6/49" lottery in France and 574 draws of the "5/36" lottery in the former USSR.[46] These lotteries were interesting because data were available for lottery draws on a daily basis, and the data included both the numbers selected by individuals and the winning numbers. These same data are collected, of course, by all lottery systems, but in the United States the raw data are difficult to obtain for research purposes.

Zilberman found that the day-to-day payout fluctuations, which according to probability theory should have been randomly distributed over time, were actually strongly related to fluctuations in the global geomagnetic field (figure 11.10). The overall odds against chance were four hundred to one.[47] The relationship was the same as we've observed before: higher payouts when the geomagnetic field was lower, leading to a *negative* correlation.

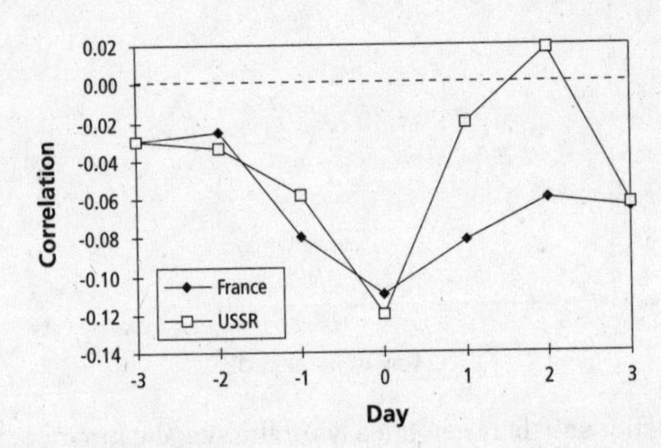

Figure 11.10. Relationship between geomagnetic field fluctuations and lottery payouts for France and the USSR in the 1980s.

Summary

The studies described in this chapter suggest that daily fluctuations in casino and lottery payouts are not due to pure chance. Some fraction of the payout rates appears to be related to daily fluctuations in the average psi ability of millions of gamblers. We know this because the payout rates fluctuate in ways that are consistent with what we independently know about environmental influences on psi performance. This provides additional support for one implication of the field-consciousness studies discussed in chapter 10: psi effects are probably more pervasive than we've thought. In fact, in some realms, the scientific controversy has been finessed for decades. Quietly, and without any fuss, psi applications are already being used.

CHAPTER 12

Applications

We didn't know how to explain it, but we weren't so much interested in explaining it as determining whether there was any practical use to it.

MAJOR GENERAL EDMUND R. THOMPSON,
ARMY ASSISTANT CHIEF OF STAFF FOR INTELLIGENCE, 1977–81

In our survey of the scientific evidence for psi, we've moved from the inner world of psi perception to the outer world of mind-matter interactions, from artificial tasks in the laboratory to psi influences in the world at large. While science has been slow to unravel the secrets of psi, the rest of the world is more pragmatic. Most people don't care how psi works, only *that* it works.

Scientists often interpret legends and other traditional accounts of applied psi as nothing but old wives' tales. Meanwhile, modern psi research is revealing that some of those "old wives" were probably more clever than we'd thought. And while the scientific debates simmer, practical applications of psi can already be found. These applications fall into five broad categories: medicine, military, detective work, technology, and business.

Medicine

Medical applications of psi can be traced to antiquity. Individuals claiming to heal with various forms of mental intention were called shamans, medicine men and women, witch doctors, wizards, psychic healers, and spiritual healers.[1] Today, thousands of conventional nurses use a form of mental healing called therapeutic touch.[2] The practice of distant prayer is also pervasive worldwide. As we saw in chapter 9, there's persuasive laboratory evidence that some of these methods really do work.

Another medical application is psychic or "intuitive" diagnosis. Perhaps the most famous twentieth-century psychic diagnostician was Edgar Cayce, who died in 1945. Detailed documentation and analyses of thousands of his readings are available in dozens of books.[3] Cayce was exceptionally good at what he did, but his ability was not unique. For instance, in 1995 D. Lawrence Burk, a physician with Duke University's Medical Center, described the case of a thirteen-year-old girl who complained of pain in her leg and back. X rays and magnetic resonance imaging were used to examine her pelvis and spine, and a tumor was detected in her left sacrum, extending into the spine. Before a biopsy was taken of the tumor, Dr. Burk phoned a medical "intuitive" he knew and gave her only the name and age of the patient. He did not mention the girl's symptoms or the results of the X rays and magnetic resonance imaging. As Burk described it, the intuitive thought for a few seconds, then replied, "There is a tumor in the pelvis working its way into the spine. This immature girl has a terminal condition."[4] The results of a biopsy, taken after this reading, revealed that the tumor was indeed cancerous.

Burk later found many reports of intuitive diagnosis in the so-called alternative medical literature, but he was more interested in whether the mainstream had taken seriously this remarkable ability. He was surprised to find, in the 1993 program of the annual meeting of the Society for Medical Decision Making, that that group had devoted a full morning session to the topic. Intuitive diagnosis was also discussed at length at a 1995 research meeting of the National Institutes of Health, Office of Alternative Medicine. This shows that in the pragmatic world of medicine, it really doesn't matter if useful information comes from magnetic resonance imaging, blood tests, or psi.

But from a scientific perspective, research on distant healing and intuitive diagnosis is still in its infancy. While many clinical techniques are being used and taught, there are still no standardized guidelines for determining who can do what, or how distant healing works, or even under what conditions it is likely to work. As with any claim about psychic ability, simply accepting an assertion about the efficacy of mental healing without having substantial evidence to back it up is fraught with risk.

The economic implications of even weakly effective psi-based distant healing, or approximately accurate psi diagnosis, are *enormous*. Imagine if distant healing were found to be effective in treating a fraction of chronic illnesses, or if psi diagnosis could enhance physicians' likelihood of correctly assessing a medical condition after using conventional diagnosis techniques. This would translate into huge savings and improved quality of life for millions of people. While the jury is still out, existing data already suggest that (at minimum) stress-related illnesses, which affect tens of millions, can be treated to some extent by distant-mental-healing techniques, even by volunteers with no special aptitude. It is equally likely that some people are really good at providing useful medical diagnoses. Obviously, no one should rely solely on these techniques, but they can be useful as adjuncts to conventional methods.

If the effects observed in the laboratory can be enhanced through training, or by selecting individuals for "talent," then it is no exaggeration to predict that untold *billions* of dollars in medical costs could be saved. Of course, those billions would be taken directly out of the coffers of the established medical community, and there are serious consequences of challenging the status quo.

Military and Intelligence Applications

Psi has had national security applications for millennia. One of the earliest essays on the art of war, written by the Chinese general Sun Tzu in 500 B.C., described how success in battle depended not only on military strategy and tactics, but on the application of *ch'i*, the life force. Soldiers trained in the proper use of *ch'i* were said to be able to influence the minds of their enemies at a distance.[5]

In more recent times, rumors persisted for decades that military and intelligence agencies were supporting research on psi phenomena.[6] Investigative journalists wrote of secret programs with exotic code names like Grill Flame and Stargate. But the rumors were always shrouded in conspiracy theories, plausible denials, and orchestrated disinformation, and very few people knew what was really going on.

It was a safe bet, however, that something was afoot, because military historians had already documented the use of remote viewing during World War II. After the war, secret British army documents revealed that the wife of the head of the Royal Air Force—her husband was known as the "man who won the Battle of Britain"—was a "sensitive." She was credited with using remote viewing to locate enemy air bases that conventional methods had not detected. Another key military leader, the American general George S. Patton, believed that he was the reincarnation of a Roman general, and General Omar Bradley agreed that Patton seemed to possess a "sixth sense."[7]

In the 1950s secret CIA-funded programs involving some psi research were code-named Projects Bluebird and Artichoke. In 1978 University of California psychologist Charles Tart surveyed fourteen psi research laboratories and found that five had been approached by government agencies interested in tracking their progress. During the Reagan administration, the House Science and Technology Subcommittee released a report containing a chapter on the "physics of consciousness." The report stated that psi research deserved Congress's attention because "general recognition of the degree of interconnectedness of minds could have far-reaching social and political implications for this nation and the world."[8]

Then, in November 1995, most aspects of the rumored, highly classified programs were declassified, and several dozen people who had worked on the programs over the preceding few decades were free to admit that some of the rumors had been correct.[9] We have already briefly discussed these projects in chapter 6, because the bulk of the U.S. government–supported research involved remote viewing.

Why was this topic supported for two decades, under the watchful eyes of highly skeptical CIA and DIA[10] contract monitors and a world-class scientific oversight committee? For one very simple reason: remote viewing works—sometimes. The "hit rate" for the military remote viewers was not wildly greater than the results observed in the clairvoyance experiments discussed in chapter 6, but when conventional investigation and intelligence techniques were at a loss to provide critical information on sensitive missions, sometimes remote viewing worked spectacularly.

For example, in September 1979 the National Security Council asked one of the most consistently accurate army remote viewers, a chief warrant officer named Joe McMoneagle, to "see" inside a large building somewhere in northern Russia.[11] A spy-satellite photo had shown some suspicious heavy-construction activity around the building, which was about a hundred yards from a large body of water. But the National Security Council had no idea what was going on inside, and it wanted to know. Without showing McMoneagle the photo, and giving him only the map coordinates of the building, the officers in charge of the text asked for his impressions. McMoneagle described a cold location, with large buildings and smokestacks near a large body of water. This was roughly correct, so he was shown the spy photo and asked what was inside the building. McMoneagle sensed that the interior was a very large, noisy, active working area, full of scaffolding, girders, and blue flashes reminiscent of arc welding lights. In a later session, he sensed that a huge submarine was apparently under construction in one part of the building. But it was too big, much larger than any submarine that either the Americans or the Russians had. McMoneagle drew a sketch of what he "saw": a long, flat deck; strangely angled missile tubes with room for eighteen or twenty missiles; a new type of drive mechanism; and a double hull.

When these results were described to members of the National Security Council, they figured that McMoneagle must be wrong, because he would be describing the largest, strangest submarine in existence, and it was supposedly being constructed in a building a hundred yards from the water. Furthermore, other intelligence sources knew absolutely nothing about it. Still, because McMoneagle had gained a reputation for accuracy in previous tasks, they asked him to view the future to find out when this supposed submarine would be launched. McMoneagle scanned the future, month by month, "watching" the future construction via remote viewing, and sensed that about four months later the Russians would blast a channel from the building to the water and launch the sub.

Sure enough, about four months later, in January 1980, spy-satellite photos showed that the largest submarine ever observed was traveling through an artificial channel from the building to the body of water. The pictures showed that it had twenty missile tubes and a large, flat deck. It was eventually named a *Typhoon* class submarine.

Scores of generals, admirals, and political leaders who had been briefed on psi results like this came away with the knowledge that remote viewing was real. This knowledge remained highly classified because remote viewing provided a strategic advantage for intelligence work. In addition, the agencies that had supported this research knew very well that the topic was politically and scientifically controversial. They had to deal with the same "giggle factor" that has caused academic and industrial scientists to be careful about publicizing their interest in psi.

Scientists who had worked on these highly classified programs, including myself, were frustrated to know firsthand the reality of high-performance psi phenomena and yet we had no way of publicly responding to skeptics. Nothing could be said about the fact that the U.S. Army had supported a secret team of remote viewers, that those viewers had participated in hundreds of remote-viewing missions, and that the DIA, CIA, Customs Service, Drug Enforcement Administration, FBI, and Secret Service had all relied on the remote-viewing team for more than a decade, sometimes with startling results.[12] Now, finally, the history of American and Soviet military- and intelligence-sponsored psi research is emerging as participants come forward to document their experiences.[13]

A related, but completely separate source of military interest in psi comes from jet fighter pilots. The complexity of modern weapons systems, severe mission requirements, and the ever-present danger of enemy fighter jets and missiles have forced the development of extremely fastidious criteria for selecting pilots. Unfortunately, in spite of rigorous selection procedures and training, not all fighter pilots are equally effective. It is estimated, for example, that about 5 percent of fighter pilots have accounted for about 40 percent of the successful engagements with hostile aircraft (i.e., "kills") in every aerial combat since World War I.[14]

While opportunity plays a role in these percentages, substantial differences remain even after the "kill opportunity" is equalized. Jet pilots and aerospace engineers would like to understand what separates the "top guns" from the less-effective pilots. In 1991 researchers B. O. Hartman and G. E. Secrist published an article in a conventional aerospace medical journal with the euphemistic title, "Situational Awareness Is More Than Exceptional Vision."[15] "Situational awareness" refers to a pilot's hypersensitivity to aircraft performance and ability to quickly anticipate and act upon changes during combat. In some instances, situational awareness surpasses hypersensitive levels, and Hartman and Secrist compared this level of performance to psi perception. For example, combat pilot L. Forrester described superior situational awareness as follows:

> There is some sixth sense that a man acquires when he has peered often enough out of a [jet fighter cockpit] into a hostile sky—hunches that come to him, sudden and compelling, enabling him to read signs that others don't even see. Such a man can extract more from a faint tangle of condensation trails, or a distant flitting dot, than he has any reason or right to do.[16]

Detective Work

When faced with long-standing unsolved crimes, police have occasionally turned to psychics for assistance. Well-documented cases of psychic detective work can be traced back to the early part of the twentieth century,[17] and psychic detective work is still popular. Author Arthur Lyons and sociologist Marcello Truzzi from Eastern Michigan University, who recently conducted a comprehensive evaluation of psychic detectives,[18] report that some police detectives with exceptionally good crime-solving abilities refer to their hunches as "the blue sense." Interestingly, one of the most accurate remote viewers ever to work on the CIA-sponsored psi research program at SRI International was a former police commissioner named Pat Price.

There are many fascinating anecdotes of cases in which psychics seemed to be instrumental in solving crimes, but any neutral observer will also acknowledge that—as with psychic spying and most other anecdotal cases—it is extremely difficult to reach any strong conclusions about individual cases. However, given the strength of the laboratory evidence for psi perception, and the evidence from dozens of successful cases of military remote viewing, it is very likely that *some* cases of psychic detective work actually are due to genuine psi.

Technology

Taking the mind-matter interactions observed in laboratory studies as their starting point, some scientists have proposed that psi-based communication and switching devices could be built.[19] In principle, such devices could be developed now because most of the engineering design problems have already been worked out. For example, the same methods used to communicate with spacecraft millions of miles from Earth (like planetary exploration satellites) would be useful in decoding the weak, noisy "signals" that psi-based technologies would probably generate. Likewise, the pattern-recognition methods used in advanced sonar and radar systems would be useful in psi-based systems that could be trained to respond to individual thought-patterns at a distance.[20]

If such devices could be developed, they might allow thought control of prosthetics for paraplegics, mentally directed deep-space and deep-sea robots, and mind-melding techniques to provide people with vast, computer-enhanced memories, lightning-fast mathematical capabilities, and supersensitive perceptions. It may also be possible to create technologically enhanced telepathic links between people.[21]

Other devices, based on presentiment, may be developed into novel early-warning systems that monitor our "presponses" to future events whose effects are "reflected" backward in time. Imagine, for example, an aircraft in which each member of the flight crew is connected to an onboard system that continuously monitors several aspects of his or her bodily state. This might include heart rate, electrodermal activity, and blood flow. Such monitors are already commonly used for astronauts.

Before the crew boards the aircraft, they would be calibrated to see how each responds to different kinds of emotional and calm events, using a method similar to the presentiment experiment described in chapter 7. Each person's idiosyncratic responses could be used to create a unique, emotional "response template." Now we inform a computer onboard the aircraft about each person's response template, and using telemetry, we have it monitor each crew member's ongoing bodily state to look for times when anyone seems to be having an emotional response.

We would expect to find a crew member occasionally producing responses that may actually reflect emotional states, such as when a flight attendant is dealing with a rude passenger. At other times, such responses may appear to reflect emotional states, but they are false alarms, such as when a flight attendant lifts a piece of luggage into an overhead rack. With appropriate programming, the computer can learn to reject such false alarms.

But what if the computer suddenly detects that every member of the crew is responding emotionally at the same time? This would not be a good sign, but it is possible that, say, unexpected flight turbulence has caused a simultaneous emotional response in the crew. Fortunately, the computer can also reject this as a false alarm, because it can use other sensors on the aircraft to detect coincidences between sudden emotional responses and changes in aircraft performance and environmental factors.

Now imagine that we have refined the presentiment-detection technique to the point where we can reliably anticipate when an emotional response is *about* to occur, before the problem even exists. If our onboard computer suddenly detected that all crew members were about to have an emotional response, *and* the aircraft was still operating normally, then the computer could alert the pilot (perhaps seconds or minutes in advance). Sometimes even a few seconds of advance warning in an aircraft can save the lives of everyone on board.

Quiet research programs examining these and other exotic technological possibilities have been under way for several years in academic and industrial laboratories. Consider, for example, the following story, which appeared in the December 10, 1995, *South China Morning Post:*

> SONY, the corporation which revolutionized the world of audio and electronics has acknowledged it is conducting research into alternative medicine, spoon bending, X-ray vision, telepathy and other forms of extra-sensory perception (ESP).
>
> The Institute of Wisdom was founded in 1989 at the instigation of Sony's founder. . . . The company believes it has proved the existence of ESP, and is already developing a diagnostic machine based on the principles of oriental medicine. . . .
>
> A sub-division of the institute, Extra-Sensory Perception Excitation Research, has worked with more than 100 possessors of ESP. In one test, subjects were presented with two black plastic containers, one of them containing platinum, the other empty. Psychic individuals were able to "see" the platinum seven times out of 10.
>
> Yoishiro Sako, a former specialist in artificial intelligence who heads the four man research team, believes commercial applications could apply to his research. "We haven't come up with such great results so far," he said, "but if we eventually discover that ki energy is based on a kind of information transmission, it would lead to a complete energy revolution. If we can understand the mechanism of telepathy, it would totally transform communication methods."

While high-technology companies in the United States and Europe have been more reluctant to publicize similar interests in psi, at least two large companies have had in-house psi research projects: Bell Laboratories[22] and Contel Technology Center.[23] At Bell Labs in the 1980s, I explored mind-matter interaction effects to see whether certain electronic circuits might be susceptible to psi influences; some aspects of this research achieved the Bell Labs imprimatur.[24]

At Contel in the early 1990s, I began to experiment with commercially available, off-the-shelf electronics to see if ordinary components were susceptible to psi influences. This was an important first step toward building psi-based devices, because unless scientists are able to demonstrate proof-of-principle with existing hardware and software, they will have no hope of obtaining funding to create speculative, custom-made microelectronics from scratch. For the physical target, I used a random-number generator on a single chip, made by AT&T Microelectronics. The intended purpose of this chip was to generate a random key for a highly secure data-encryption method, so great care was taken in its design to ensure that the numbers it produced were truly unpredictable by any ordinary means. I conducted two psi experiments using this chip, and both were successful in demonstrating mind-matter interaction influences precisely where I had predicted they would appear.[25]

I then proposed to build a prototype "thought-switch" and test it in-house to see if we could demonstrate proof-of-principle of a new psi-based technology. The project was approved and the device was built and tested in late 1990. The prototype incorporated a new type of physical detector, involving a matrix of random-number generators, and some advanced statistical and signal-processing techniques to detect the predicted psi influences.[26] The test involved ten volunteers who were asked to mentally influence the random system in strictly prescribed ways. The experiment was successful, prompting us to prepare a patent disclosure.[27] Unfortunately, immediately after we completed the prototyping tests, GTE Corporation merged with Contel, and the disruption of the merger halted our efforts on this project.

One of the ideas underlying this technology is that we know—as described in chapter 8—that a single random-number generator behaves in nonchance ways when an individual is asked to direct his or her mental intention toward it. But a random-number generator doesn't "like" to change its behavior. It is bound to operate in certain statistically prescribed ways; otherwise, it is no longer operating properly. If it is *forced* to generate numbers well beyond its normal operating conditions, say by heating it or exposing it to radiation, the device will probably be permanently damaged. This is one reason that psi effects with random-number generators are not very large—the random fluctuations it can produce are restricted by the design of the device and the statistics that govern its behavior.

But there may be a way around this restriction. By analogy, let's say that we were trying to mentally influence a gas molecule. We might stare intensely at the gas molecule (assuming we had some method of seeing it) and mentally try to "push" it from the left to the right side of a jar. This might work, but it would take an enormous amount of mental effort because molecules don't like to be pushed around. To move across the jar, the molecule has to push aside all those other gas molecules and overcome its own tendency to stay in approximately the same location. So instead, let's say that all we really care about is influencing the entire *collection* of gas molecules to drift a little to the right. Now we are not asking a given molecule to do anything it wouldn't ordinary do through random fluctuations, and the new task involves mentally influencing the statistics of the entire *system* rather than the position of an individual molecule.

The same idea can be applied to the outputs of random-number generators. The technological challenge is not in building the required physical system, but in developing analytical methods to detect that the system has actually responded to a specific mental influence and not just to random movements. In my lab at the University of Nevada, Las Vegas, I am continuing to investigate ideas for psi-based technologies, working with former "psychic spy" Joe McMoneagle to help flesh out the details of these technologies. Joe has no doubt that such devices will exist someday, and I tend to agree. But the question is whether we are clever enough to make the conceptual leap from existing principles to the future devices that Joe has sensed in his remote-viewing sessions.

Business

In science and technology, intuition is widely recognized as an essential source of innovation and discovery. Architect Buckminster Fuller once examined the diaries of great scientists and inventors, looking for common denominators. The single element he found in common was "that their diaries declared spontaneously that the most important item in connection with their great discovery of a principle that nobody else had been able to discover, was intuition."[28] The philosopher Bertrand Russell also maintained that science needs both intuition and logic, the first to generate and appreciate ideas and the second to evaluate their truth.

In case studies of scientific breakthroughs, key insights are often said to have appeared in a flash, which resembles how psychics describe their intuitive impressions. Many highly creative scientists and engineers have been drawn to the study of psychic phenomena, among them Sir Isaac Newton, Sir William Crookes, and Nikola Tesla.[29] More recently, Earl Bakken, founder of Medtronic, the first company to build heart pacemakers, John E. Fetzer, the communications pioneer, and James S. McDonnell, founder of McDonnell-Douglas, have all supported psi research because of their personal interest in psychic phenomena.

We have already mentioned SONY Corporation's interest in psi, reflecting, we believe, a greater societal openness toward psi in the Eastern world. SONY is not an isolated case. According to an article in the *Asian Wall Street Weekly*, in 1985 the Japanese government's Science and Technology Agency "decided to study 'man's spiritual activities' in its Creative Science and Technology Promotion Program," which began in 1987.[30] Tadshiro Sakimoto, president of NEC Corporation, said in this article, "The study of the sixth sense and telepathy will certainly prove a cornerstone of future modes of communications."

In the same article, Hiroo Yuhara, former head of the Japanese Post and Telecommunications Ministry's Radio Research Laboratories, echoed this sentiment by remarking, "What we know is that we can make wonderful communications equipment if we build it on theories of 'electric wave engineering'" (a euphemism for psychic effects).

In the 1970s, Douglas Dean and John Mihalasky, two scientists from the Newark College of Engineering, investigated the hypothesis that successful business executives sometimes benefit from precognition.[31] In a computer-based precognition experiment, they tested dozens of corporate executives and found significant evidence for precognitive abilities among them. The more successful the executives, in terms of profits attributable to their efforts, the more evident were their precognitive skills. This research, published in 1974 as a book called *Executive ESP,* generated enormous interest in the business community. A reviewer of this research wrote in *Time* magazine:

> Many an envious businessman has suspected that his more successful competitors are gifted with a sixth sense—an intuitive ability to foresee the future and make the tough, unexpected decisions that pay off handsomely. Now there is some evidence of a sort that suggests intuition really does pay off.[32]

Alexander M. Poniatoff, founder of the Ampex Corporation, once confessed, "In the past I would not admit to anyone, especially business people, why my decisions sometimes were contrary to any logical judgment. But now that I have become aware of others who follow intuition, I don't mind talking about it."[33] William W. Keeler, retired board chairman of Phillips Petroleum, mused that "there were too many incidents that couldn't be explained merely as coincidences. My strong feelings towards things were accurate when I would let myself go. Oil fields have been found on hunches, through precognitive dreams, and by people who didn't know anything about geology."[34]

In 1982 the *St. Louis Business Journal* tested how a psychic would fare against professional stockbrokers over a six-month period, and reported that the psychic, who had no formal training in stock market trading or analysis, outperformed eighteen of nineteen professional stockbrokers. During the testing period, the Dow Jones Industrial Average fell 8 percent, but the psychic's stocks went up 17 percent.[35]

The use of psi to enhance decision making is not overlooked in the intensely pragmatic world of Wall Street. In 1985 a vice president of Shearson Lehman Brothers provided a good summary of how results-oriented business views psi. In a full-page article in the *New York Times* business section, Chester Rothman observed that "If a psychic can better grasp the rationalities of the world than a market analyst, he might well give better business advice."[36] Successful entrepreneurs and investment analysts have confided to me that psi techniques, especially precognition, are beginning to play an increasingly important role on Wall Street. The driving force is that everyone now uses sophisticated computer models to help forecast which stocks to pick, so to maintain a slight edge in the accelerating world of stock trading, analysts and brokers need methods that outforecast the computers. Some brokers are successfully using psi to help nudge the computers' mathematical models, and the tiny forecasting advantages they gain are reportedly resulting in enormous profits. No wonder they've kept quiet about their interests!

A Natural Question

In other applications, psi has been used to guide archaeological digs and treasure-hunting expeditions, enhance gambling profits, and provide insight into historical events. So, given the tens of thousands of anecdotes about psi experiences, the thousands of scientific studies, and ongoing practical applications, why has mainstream science been so reluctant merely to admit the existence of psi? This question brings us to our next theme: *Understanding*.

THEME 3

UNDERSTANDING

In the second part of this book, we learned that an immense amount of persuasive anecdotal, scientific, and practical evidence for psi exists. Now we explore why mainstream science has been so reluctant—until very recently—to acknowledge that psi should be taken seriously.

Many scientists have assumed that the evidence isn't any good because it presents such a huge challenge to the well-accepted scientific worldview. As a result, most of the experiments summarized in this book are unknown to all but a fraction of mainstream scientists, and only a handful of researchers have had any firsthand experience in conducting psi experiments. In recent years, the few skeptics who have studied the scientific evidence in detail have significantly moderated their previous opinions, but this has not been well publicized.

Past inaccurate beliefs about psi have persisted in part because prominent skeptics have repeated the same old criticisms so often that many scientists just assume they are correct. In the next chapter we'll survey the skeptics' common assumptions, tactics, and assertions as a first step in understanding why the evidence for psi has remained more or less "invisible" to mainstream science.

CHAPTER 13

A Field Guide to Skepticism

I am attacked by two very opposite sects—the scientists and the know-nothings. Both laugh at me—calling me "the frogs' dancing-master." Yet I know that I have discovered one of the greatest forces in nature.

LUIGI GALVANI, ITALIAN PHYSICIAN (1737–1798)

This chapter does not argue against skepticism. On the contrary, it demonstrates that critical thinking is a double-edged sword: it must be applied to any claim, including the claims of skeptics. We will see that many of the skeptical arguments commonly leveled at psi experiments have been motivated by nonscientific factors, such as arrogance, advocacy, and ideology. The fact is that much of what scientists know—or think they know—about psi has been confused with arguments promoted by uncritical enthusiasts on the one hand and uncritical skeptics on the other. History shows that extremists, despite the strength of their convictions, are rarely correct. So, are all scientists who report positive evidence for psi naive or sloppy? No. Are all skeptics intolerant naysayers? No. Does psi justify the belief that angels from the Andromeda galaxy are among us? No.

Doubt

There are two ways to be fooled. One is to believe what isn't true; the other is to refuse to believe what is true.

SØREN KIERKEGAARD (1813–1855)

THE NECESSITY OF DOUBT

Skepticism, meaning doubt, is one of the hallmarks of the scientific approach. Skepticism sharpens the critical thought required to sift the wheat from the chaff, and it forces experimental methods, measurements, and ideas to pass through an extremely fine sieve before they are accepted into the "scientific worldview." A little critical thinking applied to many of the claims of New Age devotees reveals why many scientists are dubious of psi phenomena. Science requires substantial amounts of repeatable, trustworthy evidence before claims of unexpected effects can be taken seriously. Depending on the claim, providing sufficient evidence can take years, decades, or half-centuries of painstaking, detailed work. Learning how to create this evidence requires long training and experience in conventional disciplines such as experimental design, analysis, and statistics. Conducting research on controversial topics like psi requires all this plus an appreciation for interpersonal dynamics, politics, aesthetics, philosophy, and physics, combined with intellectual clarity and a strong creative streak to help break the bounds of conventional thinking.

From the lay perspective, science appears as a logical, dispassionate, analytic process. This is true sometimes, but science is also a harshly adversarial, emotional battlefield when it comes to evaluating unusual claims. Gaining acceptance for effects that are not easily accommodated by dominant theories takes an enormous amount of energy and persistence. This is why most scientists and psi researchers alike grimace upon reading breathless advertisements hawking, say, "The amazing miracle blue crystal, found deep beneath an ancient Mayan pyramid, proven by top researchers to relieve headaches and enhance psychic powers, and now available for a limited time for only $129.95!"

The *claim* about a blue crystal is not the problem. After all, if someone were to claim that a moldy piece of bread could cure all sorts of horrible diseases, he or she would be labeled a charlatan, unless the mold happened to be penicillin. The problem with many popular psi-related claims, especially claims for health-related products and devices, is that it doesn't take much digging to discover that sound, scientific evidence for the claim is entirely absent, is fabricated, or is based solely on anecdotes and testimonials.

The Danger of Uncritical Doubt

It's one thing not to see the forest for the trees, but then to go on to deny the reality of the forest is a more serious matter.

Paul Weiss

The same scientific mind-set that thrives on high precision and critical thinking is also extremely adept at forming clever rationalizations that get in the way of progress. In extreme cases, these rationalizations have prevented psi research from taking place at all. Ironically, the very same skeptics who have attempted to block psi research through the use of rhetoric and ridicule have also been responsible for perpetuating the many popular myths associated with psychic phenomena. If serious scientists are prevented from investigating claims of psi out of fear for their reputations, then who is left to conduct these investigations? Extreme skeptics? No, because the fact is that most extremists do not conduct research; they specialize in criticism. Extreme believers? No, because they are usually not interested in conducting rigorous scientific studies.

The word *extreme* is important to keep in mind. Most scientists seriously interested in psi are far more skeptical about claims of psychic phenomena than most people realize. Scientists who study psi phenomena grind their teeth at night because television shows predictably portray psi researchers as wacky "paranormal investigators" with dubious credentials. Psi researchers cringe when they see the word *parapsychologist* used in the telephone yellow pages to list psychic readers. And unfortunately, because the only thing most people know about parapsychology is its popular association with credulous "investigators" and psychic overenthusiasts, it is understandable why some skeptics have taken combative positions to fight what they see as rising tides of nonsense.

This book is intended to help illustrate that common stereotypes about psi research are overly simplistic at best and, in many cases, just plain wrong. As an example of "just plain wrong," here is one stereotype that many mainstream scientists have simply accepted as conventional wisdom. As philosopher Paul Churchland put it:

> Despite the endless pronouncements and anecdotes in the popular press, and despite a steady trickle of serious research on such things, there is no significant or trustworthy evidence that such phenomena even exist. The wide gap between popular conviction on this matter, and the actual evidence, is something that itself calls for research. For there is not a single parapsychological effect that can be repeatedly or reliably produced in any laboratory suitably equipped to perform and control the experiment. Not one.[1]

Wrong. As we've seen, there are a half-dozen psi effects that have been replicated dozens to hundreds of times in laboratories around the world. As another example, conventional wisdom often assumes that professional magicians and conjurers "know better" than to accept that some psychic phenomena are real. In fact, as parapsychologist George Hansen wrote:

> Although the public tends to view magicians as debunkers, the opposite is more the case. Birdsell (1989) polled a group of magicians and found that 82 percent gave a positive response to a question of belief in ESP. Truzzi (1983) noted a poll of German magicians that found that 72.3 percent thought psi was probably real. Many prominent magicians have expressed a belief in psychic phenomena. . . . It is simply a myth that magicians have been predominantly skeptical about the existence of psi.[2]

Skepticism About Skepticism

Why it is necessary to spend any time at all on the criticisms of psi research when we can simply refer to the previous chapters to demonstrate that there are valid experimental effects in search of answers? One answer is that very few people are aware that the standard skeptical arguments have been addressed in exquisite detail and no longer hold up. Another is that the tactics of the extreme skeptics have been more than merely annoying. The professional skeptics' aggressive public labeling of parapsychology as a "pseudoscience," implying fraud or incompetence on the part of the researchers, has been instrumental in preventing this research from taking place at all. In a commentary in the prominent journal *Nature,* skeptical British psychologist David Marks wrote:

> Parascience has all the qualities of a magical system while wearing the mantle of science. Until any significant discoveries are made, science can justifiably ignore it, but it is important to say why: parascience is a pseudo-scientific system of untested beliefs steeped in illusion, error and fraud.[3]

Such statements are pernicious because significant discoveries do not occur by themselves. Published in influential journals, these opinions have strongly affected the ability of scientists to conduct psi research. Many funding agencies, both public and private, have been reluctant to sponsor parapsychological studies because they fear being associated with what conventional wisdom has declared a "pseudoscience." Fortunately, some funding agencies know that there is a difference between popular stereotypes and serious researchers.

Skepticism Today

> **The discovery of truth is prevented more effectively, not by the false appearance of things present and which mislead into error, not directly by weakness of the reasoning powers, but by preconceived opinion, by prejudice.**
>
> Arthur Schopenhauer, German philosopher (1788–1860)

In 1993 the parapsychologist Charles Honorton, from the University of Edinburgh, considered what skeptics of psi experiments used to claim, and what they no longer claimed. He demonstrated that virtually all the skeptical arguments used to explain away psi over the years had been resolved through new experimental designs. This does not mean that the experiments conducted today are "perfect," because there is nothing perfect in the empirical sciences. But it does mean that the methods available today satisfy the most rigorous skeptical requirements for providing "exceptional evidence." As we've seen, such experiments have been conducted, and with successful results.

What Skeptics Used to Claim

Honorton pointed out that for many decades the standard skeptical assertion was that psi was impossible because it violated some ill-specified physical laws, or because the effects were not repeatable. It was also easy to claim that any successful experiments were really due to chance or fraud. Today, informed skeptics no longer claim that the outcomes of psi experiments are due to mere chance because we know that some parapsychological effects are, to use skeptical psychologist Ray Hyman's words, "astronomically significant."[4] This is a key concession because it shifts the focus of the debate away from the mere *existence* of interesting effects to their proper *interpretation*.

The concession also puts to rest the decades-long skeptical questions over the scientific legitimacy of parapsychology. It states, quite clearly, that skeptics who continue to repeat the same old assertions that parapsychology is a pseudoscience, or that there are no repeatable experiments, are uninformed not only about the state of parapsychology *but also about the current state of skepticism!*

Honorton then pointed out that informed skeptics no longer claim that there are any meaningful relationships between design flaws and experimental outcomes. This criticism was again based on the premise that psi did not exist, from which it followed that any psi effects observed in experiments must have been due to sloppy experimenters, flawed techniques, or poor measurements. The assertion implied that if a scientist performed the proper, "perfect" psi experiment, all claims for psi effects would disappear. The basic argument is flawed, of course, because all measurements contain some error. Nevertheless, the assertion is testable by comparing experimental outcomes with assessments of experimental quality. As we've seen, the meta-analyses described earlier have shown that design flaws cannot account for the cumulative success rates in psi experiments.

The skeptics are not eager to advertise their recent concessions. Over the past few decades Ray Hyman and other "professional" skeptics have tried with great creativity and diligence to explain away psi. They tried to show that the experiments were not really all that interesting, and that the apparently successful studies were due to one or another design flaws. Having failed on both counts, informed skeptics have been forced to admit that they have simply run out of plausible explanations.

It is not easy to change a lifelong, strongly held belief, even when there is strong evidence that the belief is wrong, so the publicly proclaimed skeptics are not likely ever to admit that psi per se is genuine. Nevertheless, it is important to emphasize that the focus of today's controversy has significantly shifted from the flat dismissals of the past.

WHAT SKEPTICS NOW CLAIM

Because no plausible explanations remain for the experimental results obtained with psi, today the few remaining hard-core skeptics rehash the same old polemical arguments used in past decades. The core assertion is the tired claim that after one hundred years of research, parapsychology has failed to provide convincing evidence for psi phenomena.

This argument follows a certain logic. Skeptics refuse to believe that psi experiments, which they admit are successfully demonstrating *something*, are in fact demonstrating psi itself. By acknowledging that the results are real and unexplainable on the one hand, but by stubbornly insisting that those results could not possibly be due to *psi* on the other, then of course they can claim that parapsychology is a failure. This is like a skeptic refusing to call a group of nine players who win the World Series a "baseball team." In that case, the skeptic can simply smile, shrug, and doggedly claim that yes, people do apparently go running after balls that other people occasionally hit with a bat. But still, after one hundred years there is no solid evidence that anything called a *baseball* team actually exists.

Remember that most parapsychologists do not claim to understand what "psi" is. Instead, they design experiments to test experiences that people have reported throughout history. If rigorous tests for what we have called "telepathy" result in effects that look like, sound like, and feel like the experiences reported in real life, then call it what you will, but the experiments confirm that this common experience is not an illusion.

Another way to demonstrate the purely rhetorical nature of the "century of failure" argument is to see if the same argument also applies to conventional academic psychology. After a hundred years and thousands of experiments, psychologists still argue vigorously about such elementary phenomena as conscious awareness, memory, learning, and perception. After a hundred years, psychology has not produced even the crudest model of how processes in the brain are transformed into conscious experience. If we adopt the reasoning of the skeptics, many of whom are psychologists, then conventional psychology is also a dismal failure.

AN UNUSUAL CONTROVERSY

After deftly exposing and dissolving the skeptical position, Honorton then pointed out an important difference between the controversy over psi and debates in more conventional disciplines. Most scientific debates occur within groups of researchers who test hypotheses, develop and critique other researchers' methods, and collect data to test their hypotheses. This is standard operating procedure, as witnessed by persistent debates over dozens of hot topics in all scientific disciplines. The same sort of vigorous debating is evident in the journals and at the annual meetings of the Parapsychological Association, the professional society of scientists and scholars interested in psi phenomena.

The psi controversy, however, differs in one important respect. Although the skeptics often write about the plausibility of various alternative hypotheses, they almost never test their ideas. This "armchair quarterbacking" is especially true of the current generation of psi skeptics, the vast majority of whom have contributed no original research to the topic.

Their reasoning is simple: If we start from the position that an effect cannot exist, then why should we bother to spend all the time and money required to study it? It makes more sense to use every rhetorical trick in the book to convince others that our opinion is correct, and that all the evidence to the contrary is somehow flawed. This may seem like a perfectly reasonable strategy, but it is not science. It is much closer to an argument based on faith, like a religious position.

The fact that most skeptics do not conduct counterstudies to prove their claims is often ignored. For example, in 1983 the well-known skeptic Martin Gardner wrote:

> How can the public know that for fifty years skeptical psychologists have been trying their best to replicate classic psi experiments, and with notable unsuccess [*sic*]? It is this fact more than any other that has led to parapsychology's perpetual stagnation. Positive evidence keeps coming from a tiny group of enthusiasts, while negative evidence keeps coming from a much larger group of skeptics.[5]

As Honorton points out, "Gardner does not attempt to document this assertion, nor could he. It is pure fiction. Look for the skeptics' experiments and see what you find." In addition, there is no "larger group of skeptics." Perhaps ten or fifteen skeptics have accounted for the vast bulk of the published criticisms.

Beyond the "century of failure" argument, some skeptics still stubbornly insist that parapsychology is not a "real science." One of them, Ray Hyman, wrote:

> Every science except parapsychology builds upon its previous data. The data base continually expands with each new generation but the original investigations are still included. In parapsychology, the data base expands very little because previous experiments are continually discarded and new ones take their place.[6]

If this were true, the meta-analyses described in this book would not exist. As we've seen, the early tests on thought transference gave rise to picture-drawing telepathy tests. They spawned telepathy experiments in the dream state, which later led to the ganzfeld experiments. The dice tests begot RNG experiments. All of these experimental variations evolved as researchers took stock of previous experimental outcomes and criticisms and refined their test designs and theories.

Of course, some skeptics have made important contributions to the development of progressively stronger evidence by systematically ferreting out design loopholes and by insisting upon stronger and stronger empirical evidence. But because skeptics today can no longer demonstrate *plausible* alternative explanations, all that remains are rhetoric and defense of a priori beliefs. Persisting in this stance in the face of overwhelming evidence has produced some excellent examples of minds struggling with logical contradictions. Honorton summarized his view of the state of skepticism as follows:

> There is a danger for science in encouraging self-appointed protectors who engage in polemical campaigns that distort and misrepresent serious research efforts. Such campaigns are not only counterproductive, they threaten to corrupt the spirit and function of science and raise doubts about its credibility. The distorted history, logical contradictions, and factual omissions exhibited in the arguments of the . . . critics represent neither scholarly criticism nor skepticism, but rather counteradvocacy masquerading as skepticism.[7]

Skeptical Tactics

Extreme skeptics who believe that all psi experiments are flawed have used an effective bag of rhetorical tactics to try to convince others to dismiss the evidence. These include accusations that even if psi effects are real, they are so weak that they are trivial or uninteresting; statements of frank prejudice; long lists of common, but scientifically invalid criticisms; and severely distorted descriptions of psi experiments that make psi researchers appear to be incompetent. Let's examine how some of these tactics have been used.

Accusations of Triviality

Some skeptics have reluctantly accepted that psi effects may be genuine. But then they attempted to reduce their discomfort by claiming that psi is simply too weak to be interesting. For example, the psychologist E. G. Boring wrote that ESP data were merely "an empty correlation,"[8] and psychologist S. S. Stevens asserted that "the signal-to-noise ratio for ESP is simply too low to be interesting."[9]

More recently, the skeptical British psychologist Susan Blackmore wrote, "What if my doubt is displaced and there really is extrasensory perception after all? What would this tell us about consciousness?"[10] To answer this question, Blackmore took a giant step backward to the 1950s psychological fad of behaviorism, and concluded that consciousness doesn't have any meaning at all, that it is merely an illusion. Not surprisingly then, she also concluded that psi, even if genuine, would tell us *nothing at all* about the nature of consciousness. This is a perplexing position that hardly anyone accepts anymore, not even other hard-nosed skeptics.[11]

In another example of trivializing psi, mathematician A. J. Ayer wrote in *Scientific American:*

> The only thing that is remarkable about the subject who is credited with extra-sensory perception is that he is consistently rather better at guessing cards than the ordinary run of people have shown themselves to be. The fact that he also does "better than chance" proves nothing in itself.[12]

Such an assertion is confused, because *any* form of genuine psi, even statistically "better than chance" psi, carries revolutionary potential for our understanding of the natural world. In addition, effects that are originally observed as weak may be turned into extremely strong effects after they are better understood. Consider, for example, what was known about harnessing the weak, erratic trickles of electricity 150 years ago, and compare that to the trillion-watt networks that run today's power-hungry world.

Prejudice

Ignorance more frequently begets confidence than does knowledge; it is those who know little, and not those who know much, who so positively assert that this or that problem will never be solved by science.

CHARLES DARWIN, *THE DESCENT OF MAN* (1871)

Prejudice—holding an opinion without knowledge or examination of the facts—is deeply embedded in human nature. It is much easier to follow the natural impulse to form a quick judgment and stick with it, rather than take the time and trouble to study the actual evidence. Prejudice continues to haunt psi researchers. Sometimes it is acknowledged as such, and sometimes it is not.

Philip Anderson, a prominent theoretical physicist at Princeton University, assumed that psi was incompatible with physics, and so in a 1990 editorial in *Physics Today* he wrote:

> If such results are correct, we might as well turn the National Institute of Standards and Technology into a casino and our physics classes into seances, and give back all those Nobel Prizes. . . . It is for this kind of reason that physicists, quite properly, do not take such experiments seri-

ously until they can be (1) reproduced (2) by independent, skeptical researchers (3) under maximum security conditions and (4) with totally incontrovertible statistics. Oddly enough, the parapsychologists who claim positive results invariably reject these conditions.[13]

It is clear that Anderson was simply ignorant of the evidence, and yet he still felt quite confident about his opinion. We can only imagine what Anderson thinks of well-regarded physicists who *do* take such experiments seriously.

Some critics have acknowledged that they simply do not wish to believe the evidence. For example, in 1951 the psychologist Donald O. Hebb wrote: "Why do we not accept ESP as a psychological fact? Rhine has offered us enough evidence to have convinced us on almost any other issue. . . . I cannot see what other basis my colleagues have for rejecting it. . . . My own rejection of [Rhine's] views is in a literal sense prejudice."[14]

In 1955 psychologist G. R. Price suggested that because psi was clearly impossible, fraud was the best, and really the only remaining explanation for psi effects. In a lead article in *Science,* Price began sensibly:

> Believers in psychic phenomena . . . appear to have won a decisive victory and virtually silenced opposition. . . . This victory is the result of an impressive amount of careful experimentation and intelligent argumentation. . . . Against all this evidence, almost the only defense remaining to the skeptical scientist is ignorance, ignorance concerning the work itself and concerning its implications. The typical scientist contents himself with retaining . . . some criticism that at most applies to a small fraction of the published studies. But these findings (which challenge our very concepts of space and time) are—if valid—of enormous importance . . . so they ought not to be ignored.[15]

Price then flatly asserted that because ESP was "incompatible with current scientific theory," it was more reasonable to believe that parapsychologists had cheated than that ESP might be real. Price based his argument on a famous essay on the nature of miracles by philosopher David Hume. Hume argued that since we know that people sometimes lie, but we have no independent evidence of miracles, it is more reasonable to believe that claims of miracles are based on lies than that miracles actually occur. Using this reasoning, Price concluded:

> My opinion concerning the findings of the parapsychologists is that many of them are dependent on clerical and statistical errors and unintentional use of sensory clues, and that all extrachance results not so explicable are dependent on deliberate fraud or mildly abnormal mental conditions.[16]

Another critic of the same era was skeptical British psychologist Mark Hansel, from the University of Wales. Like Price, Hansel emphasized the possibility of fraud:

> If the result could have been through a trick, the experiment must be considered unsatisfactory proof of ESP, whether or not it is finally decided that such a trick was, in fact, used. . . . [Therefore,] it is wise to adopt initially the assumption that ESP is impossible, since there is a great weight of knowledge supporting this point of view.[17]

Such opinions—that existing scientific knowledge is complete and that psi necessarily conflicts with it—have motivated skeptics to imagine all sorts of good reasons to make the psi "go away." The power of this motivation is illustrated by a 1987 report on parapsychology issued by the National Research Council.

National Research Council Report

In the mid-1980s the U.S. Army recruitment slogan was "Be all that you can be." The slogan reflected the army's desire to train soldiers to achieve enhanced performance. These highly trained warriors would be fearless and cunning, fight without fatigue, and employ a variety of enhanced, exotic, or possibly even psychic skills.

In 1984 the U.S. Army Research Institute asked the premier scientific body in the United States, the National Academy of Sciences, to evaluate a variety of training techniques and claims about enhanced human performance. These techniques included sleep learning, accelerated learning, biofeedback, neurolinguistic programming, and parapsychology. The National Academy of Sciences responded to the army's request by directing its principal operating agency, the National Research Council (NRC), to form a committee to examine the scientific evidence in these areas. Because the NRC is often asked to investigate leading-edge and controversial topics, it maintains an explicit policy of assembling balanced scientific committees. In fact, the policy requires members of its committees to affirm that they have no conflicts of interest either for or against the objects of study. This helps ensure that the scientific reviews are fair.

On December 3, 1987, the NRC convened a well-attended press conference in Washington, D.C., to announce its conclusions.[18] John A. Swets, chairman of the NRC committee, said, "Perhaps our strongest conclusions are in the area of parapsychology." The bottom line: "The Committee finds no scientific justification from research conducted over a period of 130 years for the existence of parapsychological phenomena."[19]

Whoops. Where did this come from? To help understand the disparity between the actual data and the NRC's conclusion, the board of directors of the Parapsychological Association (PA) selected three senior members of the PA to study the report in detail and respond to it. The three members were John Palmer, a psychologist at the Rhine Research Center, Durham, North Carolina; Charles Honorton, who at the time was director of the Psychophysical Research Laboratories in Princeton, New Jersey; and Jessica Utts, professor of statistics at the University of California, Davis.

After some study, the PA committee issued its report, with three main findings.[20] First, the two principal evaluators of psi research for the NRC committee, psychologists Ray Hyman and James Alcock, both had long histories of skeptical publications accusing parapsychology of not even being a legitimate science. In contrast, there were no active psi researchers on the committee. This violated the NRC's policy of assigning members to committees "with regard to appropriate balance."

Second, the NRC's report avoided mentioning studies favorable to psi research but quoted liberally from two background papers that supported the committee's position. As if this were not enough, the chairman of the NRC committee phoned one of the authors of a third commissioned background paper, Robert Rosenthal from Harvard University, and asked him to withdraw his conclusions because they were favorable to parapsychology.

And third, the NRC report was self-contradictory. The committee widely advertised its conclusion that there was no evidence for psi phenomena, yet the report itself admits that the committee members could offer no plausible alternatives to the research it surveyed. The committee failed to mention in the press conference its recommendation that the army continue to monitor psi research in the United States and the former Soviet Union. It even recommended that the army propose specific experiments to be conducted. The contrast between the NRC's advertised position and its actual position suggests that there were conflicts between reporting a fair evaluation of the data and what was politically expedient to report.

This was clearly revealed later when a newspaper reporter for *The Chronicle of Higher Education* asked the NRC committee chairman, John Swets, why he asked Rosenthal to withdraw his favorable conclusions. Swets replied: "We thought the quality of our analysis was better, and we didn't see much point in putting out mixed signals."[21] Swets explained, "I didn't feel we were obliged to represent every point of view."[22] This meant the NRC committee in effect had created a "file drawer" of ignored *positive* studies that it didn't wish to talk about. Apparently, the only acceptable views about psi for this committee were negative ones. Given the true nature of the evidence, this was bound to lead to some major contradictions.

And it did. The NRC committee commissioned ten background papers by experts in a variety of fields. One of these papers, by Dale Griffin of Stanford University, explained how difficult it is to objectively evaluate evidence when one is already publicly committed to a particular belief. According to Griffin:

> Probably the most powerful force motivating our desire to protect our beliefs—from others' attacks, from our own questioning, and from the challenge of new evidence—is commitment. . . . This drive to avoid dissonance is especially strong when the belief has led to public commitment.[23]

The Committee for the Scientific Investigation of Claims of the Paranormal (CSICOP) is an organization well known for its impassioned commitment against parapsychology. Ray Hyman was one of the original "fellows" of CSICOP, and he was an active member of its executive council at the same time he was evaluating psi research for the NRC. So the source of many contradictions in the NRC report is clear: Hyman's publicly committed position as a psychic debunker. For example, at the NRC press conference, Hyman confirmed his public stance by announcing that the "poor quality of psi research was 'a surprise to us all—we believed the work would be of much higher quality than it turned out to be.'"[24] Yet, in contrast to this public statement, the report itself actually says, ". . . the best research [in parapsychology] is of higher quality than many critics assume."[25]

In further contrast to the NRC's public assertions about "poor quality research" and "no scientific justification" was the actual paper commissioned by the NRC to review psi experiments and other studies of performance-enhancing techniques. Authored by psychologists Monica Harris and Robert Rosenthal of Harvard University, the report concluded that

> The situation for the ganzfeld domain seems reasonably clear. We feel it would be implausible to entertain the null [hypothesis] given the combined [probability] from these 28 studies. . . . When the accuracy rate expected under the null [hypothesis] is 1/4, we estimate the obtained accuracy rate to be about 1/3.[26]

In nontechnical language, Harris and Rosenthal concluded that there was persuasive evidence for something very interesting going on in the ganzfeld experiments because they found an average hit rate of about 33 percent rather than the 25 percent expected by chance (as we discussed in chapter 5). They also compared the quality of the ganzfeld experiments to the quality of experiments in four other, nonparapsychological research areas and concluded that "only the ganzfeld ESP studies regularly meet the basic requirements of sound experimental design."[27]

There is no need to belabor the point; it is clear that abject prejudice exists in science just as it does in other human endeavors. We were able to detect it fairly easily in the case of the NRC report by comparing the committee's public pronouncements with what its report actually says. Sometimes prejudice is not so easy to detect, because we usually do not stop to think that some skeptical criticisms are simply invalid.

Valid and Invalid Criticisms

It is commonly thought that all criticisms in science are equal. This is not so. In fact, criticisms must have two properties to be valid. First, the criticism must be *controlled,* meaning that it cannot also apply to well-accepted scientific disciplines.[28] In other words, we cannot use a double standard and apply one set of criticisms to fledgling topics and an entirely different set to established disciplines. If we did, nothing new could ever be accepted as legitimate. Second, a criticism must be *testable,* meaning that a critic has to specify the conditions under which the research could avoid the criticism; otherwise, the objection is just a philosophical argument that falls outside the realm of science.

A thorough examination of the usual skeptical allegations about laboratory psi research reveals that only one is both controlled and testable: have independent, successful replications been achieved? We now know that the answer is yes, so the criticisms should stop here. Skepticism dies hard, however, and surprisingly few scientists realize that all criticisms are not created equal. So let's briefly review why some other common criticisms are invalid.[29]

One popular assertion is that "Many phenomena that were once thought to be paranormal have been shown to have normal explanations." This is an invalid criticism because it is uncontrolled—the same criticism can be applied to many discoveries in other well-accepted scientific disciplines. Even if we originally thought that psi was one thing but later discovered that it was something else, that would not invalidate the *existence* of the effect; it would merely redefine how we thought about it.

Another criticism is that "Some paranormal effects have been shown to be the outcome of fraud or error," so we can safely ignore any successful results. This is invalid because if we were forced to dismiss scientific claims in all fields where there have been a few cases of experimenter fraud, we would have to throw out virtually every realm of science—since fraud exists in all human endeavors.[30]

Another favorite complaint is, "There are no theories of psi." This criticism is invalid because for the term "psi" we could substitute "consciousness," "gravity," "anesthesia," or dozens of other well-accepted concepts or phenomena. The fact that scientists do not understand some phenomena very well has not reduced scientific interest in them.

Skeptics have also charged that "Psi cannot be switched on and off, and the variables that affect it cannot be controlled." This too is an invalid criticism because there are all kinds of effects over which we have no direct control, including most of the really interesting aspects of human behavior; yet this does not disqualify them as legitimate objects of study. In any case, psi is somewhat controllable in the sense that we can cause predictable effects to appear by asking people to do something in their mind. If they do not pay attention to the task, which is how control periods are conducted in some psi experiments, then no unusual effects appear.

Some skeptics have protested that "It's impossible to distinguish between psi and chance effects even in a successful experiment without the use of statistics." This criticism is invalid because the same can be said for almost all experiments in biology, psychology, sociology, and biomedicine. Obviously, if there were some way of cleanly separating a signal from random noise before the experiment was conducted, then statistics would not have been used in the first place.

This litany of common criticisms could go on for many pages, but the point is clear. The vast majority of complaints about psi research are invalid, either because they pertain equally to conventional, well-accepted disciplines or because the complaints are untestable.

Distortions

Popular Media

Another reason why psi has been ignored by mainstream science, and decades of scientifically sound experiments have been judged controversial, can be traced to the heavily distorted portrayal of these studies in the media and in college textbooks.

The July 8, 1996, issue of *Newsweek* contained an article called "Science on the Fringe. Is There Anything to It? Evidence, Please."[31] Written by reporter Sharon Begley, this article is a good example of how widely disseminated information about psi experiments is sometimes seriously misleading. Begley's story began with the following:

> Say this about assertions that aliens have been, are or will soon be landing on Earth: at least a scenario like that of [the movie] "Independence Day" would not violate any laws of nature. In contrast, claims in other fringe realms, such as telepathy and psychokinesis, are credible only if you ignore a couple or three centuries of established science.[32]

This is a commonplace assertion, but it is worth noting that critics never specify which "laws of nature" would be violated by psi, because the assertion is groundless—the laws of nature are not fixed absolutes. They are fairly stable ideas that are always subject to expansion and refinement based on evidence from new observations. For example, after the advent of relativity and quantum mechanics, some of our new physical "laws" forced the classical concepts developed in the seventeenth century to expand radically. Have we magically reached a point at the end of the twentieth century where the present "laws" of science can be permanently chiseled into stone? I don't think so.

Begley apparently believes that aliens landing on Earth is more credible than psi. Does this make sense? In the case of aliens, the evidence is based exclusively on eyewitness stories and ambiguous photographs. Some of the stories and photos are engaging, but taking the leap of faith from this form of evidence to the actual presence of extraterrestrials is unwarranted. There are dozens of alternative possibilities, none of which involves either extraterrestrial or earthbound aliens. By comparison, the evidence for psi is based upon a century of repeated scientific evidence. The seduction of the status quo is so strong, however, that a skeptical journalist would rather believe stories about little green men than controlled observations in the laboratory.

Later in the *Newsweek* article, Begley described the ganzfeld telepathy experiments. After providing a good explanation of the basic procedure, and mentioning that the observed hit rate for Honorton's autoganzfeld studies was about 35 percent instead of the chance expected 25 percent, Begley asked:

> Was it telepathy? Some experiments failed to take into account that people hearing white noise think about water more often than sex (or so they say); if beaches appear more often as a target than a couple in bed, a high hit rate would reflect this tendency, not telepathy. Also, receivers tend to choose the first or last image shown them; unless the experimenter makes sure that the target does not appear in the first or last place more often than decoys do, the hit rate would be misleadingly high.[33]

While these criticisms are valid because they are testable, a skeptical reader might legitimately wonder, Did targets with water content *actually* appear more often than targets with sexual content? (No.) Did targets *actually* appear more often in the first or last place? (No.) Were researchers so naive as not to think of these possibilities? (No.) The implication was that the criticisms had been overlooked, but they weren't.

Begley continued:

> Skeptic Ray Hyman of the University of Oregon found that, in the Edinburgh runs, video targets that were used just once or twice had hit rates of about chance, while those appearing three or more times yielded a "telepathic" 36 percent. How come? A video clip run through a player several times may look different from one never played for the sender; a canny receiver would choose a tape that looked "used" over one that didn't.

In fact, as we saw in chapter 5, the ganzfeld system at the University of Edinburgh used *two separate video players* to address this criticism, and successful effects virtually identical to those Honorton had reported earlier were still observed. Again, the implication of the criticism is that the ganzfeld results are explainable by this potential flaw, and it is not true.

Next, Begley repeated another common criticism:

> Of the 28 studies Honorton analyzed in 1985, nine came from a lab where one-time believer Susan Blackmore of the University of the West of England had scrutinized the experiments. The results are "clearly marred," she says, by "accidental errors" in which the experimenter might have known the target and prompted the receiver to choose it.

What Begley fails to report is that after Blackmore's allegedly "marred" studies were eliminated from the meta-analysis, the overall hit rate in the remaining studies remained *exactly the same as before.*[34] In other words, Blackmore's criticism was tested and it did not explain away the ganzfeld results. It is also important to note that Blackmore never actually demonstrated that the flaw existed.

Begley continued by describing the mind-matter interaction experiments using random-number generators conducted at Robert Jahn's PEAR Lab.[35] Then she added:

> As for Jahn's results, there are a couple of puzzles. First, one of the subjects, rumored to be on Jahn's staff, is responsible for half of the successes even though he was in just 15 percent of the trials. Second, some peculiarities in how the machine behaved suggest that the experimenters might have ignored negative data. Jahn says this is virtually impossible. But other labs, using Jahn's machine, have not obtained his results.[36]

As discussed in chapter 8, analysis of the PEAR Laboratory data clearly showed that no one person's results were wildly different from anyone else's. Nor was any one person responsible for the overall results of the experiment. Again, the criticism was tested and found to be groundless.[37] The comment about "peculiarities" is pure rhetoric; since it does not mention the nature of the alleged problems, it is an untestable criticism. The assertion that other labs have not obtained Jahn's results is a commonly repeated skeptical mantra, but it is also false, as we've seen. Jahn's results are entirely consistent with a larger body of evidence collected by more than seventy investigators, and overall there is no question that replication has been achieved. If anything, Jahn's results are somewhat smaller in magnitude than those reported by others.

It is rather easy to pick apart Begley's article, because it is difficult to portray controversial topics fairly in the few paragraphs available in weekly newsmagazines. Some distortions are to be expected. But one would hope that book-length discussions by academic psychologists would be more thorough and neutral. Unfortunately, this is not always the case.

Books by Academic Psychologists

In 1985 psychologist Irvin Child, at the time the chairman of the Psychology Department at Yale University, reviewed the Maimonides dream-telepathy experiments for *American Psychologist,* a prominent journal published by the American Psychological Association.[38] Child was especially interested in comparing what actually took place in those experiments with how they were later described by skeptical psychologists.

The first book he considered was the 1980 edition of British psychologist Mark Hansel's critical book on psi research.[39] One page in the book was devoted to a description of the method and results of the dream-telepathy experiments. Hansel's strategy was to suggest possible flaws that *might* have accounted for the experimental results, without demonstrating that the flaws actually existed, and then assume that such flaws *must* have occurred because they were more believable than genuine psi. Child found that Hansel's descriptions of the methods used in the Maimonides studies were crafted in such a way as to lead unwitting readers to assume that fraud was a *likely* explanation, whereas in fact it was extremely unlikely given the controls employed by the researchers. Even other skeptics, such as Ray Hyman, agreed with Child. In a 1984 broadcast of the popular science program *Nova,* Hyman said:

> Hansel has a tendency to believe that if any experiment can be shown to be susceptible to fraud, then that immediately means it no longer can be used for evidence for psi. I do sympathize with the parapsychologists who rebut this by saying, well, that can be true of any experiment in the world, because there's always some way you can think of how fraud could have gotten into the experiment. You cannot make a perfectly 100 percent fraud-proof experiment. This would apply to all science.[40]

Child next reviewed a 1981 book by York University psychologist James Alcock. Alcock's basic theme in this and later publications is that parapsychologists are motivated by religious urges, a secular "search for the soul."[41] With this belief propelling many of his writings, Alcock tends to reject any psi experiments with positive outcomes as being flawed by their designers' religious drives. He also criticized the Maimonides experiments for not including a control group, writing that "a control group, for which no sender or no target was used, would appear essential."[42] Child responded:

> Alcock ... did not seem to recognize that the design of the Maimonides experiments was based on controls exactly parallel to those used by innumerable psychologists in other research with similar logical structure.[43]

The next book that Child looked at was by psychologists Leonard Zusne and Warren H. Jones.[44] Zusne and Jones wrote that the Maimonides researchers discovered that dreamers were not influenced telepathically unless they knew *in advance* that an attempt would be made to influence them. This led, they wrote, to the receiver's being "primed prior to going to sleep" by the experimenters "preparing the receiver through experiences that were related to the content of the picture to be telepathically transmitted during the night."[45] Child pointed out that it would be immediately apparent to anyone that such an experiment, if it were actually performed, would be catastrophically flawed. Obviously, if you prime someone with target-relevant information *before* he or she dreams, the entire experiment is worthless. But given that the dream-telepathy studies are so described, readers of Zusne and Jones's book unfamiliar with the actual experiments could reach no conclusion other than that the researchers were completely incompetent. Child responded:

> The simple fact, which anyone can easily verify, is that the account Zusne and Jones gave of the experiment is grossly inaccurate. What Zusne and Jones have done is to describe . . . some of the stimuli provided to the dreamer *the next morning*, after his dreams had been recorded and his night's sleep was over.[46]

As he discovered one flawed description after another, Child finally concluded that the books he reviewed contained "nearly incredible falsification of the facts about the experiments." But this was just the tip of an iceberg. It turns out that many introductory psychology textbooks have presented similarly flawed descriptions of psi experiments. These books, which are used in college courses, contain all the detail that most students will ever know about parapsychology If basic textbooks state or imply that psi researchers are stupid or naive, is it any wonder that future scientists and professors mistakenly assume that the evidence for psi is worthless?

Introductory Psychology Textbooks

There is no better soporific and sedative than skepticism.
FRIEDRICH NIETZSCHE

In 1991 psychologist Miguel Roig and his colleagues published a detailed analysis of the treatment of parapsychology in introductory psychology textbooks.[47] They surveyed sixty-four textbooks published between 1980 and 1989, then looked for words like *ESP* and *psychic* in the index and scanned through the chapters on research methods, sensation and perception, and states of consciousness. Of the sixty-four texts surveyed, forty-three included some mention of parapsychology. This is interesting in its own right, because it means that fully one-third of introductory psychology textbooks did not even *mention* a topic that all college students find fascinating.

A mere eight of the forty-three texts mentioned that since the 1970s parapsychologists have used the term "psi" as a neutral label for psychic phenomena. Twenty-one books mentioned the ESP card tests conducted by

J. B. Rhine and his colleagues from the 1930s to the 1960s. A few books incorrectly claimed that ESP card tests are still representative of contemporary research, whereas anyone even casually familiar with recent journal articles and books knows that such tests have hardly been used for decades. The remaining topics covered included discussions of spontaneous psychic experiences—which were uniformly explained away in terms of misunderstood sensory processes, coincidence, and self-deception; brief reviews of a few selected experiments; and alleged problems of methodology.

Most of the texts ended with a wait-and-see stance toward psychic phenomena, with thirty-five of the forty-three books mentioning lack of replication as the most serious problem. The second and third most serious problems were described as poor experimental designs and fraud. Surprisingly, only a few texts mentioned the development of experiments since the 1970s. Nine books mentioned RNG experiments, three mentioned the Maimonides dream-telepathy studies, and only one mentioned the ganzfeld-telepathy studies. Roig and his colleagues concluded that

> Much of the coverage reflects a lack of familiarity with the field of parapsychology, . . . there is an unacceptable reliance on secondary sources, most of which were written by nonparapsychologists who are critical of the field and who, at least in some cases, have been found to distort and sometimes fail to present promising lines of research. We conclude that most textbooks that cover the topic present an outdated and often grossly misleading view of parapsychology.[48]

This is unfortunate but not surprising. College textbooks reflect the status quo, and the status quo has not yet caught up with the latest developments in psi research. But what sustains the status quo? What has driven some academic psychologists to see psi research in such distorted ways?

Motivations

Skeptics are fond of claiming that believers in psi are afflicted with some sort of abnormal mental condition that prohibits them from seeing the truth. Skeptical psychologist James Alcock has suggested that one motivation for this "affliction" is psi researchers' hidden desires to justify some form of spiritual belief. This belief, according to Alcock, has biased psi research to such an extent that there *must* be something wrong with it.

But Alcock's belief about hidden spiritual motivations have produced an equally strong counterbias. This is clear in a lengthy background report that Alcock prepared for the NRC committee mentioned earlier. For forty pages, Alcock's report rips apart the mind-matter interaction studies of physicist Helmut Schmidt and Princeton University engineer Robert Jahn; then it concludes that

> *There is certainly a mystery here,* but based on the weaknesses in procedure mentioned above, there seems to be no good reason at this time to conclude that the mystery is paranormal in nature.[49]

In dismissing the mystery, Alcock missed the forest for the trees. It is true that any one or two experiments can be explained away as being due to chance or poor design, but the entire body of evidence, as discussed in chapter 8, cannot be dismissed so easily. And in contrast to Alcock's belief about what motivates psi researchers, parapsychology was formally recognized by the mainstream as a legitimate scientific discipline in 1969 when the Parapsychological Association, an international scientific society, was elected an affiliate of the American Association for the Advancement of Science (AAAS). Religious sects, New Age societies, and skeptical advocacy groups are not affiliates of the AAAS.

We may now turn the tables on Alcock and ask what motivates skeptics to spend so much time trying to dismiss the results of another scientific discipline. For Alcock, it seems that his feelings toward organized religion and his fears about genuine psi are motivations. For example, Alcock has written:

> In the name of religion human beings have committed genocide, toppled thrones, built gargantuan shrines, practiced ritual murder, forced others to conform to their way of life, eschewed the pleasures of the flesh, flagellated themselves, or given away all their possessions and become martyrs.[50]

And,

> There would, of course, be no privacy, since by extrasensory perception one could see even into people's minds. Dictators would no longer have to trust the words of their followers; they could "know" their feelings.... What would happen when two adversaries tried to harm the other via PK?[51]

Given Alcock's feelings about religion and psi, he should be suspicious about the motivations of the prominent physicist Stephen Hawking. In Hawking's widely acclaimed *A Brief History of Time,* the final paragraph reads:

> ... if we do discover a complete theory, it should in time be understandable in broad principle by everyone.... Then we shall all, philosophers, scientists, and just ordinary people, be able to take part in the discussion of the question of why it is that we and the universe exist. If we find the answer to that, it would be the ultimate triumph of human reason—for then we would know the mind of God.[52]

In other writings, Hawking has declared his skepticism about psi, so apparently his religious feelings do not interfere with his skepticism. On the other end of the spectrum, what would Alcock say about the motivations of his fellow superskeptic, Martin Gardner, who wrote:

> As for empirical tests of the power of God to answer prayer, I am among those theists who, in the spirit of Jesus' remark that only the faithless look for signs, consider such tests both futile and blasphemous. . . . Let us not tempt God.[53]

In other words, religious faith can motivate scientists both toward or against psi research. Ultimately, there are as many reasons for why people may be for or against something as there are people. Then, from the skeptical perspective, what else might account for the widespread belief in psi? Is society going crazy?

Is Society Crazy?

If there is no scientific evidence that psi exists, then strong public belief in such topics must be a sign of mass delusion. This is a common but rather peculiar skeptical position, since we could draw a parallel with, say, belief in God. There is no scientific evidence that God exists, yet there is strong public belief in God. For some reason, skeptics do not openly point to mental delusion as a reason for the widespread, "unscientific" belief in God.

But is there any evidence that society is delusional? Can paranormal experiences be attributed only to known psychological processes? This question was examined by Catholic priest Andrew Greeley, a sociologist at the University of Arizona. Greeley was interested in the results of surveys consistently indicating that the majority of the population believes in ESP. In a 1978 survey asking American adults whether they had ever experienced psychic phenomena such as ESP, 58 percent said yes; a 1979 survey of college and university professors found that about two-thirds accepted ESP;[54] a 1982 survey of elite scientists showed that more than 25 percent believed in ESP; and in a 1987 survey, 67 percent of American adults reported that they had had psychic experiences. The same surveys showed, according to Greeley, that

> People who've tasted the paranormal, whether they accept it intellectually or not, are anything but religious nuts or psychiatric cases. They are, for the most part, ordinary Americans, somewhat above the norm in education and intelligence and somewhat less than average in religious involvement.[55]

Because Greeley was surprised by this outcome, he explored it more closely by testing people who had reported profound mystical experiences such as being "bathed in light." He used the Affect Balance Scale of psychological well-being, a standard psychological test used to measure healthy personality. People reporting mystical experiences achieved top scores. Greeley reported that "The University of Chicago psychologist who developed the scale said no other factor has ever been found to correlate so highly" as reports of mystical experience.

Greeley then investigated whether prior belief in the paranormal or the mystical *caused* the experiences, or whether the experiences themselves caused the belief. He found that many widows who reported contact by their dead husbands had not previously believed in life after death. This suggests that they were not unconsciously creating hallucinations to confirm their prior beliefs.

Greeley also studied whether people who had lost a child or parent reported contact with the dead more often than people whose siblings had died. The assumption was that people who had lost family members closer to them might have had a stronger need to communicate, and hence a greater frequency of reported contacts. According to Greeley, "We were surprised: People who'd lost a child or parent were less likely to report contact with the dead than those who'd lost siblings." Such findings are incompatible with the skeptics' hypothesis that reports of paranormal experiences are due solely to hallucination, self-delusion, wish fulfillment, or other forms of mental aberrations.

Summary

Most of the commonly repeated skeptical reactions to psi research are extreme views, driven by the belief that psi is impossible. The effect on mainstream academics of repeatedly seeing skeptical dismissals of psi research—in college textbooks and in prominent scientific journals—has been diminished interest in the topic. Informed opinion, however, even among skeptics, shows that virtually all the past skeptical arguments against psi have dissolved in the face of overwhelming positive evidence, or they are based on incredibly distorted versions of the actual research.

So how can we understand the extreme reactions to the evidence for psi, and the somewhat less extreme, but still extraordinarily obstinate position taken by mainstream science? This brings us to the next topic—the remarkable power of preconceptions to determine what we can and can't see.

CHAPTER 14

Seeing Psi

It is as fatal as it is cowardly to blink facts because they are not to our taste.

JOHN TYNDALL (1820–1893)

Given the substantial historical, anecdotal, and experimental evidence for psi, why do some intelligent people positively bristle at the mere *suggestion* that the evidence for psi be taken seriously? After all, scientists studying psi simply claim that every so often they find interesting evidence for strange sorts of perceptual and energetic anomalies. They're not demanding that we also believe aliens have infiltrated the staff of the White House. Still, some people continue to insist that "there's not a shred of evidence" for psi. Why can't they see that there are thousands of shreds that, after we combine the weft of experiences and the warp of experiments, weave an immense, enchanting fabric?

The answer is contained in the odd fact that we do not perceive the world as it is, but as we wish it to be.[1] We know this through decades of conventional research in perception, cognition, decision making, intuitive judgment, and memory. Essentially, we construct mental models of a world that reflect our expectations, biases, and desires, a world that is comfortable for our egos, that does not threaten our beliefs, and that is consistent, stable, and coherent.

In other words, our minds are "story generators" that create mental simulations of what is really out there. These models inevitably perpetuate distortions, because what we perceive is influenced by the hidden persuasions of ideas, memory, motivation, and expectations. An overview of how we know this help clarify why we should be skeptical of both overly enthusiastic claims of psychic experiences and overly enthusiastic skeptical criticisms, and why controversy over the existence of psi has persisted in spite of a century of accumulating scientific evidence.

The bottom line is that if we do not expect to see psi, we won't. And because our world will not include it, we will reach the perfectly logical conclusion that it does not exist. Therefore anyone who claims that it does is just stupid, illogical, or irrational. Of course, the opposite is also true. If we expect to see psi everywhere, then our world will be saturated with psychic phenomena. Just as uncritical skepticism can turn into paranoia and cynicism, uncritical belief can turn into an obsessive preoccupation with omens, signs, and coincidences. Neither extreme is a particularly balanced or well-integrated way of dealing with life's uncertainties.

Four Stages Redux

This book opened with a listing of the four stages by which we accept new ideas. In Stage 1 the idea is flat-out impossible. By Stage 2 it is possible, but weak and uninteresting. In Stage 3 the idea is important, and the effects are strong and pervasive. In Stage 4 everyone thinks that he or she thought of it first. Later, no one remembers how contentious the whole affair was.

These same four stages are closely associated with shifts in perception and expectation. In Stage 1, expectations based on prior convictions prevent us from seeing what is out there. At this stage, because "it" can't be seen, then of course "it" is impossible. Any evidence to the contrary must be flawed, even if no flaw can be specified. The stronger our expectation, the stronger our conviction is that we are correct.

In Stage 2, after our expectations have been tweaked by repeated exposure to new experiences or to overwhelming evidence, we may begin to see "it," but only weakly, sporadically, and with strong distortions. At this stage, we sense that something interesting is going on, but because it not well understood, we can't perceive it clearly. Authorities declare that it may not amount to much, but whatever it is, it might be prudent to take it seriously.

In Stage 3, after someone shows how it *must* be there after all, either through a new theoretical development or through the unveiling of an obvious, practical application, then suddenly the idea and its implications are obvious. Moreover, if the idea is truly important, it will seem to become omnipresent. After this stage, all sorts of new unconscious tactics come into play, like retrocognitive memory distortion (revisionist history), and a whole new set of expectations arises. Inevitably, mental scaffolding begins to take shape that blocks perception of future new ideas. History shows that this cycle is repeated over and over again.

Effects of Prior Convictions

A classic experiment by psychologists J. S. Bruner and Leo Postman demonstrated that sometimes what we see—or think we see—is not really there.[2] Bruner and Postman created a deck of normal playing cards, except that some of the suit symbols were color-reversed. For example, the queen of diamonds had black-colored diamonds instead of red. The special cards were shuffled into an ordinary deck, and then as they were displayed one at a time, people were asked to identify them as fast as possible. The cards were first shown very briefly, too fast to identify them accurately. Then the display time was lengthened until all the cards could be identified. The amazing thing is that while all the cards were eventually identified with great confidence, no one noticed that there was anything out of the ordinary in the deck.

People saw a black four of hearts as either a four of spades or as a normal four of hearts with red hearts. In other words, their expectations about what playing cards *should look like* determined what they actually saw. When the researchers increased the amount of time that the cards were displayed, some people eventually began to notice that something was amiss, but they did not know exactly what was wrong. One person, while directly gazing at a red six of spades, said, "That's the six of spades but there's something wrong with it—the black spade has a red border."[3]

As the display time increased even more, people became more confused and hesitant. Eventually, most people saw what was before their eyes. But even when the cards were displayed for forty times the length of time needed to recognize normal playing cards, about 10 percent of the color-reversed playing cards were never correctly identified by any of the people!

The mental discomfort associated with seeing something that does not match our expectations is reflected in the exasperation of one participant in the experiment who, while looking at the cards, reported, "I can't make the suit out, whatever it is. It didn't even look like a card that time. I don't know what color it is now or whether it's a spade or a heart. I'm not even sure what a spade looks like. My God!"

Studies like this in the 1950s led psychologist Leon Festinger and his colleagues at Stanford University to develop the idea of *cognitive dissonance*.[4] This is the uncomfortable feeling that develops when people are confronted by "things that shouldn't ought to be, but are." If the dissonance is sufficiently strong, and is not reduced in some way, the uncomfortable feeling will grow, and that feeling can develop into anger, fear, and even hostility. A pathological example of unresolved cognitive dissonance is represented by people who blow up abortion clinics in the name of Jesus. Also, to avoid unpleasant cognitive dissonance people will often react to evidence that *disconfirms* their beliefs by actually *strengthening* their original beliefs and creating rationalizations for the disconfirming evidence.

The drive to avoid cognitive dissonance is especially strong when the belief has led to public commitment. Because the primary debunkers of psi phenomena are publicly committed to their views through their affiliation with skeptics organizations, we can better understand some of the tactics they have used to reduce their cognitive dissonance.

Reducing Cognitive Dissonance

There are three common strategies for reducing cognitive dissonance. One way is to *adopt what others believe*. Parents often see this change in their children when they begin school. Children rapidly conform to groupthink, and after a few years, they *need* this particular pair of shoes, and that particular haircut, and this video game, or they will simply *die*. Children are not just imagining their strong needs for this or that fad. Even in young children, the need to conform to social pressure can be as painful as physical pain. Likewise, a college student faced with trying to please a skeptical professor will soon come to agree that anyone who believes in all that "New Age bunk," or psi, is either mentally unstable or stupid.

A second way of dealing with cognitive dissonance is to *apply pressure* to people who hold different ideas. This explains why mavericks are often shunned by more conventional scientists and why there is almost no public funding of psi research. In totalitarian regimes, the heretics are simply tracked down and eliminated. To function without the annoying pain of cognitive dissonance, groups will use almost any means to achieve consensus.

The third way of reducing cognitive dissonance is to *make the person who holds a different opinion significantly different* from oneself. This is where disparaging labels like "heretic" and "pseudoscientist" come from. The heretic is stupid, malicious, foolish, sloppy, or evil, so his opinion does not matter. Or she has suspicious motives, or she believes in weird practices, or she looks different. The distressing history of how heretics were treated in the Middle Ages and the more recent "ethnic cleansings" of the last half-century remind us that witch-hunts are always just below the veneer of civility. The human psyche fears change and is always struggling to maintain the status quo.[5]

Vigorous struggles to promote the "one right" interpretation of the world have existed as long as human beings have held opinions. As history advances, and we forget the cost in human suffering, old controversies begin to look ridiculous. For example, an explosive controversy in the Middle Ages was whether God the Father and God the Son had the *same* nature or merely a *similar* nature. Hundreds died over that debate.[6]

Cognitive Dissonance and Psi

When we are publicly committed to a belief, it is disturbing even to consider that any evidence contradicting our position may be true—because public ridicule adds to the unpleasantness of cognitive dissonance. This is one reason that the psi controversy has persisted for so long. It also helps to explain why it is much easier to be a skeptic than it is to be a researcher investigating unusual effects. Skeptics may be overly conservative, but if they are ultimately proved wrong they can just smile and shrug it off and say, "Whoops, I guess I was wrong. Sorry!" By contrast, frontier scientists are often blindly attacked as though their findings represented a virus that must be extinguished from the existing "body" of knowledge at all costs.

Commitment stirs the fires of cognitive dissonance and makes it progressively more difficult to even casually entertain alternative hypotheses. This is as true for proponents as it is for skeptics. Cognitive dissonance is also one of the main reasons that many scientists dismiss the evidence provided by psi experiments without even examining it. In science, said the philosopher of science Thomas Kuhn, "novelty emerges only with difficulty, manifested by resistance, against a background provided by expectation. Initially only the anticipated and usual are experienced even under circumstances where an anomaly is later to be observed."[7]

This means that in the initial stages of a new discovery, when a scientific anomaly is first claimed, it literally cannot be seen by everyone. We have to change our expectations in order to see it. When one scientist claims to see something unusual, another scientist who is intrigued by the claim, but does not believe it yet, will simply fail to see the same effect.

Kuhn illustrated this bewildering state of affairs with the case of Sir William Herschel's discovery of the planet Uranus. Uranus was observed at least seventeen times by different astronomers from 1690 to 1781. None of the observations made any sense if the object was a star, which was the prevailing assumption about most lights in the sky at the time, until Herschel suggested that the "star" might have been in a planetary orbit. Then it suddenly made sense. After this shift in perception, caused by a new way of thinking about old observations, suddenly everyone was seeing planets.[8]

The same was true for studies of subliminal perception in the 1950s. Not all early experimenters could get results. No theory could account for the bizarre claim that something could be seen without being aware that it was being seen. But once computer-inspired information-processing models were developed, with their accompanying metaphors about information being processed simultaneously at different levels, then suddenly subliminal processing was acceptable and the effects were observable.[9]

The effect of shifting perceptions was observed more recently when high-temperature superconductors were unexpectedly discovered in 1986. Soon afterward, superconducting temperatures previously considered flatly impossible were being reported regularly. The same had occurred with lasers. It took decades to get the first lasers to work; then suddenly everything was "lasing." It took decades to get the first crude holographs to work, and now they are put on cereal boxes by the millions. Some of these changes were the result of advancements in understanding the basic phenomenon, but those advancements could not occur until expectations about what was *possible* had already changed.

Another famous and poignant example is the case of German meteorologist Alfred Wegener. In 1915 Wegener published a "ludicrous" theory that the earth's continents had once been a single, contiguous piece. Over millions of years, he claimed, the single continent split into several pieces, which then drifted apart into their current configuration. Wegener's theory, dubbed "continental drift," was supported by an extensive amount of carefully cataloged geological evidence. Still, his British and American colleagues laughed and called the idea impossible, and Wegener died an intellectual outcast in 1930. Today, every schoolchild is taught his theory, and by simply taking the time to examine a world map, we can now observe that Wegener's impossible theory is entirely self-evident.[10]

Expectancy Effects

I know I'm not seeing things as they are, I'm seeing things as I am.

LAUREL LEE

In attempting to understand how intelligent scientists could seriously propose criticisms of psi research that were blatantly invalid, sociologist Harry Collins showed that for controversial scientific topics where the mere *existence* of a phenomenon has been in question, scientific criticisms are almost completely determined by critics' prior expectations. That is, criticisms are often unrelated to the actual results of experiments. For example, Collins showed that in the case of the search for gravity waves (hypothetical forces that "carry" gravity), reviewers' assessment of the competency of experiments conducted by proponents and critics depended entirely on the reviewers' expectations of what effects they thought *should* have been observed.[11]

The expectancy effect has also been observed in experimental studies by Stanford University social psychologists Lee Ross and Mark Lepper. They found that precisely the *same* experimental evidence shown to a group of reviewers tended to polarize them according to their initial positions.[12] Studies conforming to the reviewers' preconceptions were seen as better designed, as more valid, and as reaching more adequate conclusions. Studies not conforming to prior expectations were seen as flawed, invalid, and reaching inadequate conclusions. Sound familiar?

This "perseverance effect" has been a major stumbling block for parapsychology. Collins and sociologist Trevor Pinch studied how conventional scientists have reacted to claims of experimental evidence for psi phenomena. In an article they wrote that focused on issues of social psychology, and in which they explicitly stated that their own position was entirely neutral with regard to the existence of psi, they received

> a spleenful letter from a well known professional magician-and-sceptic which attempts to persuade us to change our attitude to research in the paranormal and claims that: "Seriously, how men of science such as yourselves can make excuses for . . . [the proponents'] incompetence is a matter of astonishment to me. . . . I was shocked at your paper; I had expected science rather than selective reporting."[13]

Reviewer bias is not just evident in skeptics' reviews of psi research; it is endemic in all scientific controversies. This is especially true for controversies concerned with questions about morality or mortality. For example, science becomes muddled with politics when we seek answers to difficult questions such as whether herbal remedies should be used to treat cancer, or whether nuclear power is safe, or whether a particular concentration of benzene or asbestos in the workplace is tolerable.

A reviewer's judgment of a researcher's level of competency is often established on the basis of *who* produced the results rather than on independent assessments of the experimental methods. For example, results reported by "prominent professors at Princeton University" will be viewed as more credible than *identical* results reported by a junior staff member at "East Central Southwestern Community College."

Ultimately, it seems that scientific "truth," at least for controversial topics, is not determined as much by experiment, or replication, or any other method listed in the textbooks, as by purely nonscientific factors. These include rhetoric, ad hominem attack, institutional politics, and battles over limited funding. In short, scientists are human. Assuming that scientists act rationally when faced with intellectual or economic pressures is a mistake.

Sociologist Harry Collins calls one element of this problem about getting to the "truth" of controversial matters the *experimenters' regress*. This is an exasperating catch-22 that occurs when the correct outcome of an experiment is unknown. To settle the question under normal circumstances, where results are predicted by well-accepted theory, the outcome of a single experiment can be examined to see if it matches the expectation. If it does, the experiment was obviously correct. If not, it wasn't.

In cases like parapsychology, to know whether the experiment was well performed, we first need to know whether psi exists. But to know whether psi exists, we need to run the right experiment. But to run the right experiment, we need a well-accepted theory. But . . . And so on. This forms an infinite, potentially unbreakable loop. In particular, this loop can continue unresolved in spite of the application of strict scientific methods. In an attempt to break the experimenters' regress, skeptics often argue that the phenomenon does not exist. Of course, to do that they must rely on invalid, nonscientific criticisms, because there is plenty of empirical evidence to the contrary.

It is difficult to detect purely rhetorical tactics unless one is deeply familiar with both sides of a debate. As Collins put it:

> Without deep and active involvement in controversy, and/or a degree of philosophical self-consciousness about the social process of science (still very unusual outside a small group of academics) the critic may not notice how far scientific practice strays from the text book model of science.[14]

Judgment Errors

The acts of perception and cognition, which seem to be immediate and self-evident, involve absorbing huge amounts of meaningless sensory information and mentally constructing a stable and coherent model of the world. Mismatches between the world as it really is and our mental "virtual" world lead to persistent, predictable errors in judgment. These judgment errors have directly affected the scientific controversy about psi.

When a panel of expert clinicians, say psychologists, physicians, or psychiatrists, are asked to provide their best opinions about a group of patients, they are usually confident that their assessments will be accurate. After all, highly regarded clinicians have years of experience making complex judgments. They believe that their experiences in judging thousands of earlier cases have honed their intuitive abilities into a state of rarefied precision that no simple, automated procedure could ever match. They're often wrong.

Psychologist Dale Griffin of Stanford University reviewed the research on how we make intuitive judgments for the same National Research Council report that reviewed the evidence on psi.[15] Griffin's job was to remind the committee that when we make expert judgments on complex issues, it is important to use objective methods (like meta-analysis) to assess the evidence rather than to rely on personal intuitions. It's too bad that the committee did not pay close attention to Griffin's advice.

Starting in the 1950s, researchers began to study how expert intuition compared with predictions based on simpleminded mathematical rules. In such studies, a clinical panel was presented with personal information such as personality scores and tallies on various other tests, then asked to predict the likely outcomes for each person. The prediction might be for a medical assessment, or suitability for a job, or any number of other applications. The judges' predictions were compared to a simple combination of scores from the various tests, and both predictions were compared with the actual outcomes. To the dismay of the experts, not only were the mathematical predictions far superior to the experts' intuitions, but many of these studies showed that the amount of professional training and experience of the judges was not even vaguely related to their accuracy! To add insult to injury, the mathematical models were not highly sophisticated. In most cases, they were formed by simply adding up values from various test scores.[16]

A flurry of studies in the 1950s confirmed that simple mathematical predictions were *almost always better* than expert clinical intuition for diagnosing medical symptoms such as brain damage, categorizing psychiatric patients, and predicting success in college. Clinical experts were not amused.

Today, when we evaluate complex evidence provided by a body of experimental data, we use the successor to those early mathematical models: the quantitative meta-analysis. So the National Research Council experts who relied on their personal opinions to evaluate the evidence for parapsychology, however intuitively appealing their opinions may have felt, would be as perplexed as the clinical experts of the 1950s to discover that their subjective opinions were just plain wrong.

What Do We Pay Attention To?

How could experts be so wrong? One reason is that expectation biases are self-generating. We cannot pay attention to everything equally, so instead we rely on past experience and vague mental "heuristics," or guidelines, that worked fairly well on similar problems. Unfortunately, relying on subjective impressions and mental guidelines creates a cycle in which our past experience begins to divert us from paying attention to new things that might be even more predictive. After a while, since we no longer pay attention to anything other than what we have already decided is important, we tend to keep confirming what we already knew. This model- or theory-driven approach is called the *confirmation bias*.

The problem with the confirmation bias is that we end up learning only one or two ways to solve a problem, and then we keep reapplying that solution to all other problems, whether it is appropriate or not. This is especially compounded for highly experienced people, because past successes have made their theories so strong that they tend to overlook easier, simpler, more accurate, and more efficient ways of solving the problem. This is one reason that younger scientists are usually responsible for the giant, earth-shaking discoveries—they haven't learned their craft so well that they have become blind to new possibilities. Younger scientists are invariably more open to psi than older scientists.

One well-known consequence of being driven by theory is the "self-fulfilling prophecy"—the way our private theories cause others to act toward us just as our theories predict. For instance, if our theory says that people are basically kind and loving, and we expect that people will act this way, then sure enough, they will usually respond in kind, loving ways, reinforcing our original expectation. In contrast, if we assume that people are basically nasty and paranoid, they will quickly respond in ways that reinforce this negative expectation. Many people know about the power of self-fulfilling prophecy through Norman Vincent Peale's famous book, *The Power of Positive Thinking.*[17]

An experiment demonstrating the self-fulfilling prophecy was described by Harvard psychologist Robert Rosenthal in a classic book entitled *Pygmalion in the Classroom.*[18] Teachers were led to believe that some students were high achievers and others were not. In reality, the students had been assigned at random to the two categories. The teachers' expectations about high achievers led them to treat the "high achievers" differently than the other students, and subsequent achievement tests confirmed that the self-fulfilling prophecy indeed led to higher scores for the randomly selected "high achievers."

Such studies made it absolutely clear that when experimenters know how participants "should" behave, it is impossible not to send out unconscious signals. This is why scientists use the double-blind experimental design, so that their personal expectations do not contaminate the research results. And this is why we cannot fully trust fascinating psychic stories reported by groups that expect such things to occur, unless they also demonstrate that they are aware of, know how to, and *did* control for expectation biases.

An important consequence of the confirmation bias and self-fulfilling prophecy is that the more we think we already know the answer, the more difficult it is for us to judge new evidence fairly. This is precisely why scientific committees charged with evaluating the evidence in controversial fields such as psi *must* be composed of scientists who have no strong prior opinions about the topic. It is too bad that the National Research Council committee did not heed its own advice.

Because of the confirmation bias, skeptics who review a body of psi experiments are likely to select for review only the few studies that confirm their prior expectations. They will assume that all the other studies they could have reviewed would have had the same set of real or imagined problems. And they end up confirming their prior position. For example, one of skeptical psychologist Susan Blackmore's favorite arguments against parapsychology is based upon a single occasion when she thought she had reason to suspect one set of experiments. For years now, she has used that single experience to justify her doubt about all other psi experiments.[19]

Selecting the Attractive Evidence

Another way that we select subsets of evidence is through examples that attract our attention. This attractive quality, termed *salience,* involves objects or events that are brighter, louder, unique, more exciting, or more noticeable in some other way. Salience underlies many of the failures of human judgment; since we cannot pay attention to everything, whatever attracts our attention will guarantee that we are not getting the whole picture. It is precisely for this reason that meta-analysts insist on retrieving all available studies rather than just a few exemplars. To counter the biasing effect of salience, all the evidence must be collected and evaluated identically, whether it's vivid and exciting or dull and tedious.

Similar to salience is the fact that some items in memory are easier to bring to mind. As a general rule, it is much easier to store and retrieve exciting stories than dull, mathematical data. While statistical analyses and mathematical summaries are far more valid means of making decisions and evaluating evidence, a good story will attract and hold both attention and memory. Thus, when we are asked to make decisions about evidence, what usually comes to mind is one or two vivid stories, not the full set of available data. So, while the case studies retold in this book are much more exciting than the data summarized in the graphs, the cumulative data represented in the graphs *are far more evidential* for psi!

Another unconscious rule of thumb that we use to make judgments is what psychologists Daniel Kahneman and Amos Tversky have called the *representativeness heuristic*. We assume that "like goes with like." Unfortunately, this rule of thumb leads to predictable errors because we place far too much emphasis on a single case and ignore all the other cases. So not only is a single nice story easier to remember than masses of data, but the particulars of that one story carry far too much weight and make us imagine that we know more about the other cases than we do.

Moreover, we tend to think that a particular case is increasingly representative of all similar stories as that one story is fleshed out in more detail. The paradox is that as we add more and more details to any given case, it actually becomes *less representative!* This is again because it is easier to pay attention to an exciting, richly woven story than to a dry, simple story. In the future, if we hear a vaguely similar story, we will unconsciously fill in any missing details with what we remember from the one good story we already know. This quickly gives rise to stereotyped thinking and completely obscures any important nuances provided by new evidence. So, if you find yourself saying, "Oh, that's just like . . . ," well, maybe it is and maybe it's not.

We know that the representation bias is pervasive in the media's portrayal of psi research. When randomly selected scientists in the United States are asked what they know about psi research, they typically respond with stories about the Israeli psychic Uri Geller or the American magician James Randi. Among the general public, too, Geller and Randi are widely considered to be highly relevant to the scientific evaluation of psi. And yet, while the stories about these two are intriguing, nothing about the work of either Geller or Randi is described in this book. They are actually so *irrelevant* to the scientific evaluation of psi that not a single experiment involving either person is included among the thousand studies reviewed in the meta-analyses.

Hindsight Bias

In the five stages of the acceptance of new ideas, Stage 4 occurs when the idea has become so well adopted that people claim that "I knew it all along." This is not (always) just an attempt to usurp the glory; it can also reflect something called *hindsight bias*.

Many studies have shown that once people are aware of the correct answer to something, they are certain that they would have known the answer, even if they were previously uninformed. Knowledge of the correct answer allows us to build a nice story around the answer, and when the answer is "taken away" in our imagination, the story remains. The story structure that contained the right answer makes it *seem* as though we would have obviously selected the right answer.

Hindsight bias also affects recall of our confidence in the truth of an assertion. That is, if we find out that an earlier assertion was indeed true, this will increase our recalled confidence in its truth. And if we discover that the assertion was false, it will decrease our recalled confidence in its truth. Let's say we were originally very impressed by a certain telepathy experiment. Later, we heard a rumor somewhere that this study contained a flaw (real or imagined). Hindsight bias will covertly reconstruct our memory so that we begin to recall that we were actually not at all impressed by the experiment in the first place.

Hindsight bias also occurs if we are repeatedly exposed to an assertion. Regardless of whether the assertion is true, repetition will improve our memory of it and, as a result, falsely boost our confidence in its truth! This means that incessant television shows with stories about angels and aliens will boost our confidence in those ideas, completely independently of whether those stories are true.

Mediated Evidence

Most of what we know about the world at large, especially about science, and even more so about psi research, comes not through our personal experience but through heavily refined, preprocessed, "mediated" information on television and in magazines and books. All the informational and motivational biases already mentioned operate on this mediated information just as they would have operated on the evidence if we had seen it with our own eyes.

The difference is that we've learned (it is hoped) that preprocessed information invariably presents only part of the story. Someone had to decide what to present and what to leave out, and this means we should always be wary of scientific evidence presented in the brief formats available on television shows. When a program host says, "Here's the evidence for psychic phenomena. Now *you* decide what to believe," this sounds great but it's actually a ridiculous assignment. We haven't been shown all the evidence, nor do we have any guidelines about how to evaluate the evidence. All we saw were a few selected bits that looked good on TV.

To overcome the suspected biases in any source of mediated or predigested information, we should look for multiple sources of similar information and see if the evidence converges. Of course, this takes effort, usually more effort than most people are willing to spend. In addition, informational and motivational biases provide very reasonable-feeling excuses to ignore any mediated evidence that contradicts our beliefs. If we don't like what the TV show is saying, we just flip the channel.

The television medium in particular is designed to manipulate our attentional biases by playing the commercials a little louder, by making programs faster-paced and brighter, and by escalating the number of emotionally stimulating scenes. Television rarely presents information about what actual data mean, or how data were collected, or how to understand the analysis and interpretation of the evidence. Television shows are forced to bypass most of the caveats and alternative explanations that form the cautious side of science, because the alternatives are not always so simple to convey.

Because most of what people know about "the paranormal" comes through television programs and movies, those who would like to know what to believe often assess the quality of the evidence through the credibility of the show. A calm, sober presentation of evidence on the PBS science program *Nova* should carry more evidential weight than a sensational ghost story on a "tabloid" program. This sounds reasonable, except for something called the "sleeper effect."[20]

The sleeper effect is a memory distortion whereby information becomes separated from its source. Let's say that we see a silly ghost story on a sensational, "tabloid" show. Later, we see a scientific psi experiment described on a sober science program. Initially, we will perceive the ghost story as less credible than the experiment, but after a while the information from the tabloid show and the science show will become mixed up in memory. Soon the ghost story and the psi experiment will be remembered as *equally evidential,* because the sources of the information have been forgotten. If our prior opinion says that ghosts are silly, then the psi experiment must also be silly—because we will assume that they both came from the same source.

Beyond the Perceptual Filters

Beyond expectation, hindsight bias, and cognitive dissonance, psychotherapists have identified many other ways that the mind consciously and unconsciously protects itself from seeing what it does not wish to see. These mental protection schemes, which are the emotional first cousins of the more intellectually motivated perceptual and cognitive biases, have been labeled suppression, reaction formation, repression, identification, dissociation, and projection.[21] Let's examine them briefly.

Suppression

Suppression refers to a conscious avoidance of something we may wish to do or say. Say you're in an meeting where your boss is about to give a presentation to some important clients. You notice to your horror that your boss's toupee has slipped, and it now bears a striking resemblance to a squirrel perched on his head. You have an overriding impulse to laugh, but you suppress your impulse, because laughing would cause a public embarrassment that might put your job in jeopardy.

People who do not control their impulses, such as when they are drunk, are perceived as impulsive and antisocial. For the wheels of civilization to turn smoothly, some impulses *must* be suppressed. Little children speak whatever comes to mind and get away with it, but adults who do this quickly find that they have no friends. Children are taught very early on not to speak about what adults have called "psychic experiences," because the social order in the Western world does not know what to do with children who are "violating natural law."

Reaction Formation

Defense mechanisms can be consciously applied, as in suppression, or applied in the unconscious. The deeper in the unconscious the defense is, the more powerful are its effects, and the more difficult it is to recognize the defense as being part of us. Reaction formation is usually just below the level of awareness, but it can be recognized as an unrealistically enthusiastic or wildly negative response to something.

For example, say you really wanted to get a promotion, and you thought that you were the obvious choice, but another office worker, call him Bob, was promoted instead. When you learn of this shocking miscarriage of justice, reaction formation will immediately protect your ego by making Bob out to be a nasty, backbiting, stupid, ugly person.

The same reaction can occur in the opposite direction. Say that you were taught as a child that only bad, evil people become angry. This message may have been deeply engrained throughout your childhood, and it has become part of the way you perceive the world. Thus, when Bob is given the promotion that you worked hard for, your initial impulse to become angry may be transformed, because you are not allowed to become angry. Reaction formation would cause you to become wildly enthusiastic about the wisdom behind Bob's promotion, and to tell everyone what a wonderful guy he is.

Typically, when someone complains a little too much, or is inappropriately over-enthusiastic, it may be a sign of reaction formation, in which case the person's real feelings may be the opposite. As the queen responded after Hamlet asked her how she liked the play, "The lady doth protest too much, methinks." The more vigorous skeptical attacks on parapsychology are reminiscent of reaction formation. Likewise, some equally emotional attacks on psi research by psychic *enthusiasts* (who resent the intrusion of science into their private domain) suggest the presence of underlying defense mechanisms.

Repression

When a feeling or desire is completely blocked from awareness, this is a form of repression. In contrast to suppression, which is a conscious blocking of inner impulses, repression is blocked below conscious awareness. Because repression is hidden from awareness, it is usually inferred from unconscious changes in behavior. For example, suppose your spouse notices that you never call your siblings on the phone. He asks you if something happened that has caused you to ignore them. You honestly cannot think of any reason that you don't call them, so you presume it is because you have been too busy. And yet your spouse notices that you are absent-mindedly clenching your jaw and twisting your fingers as you answer the question. An inference can be made that you have repressed something about your siblings.

Repression also occurs for things that run counter to what we have been taught. Unfortunately, one of the outcomes of going to school is that the natural curiosity we are born with is repeatedly suppressed. We learn not to ask too many questions, not to wonder out loud about certain taboo topics, and to overcome our creative impulses for the sake of conforming to social pressure. In adulthood, these rules may become deeply embedded in the psyche, and are then *repressed.*

If, as an adult, you have a spontaneous, psychic experience, repression may quickly set in. This protects you from bad memories about how "only crazy people get psychic impressions," and thus the experience will disappear from awareness almost as fast as it arises. An experience so dramatic that it doesn't get repressed may weigh heavily as a secret event in your life, rarely admitted to anyone. This practically guarantees that radically new ideas rarely rise to the surface in properly socialized folks. It also guarantees that those who do suggest new ideas are quickly labeled "wacky" or "heretical" and regarded with suspicion. It further explains why the evidence for psi is more or less invisible to the orthodox.

Identification and Introjection

We all carry ideas about who we are, or who we have been taught to believe we are. If we have an inappropriate reaction to something, like a sudden outburst of anger, we might be surprised and think, "Hey, that isn't me." When this occurs, it reflects the fact that not only is our perception of the world a construction, but also our sense of who we think we are. That sense of ourselves is formed from identifying with role models during our developmental years, and with role models we admire (or fear) as adults. Sometimes you may find yourself thinking or saying certain things and suddenly realize that it wasn't "you" but rather a tape playing from something one of your parents said repeatedly when you were growing up.

If you forget that the mental tape player is not you but your parent, you may begin to identify with the messages on the tape. You will *introject* an image, or simulation, of that person into you. If you are unlucky, the internal tape recorder may be constantly replaying messages like "you will never amount to anything." Unconsciously *identifying* with these messages will lead to the unshakable belief that in fact you never did amount to anything, regardless of what you may have actually accomplished.

Say that one of your favorite college professors insisted that psi is impossible because it contradicts a dozen inviolate Laws of Nature. Say that he or she (but usually he) gave lots of wise-sounding reasons for dismissing the "obviously sloppy" research promoted by psychic researchers who were clearly motivated by secret religious cults. It would take years, or perhaps never, to no longer identify with those skeptical messages.

Dissociation

If you've learned that some desires or emotions are taboo, and yet you still have them, then parts of your personality may split off and dissociate from the rest of "you." In the extreme case of multiple-personality syndrome, these split-off personalities can act like separate people in one body.

For example, if you were brought up in a staunchly atheistic family, religious thoughts or feelings may never have been discussed. If mentioned at all, the topic was immediately derided as superstitious nonsense. As an adult, you may be faced with the uncomfortable situation of having a strong psychological block against something that seems to be a deeply instinctive part of the human psyche. To accommodate both in the same person, the mind compartmentalizes these conflicting desires and beliefs into an "ordinary you" and a "secret you" who yearns for religious experiences.

A mild form of this compartmentalization can be seen in scientists who for six days each week immerse themselves in a purely materialistic, scientific mode that provides no outlet for spiritual ideas. While playing the role of Dr. Scientist, they are essentially atheists, or at least agnostics. And yet on the weekend they attend religious services and sincerely pray that everyone in their family remains safe and sound for another week. Maintaining these conflicting attitudes can be mentally painful, especially because it is taboo to mix science and religion. So the mind compartmentalizes the two "you's" into personality segments that do not overlap. This dissociation also occurs in scientists who publicly and vigorously deny the existence of psi while harboring a couple of secret psi experiences that they have not admitted to anyone.

Projection

If we unconsciously deny that inner feelings or beliefs are from us, we may *project* them onto others. Let's say that "Mary" was taught that it was inappropriate ever to tell a lie. And yet she also discovered that she could manipulate people quite easily by telling lies. Faced with the conflict between what she has been taught and what she does, Mary may perceive that she is surrounded by inveterate liars ("they are lying, not me").

Because our perception of the world, including our perception of other people, is a mental construction, we are always projecting to some extent. As psychologist Charles Tart says, "Watch out for the tendency to assume that anyone who doesn't confirm your perception (projection) of him is lying!"[22] What then are we to make of extreme skeptics who insist that the *only* rational explanation for psi is fraud, collusion, or mushy-minded thinking? Or of extreme enthusiasts who do not see *any* value to constructive criticism of psi research, and who view all critics as malicious, evil rationalists? Could this be projection?

Perception and Belief

All this leads us to predict that a person's level of commitment to the current scientific worldview will determine his or her beliefs about psi. Because perception is linked so closely to one's adopted view of reality, people who do not wish to "see" psi will in fact not see it. Nor will they view any evidence for psi, scientific or otherwise, as valid. This effect should be strongest in people who are committed to a particular view, motivated to maintain it, and clever enough to create good rationalizations for ignoring conflicting evidence.

We can indirectly test this prediction by examining belief in psi among four groups of people: the general public, college professors, heads of divisions of the American Association for the Advancement of Science, and members of the National Academy of Sciences. We would predict that belief in psychic phenomena will decrease as the degree of commitment to and belief in orthodox science increases. Sure enough, as shown in figure 14.1, belief (measured in surveys by questions such as "Do you believe in the certainty or in the possibility of psychic phenomena?") drops from about 68 percent of the general public to only about 6 percent of members of the National Academy of Sciences.

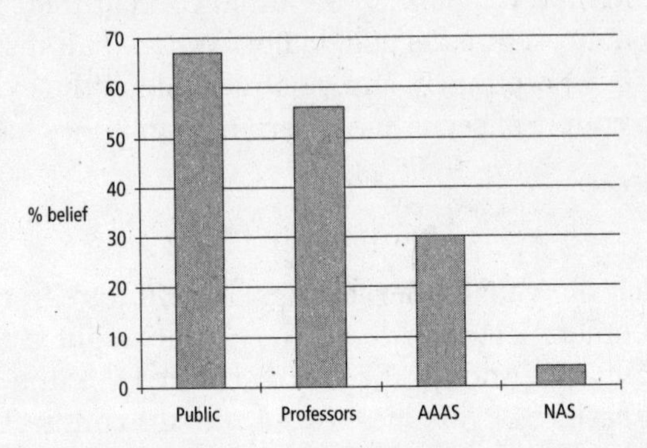

Figure 14.1. Percentage belief in psychic phenomena by degree of commitment to the scientific status quo.[23]

It might be argued that this drop in belief is related to the fact that prominent scientists know more about perceptual and memory biases, and about wishful thinking and self-delusion, and this is why they are so skeptical. But another explanation is that the expectations of the scientific elite actually put them more at risk for being swayed by perceptual biases than the general public. After all, the scientific elite have lifelong careers and their credibility on the line. They are strongly motivated to maintain certain belief systems. By contrast, most members of the general public do not know or care about the expectations of science. So if Joe Sixpack and Dr. Scientist both witness a remarkable feat of clairvoyance, we can predict that later, when we ask Joe what he saw, he will describe the incident in matter-of-fact terms. In contrast, when we ask Dr. Scientist what he saw, he may become angry or confused, or deny having seen anything unusual at all.

Given that we see what we wish to see, who is more likely to report genuine psi experiences? Probably not people who enthusiastically subscribe to New Age beliefs, because they see psi everywhere, whether it is really there or not. Probably not confirmed skeptics, because they never see psi anywhere, whether it is really there or not. And probably not the scientific elite, because they are motivated not to see psi, or at least not to publicly *admit* that they have had such experiences. This is why the strongest *neutral* evidence for the existence of psi is the cumulative results of experimental studies, evaluated through a technique like meta-analysis.

Once we realize that preconceptions make it difficult to see the scientific evidence for psi, another question naturally arises: where did these preconceptions come from? The next chapter explores the origins of these assumptions, and discusses why many of those assumptions are no longer justified.

CHAPTER 15

Metaphysics

I was thrown out of N.Y.U. my freshman year . . . for cheating on my metaphysics final. I looked within the soul of the boy sitting next to me.

WOODY ALLEN

Why has mainstream science so vigorously resisted the experimental evidence for psi?

In the preceding chapter, we saw that expectations allow us to see the world only in certain ways, and the deeply embedded expectations of modern science do not allow some people to see psi. Where did these expectations come from? They arose from the metaphysics of science, that branch of philosophy concerned with the nature of reality. Metaphysical assumptions often go unexamined, and if they're thought about at all, they're usually accepted as self-evident. Most of the time these assumptions work perfectly fine, so it doesn't matter if they are taken for granted.

But a serious problem has arisen. Most of the fundamental assumptions underlying classical science have been severely challenged in recent years. As the old assumptions dissolve because of advancements in many disciplines, new assumptions are carrying us toward a conception of the world that is entirely compatible with psi. Few scientists have paid close attention to this dramatic shift in scientific fundamentals, and the general public has heard almost nothing about it.

Many scientists working deeply within their own disciplines imagine that once the key mysteries in their realm are better understood, we will understand just about everything that is worth knowing. Many geneticists and molecular biologists believe that after we get a grip on the remaining mysteries of DNA, we will finally be on the road to the golden age. And after the human genome has been fully mapped, we will really understand human behavior at its most fundamental level. Likewise, many neuroscientists fully expect that once we have unraveled the electrochemical complexities of the brain, well *then* we will finally understand the nature of consciousness itself.[1]

Some scientists argue that surely the metaphysics underlying modern science is still working quite well. Science is showing no obvious signs of collapsing, as evidenced by technologies that have made yesterday's supercomputers available in today's homes as desktop PCs. So why would we even dream of reexamining the basic assumptions of science? The answer is in the meaning of scientific anomalies like psi. When the evidence for an anomaly becomes overwhelming, and the anomaly cannot be easily accommodated by the existing scientific worldview, this is a very important sign that either our assumptions about reality are wrong or our assumptions about how we come to understand things are wrong. Or perhaps both are wrong. Assumptions at these fundamental levels act as extremely powerful drivers of expectation and belief, and as we've seen, we only see what we expect to see.

That scientific assumptions evolve should come as no surprise. One of the most profitable consequences of science as an "open system" of knowledge, as opposed to rigid dogma, is that the future Laws of Nature will bear as much resemblance to the "laws" we know today as the cellular telephone does to smoke signals. Both sets of laws attempt to deal with and explain the same world, but the latter set is much more sophisticated and comprehensive than the former.

Going Out of Our Mind

> **Do you believe that the sciences would have arisen and grown up if the sorcerers, alchemists, astrologers and witches had not been their forerunners; those who, with their promisings and foreshadowings, had first to create a thirst, a hunger, and a taste for hidden and forbidden powers? Yea, [and] that infinitely more had to be promised than could ever be fulfilled?**
>
> FRIEDRICH NIETZSCHE

In a nutshell, psi has not been readily accepted by orthodox scientists today because our predecessors, about three hundred years ago, found it expedient to "go out of their mind." Their choice made a lot of sense at the time, but the consequences now are forcing science and society to rethink some basic assumptions. What were these assumptions, why did they arise, and what do they have to do with understanding psi? The answers involve reinterpreting the context of psi research. Parapsychology is not a misguided search for bizarre mysteries, or a thinly veiled religious search for the soul. Instead, psi research is the study of an ancient and still completely unresolved question: Is the mind causal, or is it caused? Are we zombies with "nothing" inside, or are we self-motivated creatures free to exercise our wills?[2]

Medieval Times

In the medieval Western world, all knowledge about nature was revealed solely through the literal word of theological scripture. The world was intensely personal, organic, capricious, meaningful, and teeming with supernatural causes. People found themselves buffeted by unknown, unseen causes attributed to disembodied spirits, or to divine agencies, or to God.[3]

Beginning in the sixteenth and seventeenth centuries, and attributed primarily to Polish astronomer Nicolaus Copernicus, French philosopher René Descartes, Italian astronomer Galileo Galilei, and English physicist Sir Isaac Newton, a new way of understanding the world developed.[4] One reason that this challenge was not instantly squashed by ecclesiastical authority, as it had been many times in the past, was that society desperately needed a major change. Growing economic pressures and stubborn power struggles between monarchs and the church promoted this change, as did the Black Death, or bubonic plague. The plague had ravaged most of Asia and Europe during the fourteenth and fifteenth centuries, killing tens of millions of people. Divine intervention was powerless to stop the scourge, whole villages were decimated, and one-third of Europe's population was wiped out. Many people felt that there had to be a better way.

One of the central ideas allowed to develop under these pressures was Copernicus's heliocentric (sun-centered) theory of planetary motion, which dethroned the earth as the center of the universe. This theory also challenged the common belief that human beings were at the center of the universe. Another key idea was Descartes's distinction between matter and mind—matter being characterized by involuntary activity and mind by voluntary activity. Descartes's idea, called "dualism," cracked the previously unified world into two.

Galileo proposed a similar distinction, except that he added that objects had both primary and secondary qualities. Primary qualities were objectively measurable qualities such as weight, motion, and size. Secondary qualities included color, taste, and heat, which were perceived and existed only in the mind of the beholder. Primary qualities were considered to be reliable and consistent and could thus be used as an empirical basis for science. Galileo's distinction eventually gave rise to the idea that primary qualities were more "real" than secondary qualities.

And then there was Sir Isaac Newton, who in 1687 published his monumental *Philosophiae Naturalis Principia Mathematica*. This work provided the first adequate descriptions of gravity, the laws of motion, fluid mechanics, the motions of the planets, the nature of light, and the phenomena of tides. Out of these ideas arose the first guiding principles of what is now considered to be classical science. Perhaps most important were the ideas of *determinism* and *materialism:* the universe operates according to a uniform set of impersonal rules of cause and effect, and the universe is composed of material objects. Closely related to materialism was the notion of *reductionism,* the assumption that physical objects and systems could be understood in terms of their parts.

The goal of the new science was to reduce objects to their elemental parts and to discover the cause-and-effect rules that governed them. Other ideas soon began to expand the power of these assumptions, including concepts such as what is real is measurable, called *positivism;* there is an objective, real universe separate from and independent of the observer, called *realism;* everything is ultimately made up of little particles, called *atomism;* particles interact like colliding billiard balls, or like gears in a clock, called *mechanism;* once little particles are set on their way, we can in principle predict what they are going to do indefinitely in the future, called *determinism;* and everything interacts only with its closest neighbors, and there is no action at a distance, called *localism.*

These innocent-sounding proposals were, of course, vigorously attacked by the church, because theologians feared that such ideas would eventually undermine their authority. It turns out they had good reason to be afraid. In the intervening three centuries, the church lost basically all its previously held authority to state the "truth" about nature, and science gained unchallenged authority, not only for the physical world, but for most explanations having to do with the mental world as well.

This change in worldview proved to be extremely influential both within and outside of science. It reshaped the way people thought about themselves, about society, government, art, music, and just about everything else. This is when modern liberal-democratic concepts arose, and when the nation-state was "invented." It marked the end of the Middle Ages. But this was also when existential philosophy was born. The profound impact of the new secular view of nature was captured by the nineteenth-century German philosopher Nietzsche in his famous phrase: "God is dead."[5]

The Disenchantment of Nature

Figure 15.1 illustrates how the personal, organic, purposeful world of medieval scholastic authority was split into two worlds after Copernicus, Descartes, Galileo, and Newton. The side of the world claimed by science involved concepts like matter, objective measurement, impersonal, value-free, and eventually, "reality." The other side of the world was claimed by philosophy and religion, and it included concepts like subjective, personal, purposeful, value, and eventually "illusion," in the sense of being less real than the hard-core world of science.

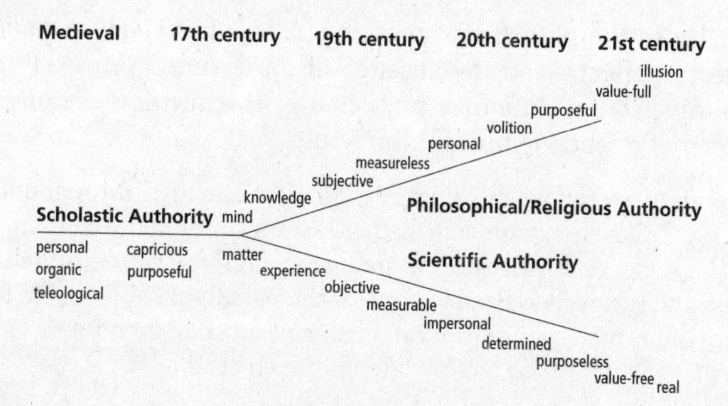

Figure 15.1. The progressive disenchantment of Nature.

The long-held worldview of an organic, personal, holistic place of belonging and meaning was increasingly fractured into a new worldview that was inorganic, impersonal, isolated, and without meaning. The medieval metaphor was of the universe as a "great organism." After the scientific revolution, that metaphor was replaced by one of a "great machine."

As science attained unprecedented power through its ability to predict and control certain limited aspects of nature, it also began to overshadow our understanding of ethics and values. History has shown that decisions affecting millions were made on the basis of industrial expediency, technological imperatives, and economic pressures. Just as the absolute power held by the church for centuries had been seductive, the growing power of science had seduced as well.[6] These seductions were not new, of course, but through science the power of individual whims, which once affected only localized fiefdoms, could now extend to the entire world.

A New Worldview

The new assumptions about the proper way of studying nature proved to be wildly successful, and led to an explosion in our understanding of the universe. In fact, the new worldview was so successful in describing the physical world, and in allowing us to create televisions and jet aircraft, that today most people take these assumptions as unquestionably self-evident. That is, we assume that because we can hold a two-inch color TV set in the palm of our hand, then the principles of determinism, positivism, reductionism, and so on, can explain *everything*.

This is a reasonable assumption. But keep in mind that Lord Kelvin (William Thomson Kelvin), a prominent British mathematician and physicist of the late nineteenth century, was so satisfied with the astonishing rate of progress in physics in his day that he confidently predicted that the rest of physics would soon consist of little more than mopping up a few minor problems on the horizon. Little did he know that those minor problems, which included the nature of something called "blackbody radiation," the "photoelectric effect," and the absence of a presumed universal "aether," would completely revolutionize physics over the next few decades. As the philosopher of science Thomas Kuhn wrote:

> Mopping-up operations are what engage most scientists throughout their careers. . . . Closely examined, whether historically or in the contemporary laboratory, that enterprise seems an attempt to force nature into the preformed and relatively inflexible box that the paradigm supplies. No part of the aim of normal science is to call forth new sorts of phenomena; indeed those that will not fit the box are often not seen at all.[7]

The problem is that over time, it was completely forgotten that those guiding principles—the metaphysical assumptions underlying science—were never intended to describe *everything*, especially not the workings of the mental world. But because they were so successful, the mind as described by modern science and the mind that we directly experience have diverged to such a degree that they are almost entirely incompatible. Many scientists were not particularly pleased with the direction that science had taken. As William James wrote in the late nineteenth century:

> This systematic denial on science's part of personality as a condition of events, this rigorous belief that in its own essential and innermost nature our world is a strictly impersonal world, may, conceivably, as the whirligig of time goes round, prove to be the very defect that our descendants will be most surprised at in our boasted science, the omission that to their eyes will most tend to make *it* look perspectiveless and short.[8]

Going Way Out of Our Mind

About forty years ago, the growing incompatibility between the inner world of personal experience and the outer world described by science resulted in the aberration called "behaviorism." This was the main approach to psychology from the 1930s to the 1950s. Popularized by Harvard psychologist B. F. Skinner, behaviorism taught that the mind, our experiences, and our personal awareness were meaningless illusions. This viewpoint, which dominated psychology for decades, argued that the concept of mental "autonomy"—meaning the mental ability to initiate, originate, or create—was equivalent to the outdated and superstitious notion of "miraculous." According to Skinner, "A scientific analysis of behavior dispossesses autonomous man and turns the control he has been said to exert over to the environment."[9]

In other words, science had evolved into the absurd position of the mind denying its own existence. Science had effectively lost its mind. Under this worldview, as the physicist Steven Weinberg put it, "The more the universe seems comprehensible, the more it also seems pointless."[10] One of the problems with the classical scientific worldview is that people want to *believe* that they are more than mere machines, but given the powerful arguments provided by science, "it has been extremely difficult to state these convictions and feelings in an intellectually defensible way."[11] Despite our wishes, if it turns out that (classical) science is correct, and we really are walking, talking zombies, then humankind will be "compelled either to surrender what we call its humanity by adjusting to the real world or to live some kind of tragic existence in a universe alien to the deepest needs of its nature."[12]

We can't really blame the behaviorists for losing their minds. Mind and matter certainly *seem* to be very different creatures. The *mind* thinks, it isn't located in space, it's concerned with values, it has free will, it's driven by purpose, it's private, and everything we personally know is through the mind's direct, conscious experience. By contrast, as far as we know *matter* doesn't think; it's localized in space; it's value-free, determined, purposeless, objective; and all knowledge about it is inferred by our mind.

So how can we have two such different things in the world? After five or six thousand years of mulling over this question, philosophers have come up with three general possibilities: dualism, materialistic monism, and transcendental monism.[13]

Dualism says that mind and matter are both primary: neither causes the other; they both just exist. Matter-energy questions are studied with the current tools of science, but mind-spirit knowledge must be explored in ways more appropriate to it. They are two complementary kinds of knowledge, and two quite different kinds of basic components in the universe. This position is held by some philosophers and scientists.

Materialistic monism says that matter causes mind, that the mind is essentially a function of the activity of matter in the brain. The basic stuff of the universe is matter and energy. We learn about reality from studying the measurable world. Whatever we learn about the nature of the mind must ultimately be explained as the operation of the physical brain. This is a popular opinion among neuroscientists.

Transcendental monism says that the mind is primary, and in some sense causes matter. The ultimate stuff of the universe is consciousness. The physical world is to the greater mind as a dream is to the individual mind. Consciousness is not the end-product of material evolution; rather consciousness was here first.[14] This idea is popular with those who are attracted to Eastern philosophical views.

Problems with the Usual Views

Unfortunately, there are problems with each of these ideas. If mind and matter are fundamentally different, as *dualism* maintains, then how do they interact? How does spatially bound matter interact with something that is nonspatial? How does purposeless matter interact with purposeful mind, or value-free with value-laden? Also, why should everything we know be explainable in physical terms except for this one tiny piece of the universe inside our heads called the mind? And, "what sort of chemical process can lead to the springing into existence of something nonphysical? No enzyme can catalyze the production of a spook!"[15]

Materialistic monism attempts to avoid the problems of dualism by assuming that experience is not made of spooky minds but of brains. In doing so, however, it creates another problem: if the mind is an evolved form of matter, then it presumably exists because it offers some survival value. But no one has the slightest idea what this survival value might be, because the brain as an organic computer seems to work perfectly well without requiring conscious awareness. We also know that a vast amount of mental processing and decision making goes on without conscious awareness. These facts have given rise to vigorous debates about "zombies." That is, a seriously debated contemporary theory of consciousness—meaning self-awareness—is that it is an illusion, and that *really* nothing is happening inside the head. This seems to be a throwback to outdated ideas of behaviorism, because without conscious awareness no one would be worrying much about mind and matter in the first place. Nevertheless, such discussions continue to fill hundreds of pages in scholarly journals.

Finally, *transcendental monism* doesn't offer much insight into the problem because it explains the mystery of mind and matter by referring back to the mystery of mind. So, after thousands of years of the best minds thinking about what a mind is, we have the concepts of dualism, which doesn't make much sense, materialistic monism, which doesn't fit our personal experience, and transcendental monism, which has a problem of circular reasoning.[16]

Does it make any difference which interpretation is correct? In practical terms, it hardly seems to matter—we would still have televised sports events, barbecues, and movies. But from the scientific and theoretical perspective, when two fundamental elements of the universe—mind and matter—are staring us in the face and no one can explain how they coexist, this gives us a strong signal that some of our usual assumptions about the nature of reality are probably very wrong.

There is also an important pragmatic reason for being concerned about how we think about mind and matter. The scientific models built up over the past few centuries imply that human beings are basically nothing but machines. For example, the late Carl Sagan wrote, "The workings of the brain, what we sometimes call mind, are a consequence of its anatomy and its physiology and nothing more."[17] And Marvin Minsky, the MIT pioneer in artificial intelligence, wrote, "What is the brain but a computer made of meat?"

Many people believe that Sagan and Minsky were right, because if you watch the nightly news on television you'll certainly see many people being treated like machines. When a machine is worn out, we toss it in the trash. Machines don't have any intrinsic meaning, and they certainly don't have feelings, or values, or ethics. This is a rather depressing picture, but the scientific worldview is the closest thing we have to an accurate account of reality, so perhaps that's just the way it is.

The Rise of "Nothing But-ism"

Of course, "that's just the way it is," *assuming* that the scientific worldview is both correct and complete, because it's that worldview that gave rise to the idea that the mind is *nothing but* a machine in the first place. The principle of "nothing but-ism" is an inevitable consequence of the classical scientific worldview. It means that chemistry is nothing but physics, biology is nothing but chemistry, psychology is nothing but biology, and so on, up the line. This view is directly based on the assumptions of materialism, mechanism, and reductionism.

One of the most important features of nothing but-ism is that causation flows strictly "upward," starting from physics. That is, chemistry is caused by physics, biology is caused by chemistry, and so on. This is why the mind is seen as nothing but a computer made of meat. All the action starts at fundamental physics, and by the time the activity reaches the mind, it has all been determined through the operations of physics, chemistry, molecular biology, anatomy, and so on.

Are the Classical Assumptions Correct?

Fifty years ago it would have been pointless to argue with the basic assumptions of science. They were simply too successful in explaining just about everything in the physical world. Today, however, something very odd is happening. New advancements in science are beginning to dissolve every single one of those assumptions.[18] For example, we can no longer assume that positivism (what is real is measurable) is always valid, because statistical mechanics and the Heisenberg Uncertainty Principle taught us that not all aspects of the world are directly observable or measurable. How did this come about? Psychologist Ken Wilber has suggested a possible reason:

> When the universe is severed into a subject vs. an object, into one state which sees vs. one state which is seen, something always gets left out. In this condition, the universe "will always partially elude itself." No observ-

ing system can observe itself observing. The seer cannot see itself seeing. Every eye has a blind spot. And it is for precisely this reason that at the basis of all such dualistic attempts we find only: Uncertainty, Incompleteness![19]

So positivism was undermined as soon as Descartes split the world in two, although he didn't know it yet. Materialism was cast into doubt by Einstein's equivalence of matter and energy. Later, general and special relativity, quantum mechanics, chaos theory, and dissipative systems theory all directly undermined the assumptions of positivism, mechanism, determinism, and realism.[20] The assumptions were also challenged by new ideas in developmental biology, psychology, sociology, and medicine. Classical scientific assumptions simply do not account for how mind-body interactions, biofeedback, or the placebo effect works.

A Collapsing Theoretical Network

Scientific hypotheses are embedded in a theoretical network consisting of assumptions about what it means to observe a phenomenon, the "basic laws" of the observer's scientific discipline, the accepted scientific techniques used in that discipline, and assumptions about the basic nature of reality. As these networks develop over time, they establish the power behind commonsense beliefs and can withstand all sorts of insults by the occasional odd observation. But when observations begin *repeatedly* to violate commonsense beliefs, that means that something in the theoretical network is incomplete or even false.

If we have sound evidence suggesting that something in the network is false, there is no way to tell precisely where the false part is, so we have to reconsider all the parts, including the hidden assumptions. As philosopher Patricia Churchland has argued, "Even our . . . convictions about what it is to acquire knowledge and about the nature of explanation, justification, and confirmation—about the nature of the scientific enterprise itself—are subject to revisions and correction."[21]

In other words, if the placebo effect, or intuition, or out-of-body experiences, or psi, all seem a little odd, that may be because of mismatches between what the theoretical network predicts and the true nature of the world. The easy thing to do is to disregard the anomalous observations. Some skeptics have been highly motivated to take the easy route.

Rising Doubts

To demonstrate how much we take for granted, even with our basic assumptions about straightforward, rational logic, imagine the following scenario: We are passing farm animals through a gate that only lets horses through and rejects all cows. Now we take the horses that made it through the first gate and pass them through a second gate that only lets through black animals and rejects all white animals.

Logic tells us that only animals that are both horses and black can pass through both gates. To our surprise, about half of the animals that make it through both gates turn out to be white cows! This seems completely ridiculous, and would be instantly dismissed as a mistake, except that the world at the quantum level actually does operate this way. If we assumed that elementary particles were like tiny versions of black and white horses and cows, we would be wrong![22] If we assumed that psi was like another version of ordinary perception, we might be wrong!

Consider a profound mystery in biology that is not accounted for by classical assumptions: The average neuron consists of about 80 percent water and about 100,000 molecules. The brain contains about 10 billion cells, hence about 10^{15} molecules. Each nerve cell in the brain receives an average of 10,000 connections from other brain cells, and the molecules within each cell are renewed about 10,000 times in a lifetime. We lose about 1,000 cells a day, so the total brain cell population is decimated by about 10 million cells, losing in the process some 100 billion cross-linkages.

"And yet," as P. A. Weiss writes, "despite that ceaseless change of detail in that vast population of elements, our basic patterns of behavior, our memories, our sense of integral existence as an individual, have retained their unitary continuity of pattern."[23] *All* of the material used to express that pattern has disappeared, and yet the pattern still exists. What holds the pattern, if not matter? This question is not easily answered by the assumptions of a mechanistic, purely materialistic science.

The British philosopher and mathematician Alfred North Whitehead argued that even evolutionary philosophy, itself founded on classical principles, is inconsistent with the assumption of materialism:

> The aboriginal stuff, or material, from which a materialistic philosophy starts, is incapable of evolution. This material is in itself the ultimate substance. Evolution, on the materialistic theory, is reduced to the role of being another word for the description of the changes of the external relations between portions of matter. There is nothing to evolve, because one set of external relations is as good as any other set of external relations. There can be merely only change, purposeless and unprogressive.... the doctrine thus cries aloud for a conception of *organism* as fundamental to nature.[24]

Bidirectional Causation

Challenges to the classical assumptions, which are appearing in many disciplines, suggest a new concept of causation, proposed most recently by Nobel laureate Roger Sperry. In Sperry's view, causation "flows up" as in standard reductionism, but it also "flows down." The emerging view says that we cannot predict all known chemical properties based on what is known about physics, we cannot predict all biological properties based upon chemistry, and so on. According to Sperry, science cannot claim to be complete until it recognizes "inner conscious awareness as a causal reality."[25]

Until very recently, such a suggestion would have been considered a serious heresy because it overrides the assumption of strict reductionism. Unfortunately for orthodoxy, the accumulating evidence is demonstrating that downward causation also exists. Perhaps the field best known for studying this form of causation is "psychoneuroimmunology," the study of mind-body interaction at the biochemical level.

Under a strict reductionist viewpoint, the idea of mind-body interactions does not make sense, because under that view we cannot reduce chemistry and physics to anything but matter. This is why most medical researchers in the 1950s considered psychoneuroimmunology a ridiculous fantasy.

Example of Downward Causation

Let's say you see a car moving along the street. Someone asks, Why is that car moving? From the physical level of the scientific hierarchy, the causal explanation goes something like this: Exploding bits of gasoline in the cylinders of an engine create hot gases. These cause pistons to move, which in turn impart torque to a drive shaft. This causes the shaft to move, which causes the wheels to turn, and so on. A chemist might say that the car is moving because of the action of certain molecular bonds breaking during the combustion process. A neuroscientist might say that the car is moving because someone's leg muscle contracted in a specific way, thus pressing the accelerator pedal. A psychologist might talk about how the driver wanted to go somewhere. A sociologist might talk about the creation of suburbs. And so on.

Where does the ultimate cause lie? There is none. It is distributed everywhere at every level of the explanatory hierarchy, all at once. One view sees causation flowing "up" from the exploding droplets of gasoline and another sees causation flowing "down" from the driver's volition. Neither view is more correct; instead, these causal flows are truly complementary descriptions of the same event.

But notice something very interesting here. Downward causation would *appear* to be teleological, that is *goal-directed* or purposeful, to the levels below the "source" of the causation. Thus, if we were the car's tires in the example just discussed, the "primal cause" that would make us move would be outside our scope of understanding. However, a few "maverick" tires among us might profess faith in some sort of "higher power," directed by a great and divine "driver" who had the supernatural quality known as "volition." We could never understand all the complexities of the driver's volition, or even conceive of engines, gasoline, and sociological causes, but we might come up with a few fancy mythologies that tried to explain those observations from the only perspective we could understand.

Downward Causation and Psi

What does the idea of causation flowing up and down a hierarchical model of the world have to do with psi? It allows for the conventional but extremely powerful reductionistic explanations of many natural phenomena. And it allows for all the methodology and rigor of conventional science. But it also allows for the existence of effects that appear to be driven by higher purposes. It allows us to expect that events at higher levels of the hierarchy can cause effects at lower levels. It provides a way of thinking of how the placebo effect can work, how deep hypnosis can significantly alter body chemistry, and why something as ephemeral as wishing might produce meaningful coincidences in the objective world.

This bicausal model also begins to heal or "re-member" the undifferentiated nature of the world. The hierarchies mentioned so far are only a thin slice of a continuous spectrum of hierarchies. A slightly more comprehensive model might place quantum or subquantum physics at the bottom and "spirit" or "superspirit" at the top. Causation would flow up and down the entire hierarchy. There may even be parallel hierarchies that do not fit within science, such as hierarchies of values and meaning. Perhaps causation sometimes flows "sideways" between these hierarchies.

It is important to emphasize that at the extreme ends of the hierarchy, the world is completely undifferentiated. It doesn't even make sense to think in terms of commonsense reality at the extremes. For example, imagine one of these hierarchical models as a model of how you fit into the world. Certain parts of yourself—the ultimate energy that constitutes your body at the lower end of the hierarchy, and the ultimate "spirit" that constitutes your "essence" at the upper end—would be completely indistinguishable from anyone else's body or spirit.

Is there any evidence that downward causation really exists in the "higher" levels of this hierarchy? To identify such evidence we would need to look for appearances of teleological effects. At the level of mind, there are the obvious examples of psychoneuroimmunology, hypnosis, and the placebo effect. Some forms of psi, and perhaps spontaneous remissions and miraculous healing, may belong here as well. At a higher level, say sociology, perhaps mass riots and our "field-consciousness" effects provide evidence. Biologist Rupert Sheldrake's idea of morphogenetic fields as well as certain anomalous societal effects such as the UFO phenomenon probably belong here.[26]

Higher still, at a global level, the concept of Earth as an organism, the Gaia hypothesis, is a candidate.[27] At the cosmological level, some of the so-called neoastrological effects, such as Michel Gauquelin's "Mars effect" or

Jung's archetypes, may reflect higher-level "order."[28] At the "spiritual" levels, perhaps we see signs of something lurking in the so-called perennial philosophy and in cross-cultural spiritual traditions.[29]

Yet Another Paradox

Biophysicist Harold Morowitz pointed out an interesting cyclic paradox about hierarchies in science.[30] If we strictly follow the principles of classical reductionism in an attempt to understand this mysterious property called "conscious awareness," we will soon discover that the ephemeral mind is associated with a physical lump of tissue called the brain.

If we continue our reductionistic approach by closely examining brain anatomy we will find that it is part of a central nervous system. Then as we study this nervous system, we find that it is composed of billions of neurons. And so we study neurons and discover that they are cells with certain interesting inner structures. Continuing our search, we discover biological molecules, and then biochemistry, and then elementary particles, and subparticles, and forces, and fields, and eventually, at the bottom of the known hierarchy, the zero-point field of quantum mechanics. This reductionist search is illustrated in figure 15.2.

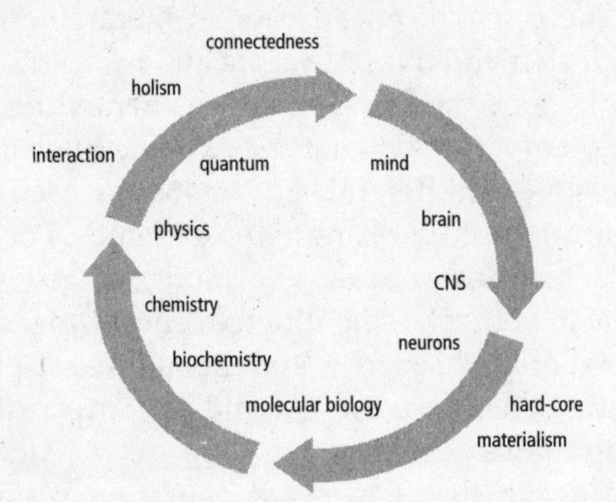

Figure 15.2. A cyclic paradox.

The paradox is that the best scientific interpretations we have of the *living* part of this cycle—the mind, cells, and neurons—are hard-core materialistic interpretations. This electrical event causes that synapse to fire, and this network of cells creates that response, and so on. In other words, there is no place for anything as abstract as "mind" in here. But by the time we cycle around to what we might *think* would be the hardest of the hard core, the hard physical matter of subelementary particles, we begin to run into descriptions couched in terms of holistic, mushy-minded interactions. Concepts like observation—a property of the mind—can no longer be ignored.

In other words, there's an unexpected role reversal: biologists begin to sound like hard-core materialists, and physicists begin to sound like mystics. Biologists studying living systems have enjoyed great success standing apart from their object of study, while physicists studying dead matter have been forced to adopt the idea that they are inseparable from their object of study! As psychologist Ken Wilber put it:

> Most branches of science remain today thoroughly and solidly dualistic, hotly pursuing as they are the "objective facts," but some of the "purer" forms of science, such as physics and mathematics, and some of the emergent sciences, such as systems theory and ecology, have dealt lethal blows to several long-cherished dualisms. . . . Nevertheless, all of these forms of science are relatively recent inventions, being hardly 300 years old, and thus it is only in recent history that we have started to see the elimination of the dualisms that have plagued Western thought for 25 centuries. There is no doubt that all sciences began as pure dualisms—some, however . . . pursued their dualisms to the "annihilating edge," and for those scientists involved, there awaited the shock of their lives.[31]

The shock was that reductionism *did not hold true* when we got closer and closer to the ultimate constituents of matter. In those realms, we were unable to maintain the subject–object distinctions required by the assumptions of classical science, and holistic and mentalist concepts began to take over.

Where Does Psi Fit In?

When modern science began about three hundred years ago, one of the consequences of separating mind and matter was that science slowly lost its mind. This split became painfully obvious about seventy-five years ago when psychotherapy began to intensely *embrace* the value of personal experience and behaviorism began to intensely *deny* the value of personal experience.

Parapsychology fits in this picture by straddling the edge separating the mind-oriented disciplines such as clinical and transpersonal psychology and the matter-oriented disciplines such as neuroscience and cognitive science. Parapsychology explicitly studies the interactions between consciousness and the physical world. It assumes that downward causation exists in some form, and it assumes that scientific methods can be used to study this middle realm in a rigorous way.

Thus, the persistent controversy over psi can be traced back to the founding assumptions of modern science.[32] These assumptions have led many scientists to believe that the mind is a machine, and as far as we can tell, machines don't have psi. The problem is that most of the classical assumptions that originally spawned the idea of mind-as-a-machine have dissolved in the wake of new discoveries. The old assumptions are transforming into new concepts that must take into account factors such nonlocality, quantum logic, systems theory, downward causation, and an active role for the mind. The new concepts will not be easy to accept, for as Sir James Jeans put it in 1948, musing about the perplexing implications of quantum theory, "The concepts which now prove to be fundamental to our understanding of nature . . . seem to my mind to be structures of pure thought, . . . the universe begins to look more like a great thought than a great machine."[33] Is the universe a great thought? Is the universe conscious?

Today, some scientists believe that we are once again ready for a minor mopping-up operation. We're just about ready to close the book on the ultimate theory of everything. But just as there were a few clouds on the horizon at the close of nineteenth century, they are again approaching as the twenty-first century dawns (figure 15.3). The problem this time is that the clouds are immense and are gathering quickly, and they portend a major storm.

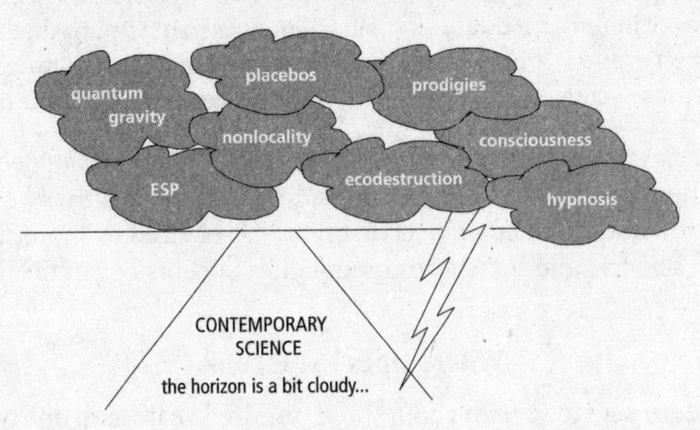

Figure 15.3. A few clouds have appeared on the horizon of 21st century science.

Raising Consciousness

One consequence of the changes under way in our metaphysical assumptions is that previously taboo topics, like the nature of human consciousness, are rapidly becoming mainstream. Consciousness, primarily referring to self-reflective awareness, is not merely a scientific curiosity. It is literally the source of everything we know.[34] Without consciousness, it is doubtful that any aspect of our modern civilization would have developed, and even if it had, we wouldn't know it!

As Nobel laureate Erwin Schrödinger described it, "Consciousness is that by which this world first becomes manifest, by which indeed, we can quite calmly say, it first becomes present; that the world *consists* of the elements of consciousness."[35] And yet, we know virtually nothing about what consciousness is, or how it works, or even what it is for. According to Nobel laureate Eugene Wigner, "We have at present not even the vaguest idea how to connect the physio-chemical processes with the state of the mind."[36] Perhaps physicist Nick Herbert put it best:

> Science's biggest mystery is the nature of consciousness. It is not that we possess bad or imperfect theories of human awareness; we simply have no such theories at all. About all we know about consciousness is that it has something to do with the head, rather than the foot.[37]

One way of gauging the accelerating interest in consciousness is to examine how many books have been published on this topic, and when. Figure 15.4 shows the percentage of all books published between 1800 and 1990 with the words *psychology* and *consciousness* in their titles.[38] The bars in figure 15.4 add up to 100 percent for each category. We see that more than 50 percent of all books ever published with "consciousness" in the title have appeared since the 1980s. Books with "psychology" in the title show a similar rise, but not as dramatic an increase as "consciousness" (because books on psychology have been published over a longer timescale).

As interest in the topic of consciousness heats up, there has been a corresponding increase in the number of books published on parapsychological topics.[39] Figure 15.5 shows that more than 50 percent of all books ever published on parapsychology have appeared since the 1970s. The dip in parapsychology and psychology books published in the 1930s, 1940s, and 1950s is probably related both to the rise in behaviorist psychology and to limitations on publishing new books during World War II.

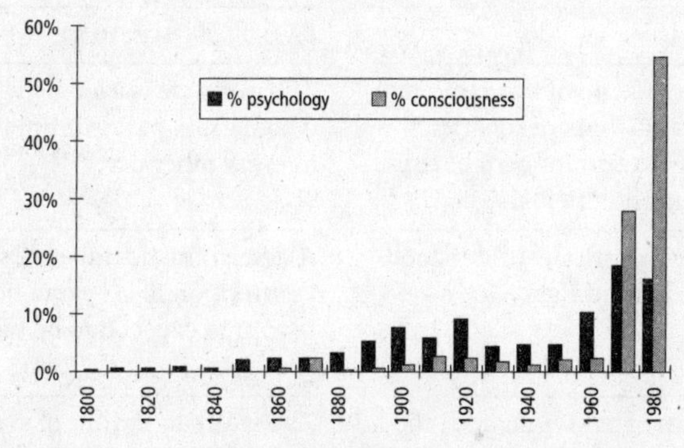

Figure 15.4. Percentage of books published on psychology and consciousness from 1800 to 1980.

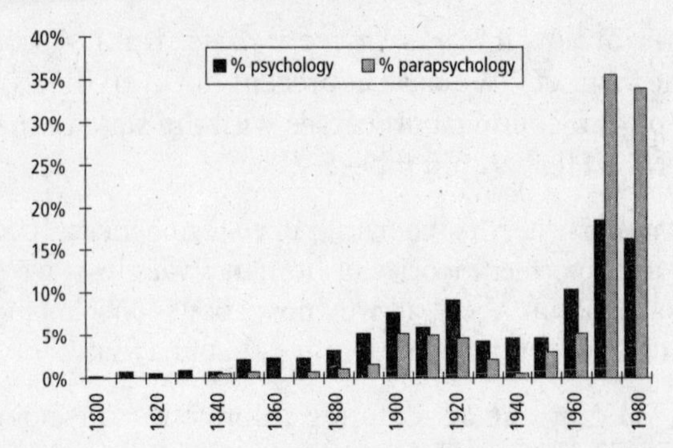

Figure 15.5. Percentage of books published on psychology and parapsychology from 1800 to 1980.

The New Metaphysics

We know that the assumptions of classical science are not adequate to understand psi, just as they are not adequate to understand consciousness. As mathematician Sir Roger Penrose has pointed out, if a classical world is not something that consciousness could be a part of, then "our minds must be in some way dependent upon specific deviations from classical physics."[40] To help clarify the differences between the classical and the newly evolving assumptions in science, philosopher of science Willis Harman proposed the comparisons shown in figure 15.6.[41]

Orthodox "Separateness" Science	**Proposed "Wholeness" Science**
BASIC ASSUMPTION:	BASIC ASSUMPTION:
The universe is made up of fundamental particles and quanta that are separate from one another except for certain connections made through fields.	The universe is a single whole within which every part is intimately connected to every other part.
The universe is scientifically understood to be ultimately deterministic.	A deterministic universe stems from the assumption of "separateness"; there is no reason to expect it to be borne out in experience.
Nonnormal states of consciousness, dissociation, and so on, are to be studied in the context of the pathological. Consciousness is a by-product of material evolution and is an epiphenomenon with no intrinsic meaning or purpose.	The entire spectrum of states of consciousness, including religious experiences and mystical states, has been at the heart of all cultures. These states of consciousness may be an important investigative tool, a "window" to other dimensions of reality.

Orthodox "Separateness" Science BASIC ASSUMPTION:	Proposed "Wholeness" Science BASIC ASSUMPTION:
Commonly reported experiences known as "meaningful coincidences," synchronistic, and psychic, must ultimately have a physical or psychological explanation or be merely coincidence or fraud.	The question is not "how can we explain telepathy?" but rather, "how can we explain why our minds are not cluttered by all that information in other minds?" Not "how can we explain psychokinesis?" but rather, "how can we understand why our minds have such a limited effect in the physical world?"
There is no evidence for "drives" or "purposes" in evolution. What appears as a survival instinct is merely the result of natural selection; any organisms that did not have such a drive were selected out. There is no scientific evidence for anything in the universe resembling "purpose" or "design." The biological sciences use the term "teleology" for convenience, but what it really means is that those structures and behaviors were ones that contributed to survival.	Human beings are part of the whole and there is no justification for assuming that "drives" such as survival, belongingness, achievement, and self-actualization are not also characteristics of the whole. Similarly, since we experience "purpose" and "values," there is no justification for assuming these are not also characteristics of the whole. The universe may be genuinely, and not just apparently, purposeful and goal-oriented.
A scientific explanation of a phenomenon consists in relating the phenomenon to increasingly general, fundamental, and invariant scientific laws. Ultimate scientific explanations are in terms of the motions and interactions of fundamental particles and forces.	There is no reason to assume that scientific laws are invariant; it seems more plausible that they too evolve. Hence, extrapolation to the big bang may be suspect. Evidence points to consciousness either evolving along with, or being prior to, the material world.
The truest information about objective reality is obtained through the observer being as detached as possible. A clear separation can be maintained between subjective and objective knowledge.	There is an ultimate limit to objectivity, in that some "observer effect" is inevitable in any observation. Understanding comes not from detachment, objectivity, and analysis but from identifying with the observed, becoming one with it.
All scientific knowledge is ultimately based on data obtained through the physical senses. Such information is ultimately quantifiable.	Reality is contacted through physical sense data and through inner, deep, intuitive knowing. Our encounter with reality is not limited to being aware of messages from our physical senses, but includes aesthetic, spiritual, and mystical senses.

Figure 15.6. Comparisons of the classical and evolving scientific worldviews.

Harman's comparisons reveal that the new metaphysics is shifting toward—for want of a better term—a "mystical" worldview. Some scientists will be suspicious of this interpretation, and yet what else are we to make of the writings of Nobel laureate physicist Erwin Schrödinger:

> I have . . . no hesitation in declaring quite bluntly that the acceptance of a really existing material world, as the explanation of the fact that we all find in the end that we are empirically in the same environment, is mystical and metaphysical.[42]

Or of Albert Einstein, who wrote:

> The most beautiful and most profound emotion we can experience is the sensation of the mystical. It is the sower of all true science. He to whom this emotion is a stranger, who can no longer wonder and stand rapt in awe, is as good as dead.[43]

Some of the suspicions that scientists have about the concept of the mystical almost certainly derive from its close association with religious doctrine. But that is not what Schrödinger, Einstein, James, and dozens of other eminent scientists meant. They were talking about the nature and experience of *interconnectedness.*

Interconnectedness

Underlying the isolated world of ordinary objects and human experience is another reality, an interconnected world of intermingling relationships and possibilities. This underlying reality is more fundamental—in the sense of being the ground state from which everything originates—than the transient forms and dynamic relationships of familiar experience.

This basic theme is as ubiquitous in physics as it is in ecology, economics, complexity theory, systems theory, social psychology, psychotherapy, philosophy, and theology. For example, in systems theory, we find statements like:

> What [physicists] found is very much what an ecology portrays: a web of cause and effect that has coherence, hidden order, inseparability, and subtle connectively.[44]

And in physics:

> Bell's theorem shows that although the world's phenomena seem strictly local, the reality beneath this phenomenal surface must be superluminal. The world's deep reality is maintained by an invisible quantum connection whose ubiquitous influence is unmediated, unmitigated, and immediate.[45]

And in philosophy:

> The farther and more deeply we penetrate into matter, by means of increasingly powerful methods, the more we are confounded by the interdependence of its parts. Each element of the cosmos is positively woven from all the others. . . . All around us, as far as the eye can see, the universe holds together, and only one way of considering it is really possible, that is, to take it as a whole, in one piece.[46]

And in theology:

> The Buddha compared the universe to a vast net woven of a countless variety of brilliant jewels, each with a countless number of facets. Each jewel reflects in itself every other jewel in the net and is, in fact, one with every other jewel. . . . Everything is inextricably interrelated: We come to realize that we are responsible for everything we do, say, or think, responsible in fact for ourselves, everyone and everything else, and the entire universe.[47]

Indeed, the name "uni-verse" suggests a connected whole, not a set of isolated fragments. It would ordinarily seem that the relationship between the motion of my fingers while typing and the display of the words on the computer screen is closer than, say, "between either of them and the price of yak milk in Tibet."[48]

But in the long run, the motion of my fingers, the computer display, and the price of yak milk are indeed interrelated. We do not normally see or pay attention to these interrelations, but the fact is that everything does interact with everything else, and how much of an interaction is only a matter of degree. Some may argue that the *effects* of these interactions are so minuscule that for all practical purposes they can be ignored. But we should remember that what is considered to be "practical" is determined by events that occur on human time- and size-scales. Whether we notice it or not, the interactions are still there. For example, what difference does it make if Wegener's continental drift theory is correct? On the human timescale it seems completely inconsequential. That is, unless we happen to live in an earthquake zone, where it suddenly becomes critical to understand the minuscule movements of huge chunks of the earth over very long periods of time.

Deep Interconnectedness

The interconnectedness revealed by modern science and described in ancient doctrines suggests a richly connected network of physical variables interacting like a shimmering weaver's loom. But a more extended view of interconnectedness, especially in light of psi, quantum field theory, and general relativity, goes far beyond this physical metaphor. As both modern physics and ancient Buddhist doctrine suggest, "deep" interconnectedness embraces everything, unbound by the usual limitations of time and space.

Recognition of the effects of mind in the act of observation forced the founders of quantum theory to reconsider the commonly held assumption of strict separations between mind and matter. This is the reconsideration that eventually caused Sir James Jeans to conclude that "The universe begins to look more like a great thought than like a great machine"[49] and Sir Arthur Eddington to agree that "The stuff of the world is mind-stuff."[50]

Today, physicists and philosophers interested in the implications of modern physics are explicitly putting consciousness back into the interconnectedness soup. For example, physicist David Peat has written, "The universe appears as a single, undivided whole whose patterns and forms emerge out of a ground, are sustained for a time, and then die back into the field. . . . Consciousness too can be considered to arise out of a deeper ground that is common to both matter and mind."[51]

Dualistic Conundrums

Observations about deep interconnectedness raise the possibility that the debate over how mind and matter interact may have been misconceived. That is, we are probably not dealing with interactions between two dissimilar entities, but with a single phenomenon. Likewise, the puzzling dualisms of subjective versus objective, inner versus outer, mind versus body, all dissolve into illusions created and sustained by the nature of language. These illusions are certainly compelling, but as systems theorist Sally Goerner wrote:

> Interactive causality requires how, when, where and how-much questions instead of either/or questions. If you try to explore a deeply interactive system with either/or thinking, you are likely to get a double-bind answer and this results in both anger and confusion.[52]

Double-bind arguments are quite common in science and philosophy and often reflect deep misunderstandings. The conundrum presented by how two significantly different entities (mind and matter) interact suggests that they are in fact one and the same. "The same" is not quite captured by a metaphor like "two sides of the same coin," for this perpetuates a false distinction. But it is also incorrect to imagine that mind and matter are *literally* the same, since our experience of them is so different. Still, the metaphor of two sides of a coin is close enough for our present purposes, provided we do not forget that it is, after all, a metaphor.

As with the idea of interconnectedness, the idea that mind and matter are part of a common whole has been observed across many scientific and scholarly disciplines.[53] For example, in physics we find statements like:

A human being is part of a whole, called by us the "Universe," a part limited in time and space. He experiences himself, his thoughts and feelings, as something separated from the rest—a kind of optical delusion of his consciousness. (Albert Einstein)[54]

In biology:

Our growing scientific knowledge . . . points unmistakably to the idea of a pervasive mind intertwined with and inseparable from the material universe. This thought may sound crazy, but such thinking is not only millennia old in the Eastern philosophies but arose again and again among the monumental generation of [quantum theory] physicists in the first half of this century. (George Wald)[55]

In systems theory:

Our bodies and minds are much more sensitively coupled and deeply integrated into a larger process than classical science imagined. We each contain more information than was dreamt of in mechanistic philosophy. . . . Mystical experiences may be a form of knowing arising from our deep evolutionary entwinement with the world. (Sally Goerner)[56]

In philosophy:

Our ordinary conception of the world as a complex of things extended in space and succeeding one another in time is only a conventional map of the universe—it is not real. It is not real because this picture painted by symbolic-map knowledge depends upon the splitting of the universe into separate things seen in space-time, on the one hand, and the seer of these things on the other. In order for this to occur, the universe necessarily has to split itself into observer vs. observed. (Ken Wilber)[57]

And in religion:

That they all may be one; as thou, Father, art in me, and I in thee, that they also may be one in us. (John 17:21, KJV)

But if all this is really true, and mind and matter are something like two sides of the same coin, then surely there should be *substantial* evidence of psi effects, both for mind-matter interactions and for distant perception. There shouldn't be just spontaneous, sporadic effects, but pervasive, persistent, ubiquitous effects. As the philosopher C. D. Broad wrote:

If paranormal cognition and paranormal causation are facts, then it is quite likely that they are not confined to those very rare occasions on which they either manifest themselves sporadically in a spectacular way,

> or to those very special conditions in which their presence can be experientially established. They may well be continually operating in the background of our normal lives.[58]

As we've seen, there is indeed evidence of psi operating "in the background": spontaneous psi experiences are recorded in the tens of thousands; many people who've paid attention to odd coincidences in their lives report dozens of profoundly synchronistic episodes; thousands of controlled laboratory experiments have measured psi; we've recently found new hints of psi in the real world; and psi applications are being used every day. There are probably many other effects that we haven't named yet that go completely unnoticed.

Mystical Mumbo Jumbo?

But hold on a minute, couldn't all this talk about metaphysics be just "mystical mumbo jumbo"? I don't think so. Much has been written about the mystical roots of modern science, and how the goals of science and mysticism are strikingly similar. Both attempt to understand the world by searching for unity in the apparent diversity of nature.

In spite of much being written on this topic, many contemporary scientists—when presented with evidence that virtually *all* the founders of quantum theory thought deeply about the relationship between science and mysticism—adopt facial expressions that combine awe with contempt.[59] But whether we like it or not, the fact is that science and mysticism both sprang from the same primeval urge to understand the world around us. As a result, there are more similarities between the two than most people know. Reneé Weber, a philosopher at Rutgers University, expressed the relationship as follows:

> In the beginning there was wonder and awe. These inspired the search with which science and religion began. Originally they were one, untroubled by the modern separation that would develop to decree that they become distinct domains with uncrossable borders. In that separation, the sense of wonder became science, the sense of awe, mysticism. . . . To this day, science seeks the boundaries of nature, mysticism its unboundedness, science the droplet of the ocean, mysticism the wave. . . . They share the search for reality because, in their own way, both science and mysticism look for the basic truth about matter and the source of matter.[60]

Weber suggests that the similarity between science and mysticism is a common search for *unity*. Theoretical physicists seek this unity from the "outside," expressing their ideas in scientifically elegant proposals for a "Grand Unified Theory" and a physical "Theory of Everything." Mystics seek this unity from the "inside," through the direct experience of oneness with the universe. Weber raises the unexpected possibility that mysticism may in a sense be more committed to the spirit of scientific exploration than science itself:

> It is mysticism, not science, which pursues the Grand Unified Theory with ruthless logic—the one that includes the questioner within its answer. Although the scientist wants to unify everything in one ultimate equation, he does not want to unify consistently, since he wants to leave himself outside that equation. Of course, with the advent of quantum mechanics, that is far less possible than it was in classical physics. Now observer and observed are admitted to constitute a unit. But the full meaning of this has not yet caught up with most of the community of scientists who, despite quantum mechanics, believe they can stand aloof from what they work on.[61]

The Big Picture

After our lengthy detour through the metaphysics of science, we should now have a better appreciation for how psi fits into the bigger picture. Psi is our experience of the invisible interconnections that bind the universe together. Psi research is at the core of the new metaphysical foundations of science, and psi experiments delve directly into holistic realms previously described only by mystics and mythology. Where these explorations will ultimately lead is unknown. One thing we do know is that the direction we are going is consistent with developments in many other scientific disciplines. The next chapter explores how some of these theories are converging and suggests what's ahead.

THEME 4
IMPLICATIONS

When a scientist states that something is possible, he is almost certainly right; when he states that something is impossible, he is very probably wrong.

ARTHUR C. CLARKE

So far, we've learned that the effects observed in a thousand psi experiments are not due to chance, selective reporting, variations in experimental quality, or design flaws. They've been independently replicated by competent, conventionally trained scientists at well-known academic, industrial, and government-supported laboratories worldwide for more than a century, and the effects are consistent with human experiences reported throughout history and across all cultures. We've also learned that one of the main reasons this evidence is largely unknown is that psi effects do not fit the preconceptions underlying conventional scientific theories. In the last chapter, we saw that a new metaphysics of science is now emerging that provides new expectations and perspectives about the nature of reality.

The theme in the last two chapters is the *implications* of psi. We'll begin by considering how leading-edge theoretical developments are converging toward a scientific explanation of psi; then we'll look at what psi implies and what the future may bring.

CHAPTER 16

Theory

I have yet to see any problem, however complicated, which when you looked at it in the right way, did not become still more complicated.

PAUL ANDERSON

One of the most shocking events in twentieth-century science—an event so outrageous that its repercussions are still barely understood—was quantum theory's prediction and subsequent verification of *nonlocality*.[1] This idea challenged long-held classical assumptions that objects separated in space are strictly isolated. Instead, nonlocality shows that physical objects that appear to be separate are really connected in ways that transcend the limitations of space and time. This may seem like a stark violation of common sense, but that is what the theory predicts and the experiments show.

Even more shocking than the demonstration of nonlocality was the fact that overturning centuries of commonsense assumptions took only a handful of experiments. The experimental coup de grâce was a study in 1982 showing results that were "five standard deviations larger than the prediction of [hidden variables theory]," the countertheory to the quantum prediction.[2]

An experimental result of "five standard deviations" greater than some alternative hypothesis is equivalent to odds against chance of about 3.5 million to 1. As we've seen, some individual psi experiments have produced results with odds against chance greater than a *billion* to one. And the odds after combining thousands of psi experiments are astronomically beyond that. So why was nonlocality accepted on the basis of a few studies, but psi is not? The answer is that quantum theory had *predicted* nonlocality, and so far, hardly anything predicts psi.

This is not to say that there are no theories of psi, for actually there are many. They range from serious speculations in physics about the possibility of "advanced" electromagnetic waves carrying precognitive information, to how enhancements to quantum mechanics would allow an observer to mentally alter the physical probabilities of events.[3] There are psychological speculations about how some aspects of the world may be driven by mental concepts like goals and purpose. There are theories based on Eastern philosophical concepts in which the world is primarily composed of Mind, which gives rise to matter.

And there are dozens of other theories, including ideas based on the evolutionary value of psi, on teleological (purpose-driven) concepts, and on metaphysical, occult, religious, mystical, holographic, and other ideas. Some theories are domain-specific, in the sense that they attempt to explain effects observed in certain experiments without worrying about explaining everything. Others try to be all-encompassing, to explain the big picture without considering the details. A comprehensive survey of existing theories, none of which is completely satisfactory, is beyond the scope of this book. But we can consider what an adequate theory must look like.

Toward an Adequate Theory

The whole of science consists of data that, at one time
or another, were inexplicable.
BRENDAN O'REGAN

What would a good theory of psi look like? First, the theory must be compatible with what is already known with high confidence in physics, psychology, and neuroscience. To ignore well-established principles in these disciplines would guarantee that no one need take the theory of psi seriously. An adequate theory of psi, however, will almost certainly have to expand upon *and synthesize* aspects of certain puzzles in existing physical, psychological, and neurological theories. This means that the existing theories in these disciplines will eventually be seen as special cases, applicable only to certain, limited conditions rather than explaining all conditions. Cross-disciplinary theories are exceptionally difficult to develop, but that's probably what psi will require. Systems theorist Irvin Laszlo has made a good beginning.[4]

INFORMATION ACQUISITION

The theory will have to explain how information can be obtained at great distances, unbound by the usual limitations of space or time. Here we must point out that the existing laboratory data certainly *suggest* that psi effects are completely independent of space and time, but there is not enough evidence yet to state this with certainty.

Such a theory must also explain, not only *how* one can get information from a distance in space or time, but also how one can get *particular* information. Because the evidence suggests that we can get specific, meaningful information from anywhere or "any-when" that we can clearly specify, the theory must account for why we are not overwhelmed with information all the time. As suggested earlier, perhaps the same brain processes that prevent us from being overwhelmed by sensory input also filter out the meaningless chatter from elsewhere in the universe. Perhaps the same unconscious perceptual filters that alert us to our name being called across a crowded room allow us to become aware of events that are meaningful to us on the other side of the planet.

In principle, clairvoyance could be used to "view" events on other planets in other galaxies. But if the viewer did see some sort of alien intelligence, unless that intelligence was extremely similar to human intelligence, it would be perceived only with severe distortions. In fact, we could predict based on what we know about the reality-shaping power of expectations, that what is perceived with clairvoyance will be driven almost entirely by prior beliefs. If a viewer expects to see aliens as angels, she almost certainly will. If she expects to see demons, her perceptions will comply. Consider too that many people have trouble assimilating into other human cultures on this planet. Imagine the difficulties we would have, with our imperfect psi perceptions, trying to perceive accurately what is going on in a truly alien culture.

Randomness

The theory must explain how random processes can be tweaked by mental intention. For example, while the results of random-number-generator (RNG) experiments and the mass-consciousness studies using RNGs suggest that the random bits are being mentally *forced* to change from purely random to more orderly, in reality this is not so clear. A highly accurate model of the RNG results, proposed by physicist Edwin May and his colleagues, is based on the idea that RNG effects are "caused" by precognition rather than by any form of microforce.[5] Other theorists familiar with the RNG studies agree that what seems to be happening in mind-matter interaction phenomena is better described in terms of exchanges of *information* rather than by the application of conventional forces.

Thinking in terms of information, and especially in terms of *meaningful* information, which seems to be a necessary next step, immediately shifts the type of theory from the hard, concrete world of forces and particles to the more abstract world of ideas. Science is already comfortable with this, since quantum theory is a purely mathematical, abstract theory and is also one of the most successful physical theories in history.

A theory specifically involving *meaning* will be more of a stranger to the hard sciences, because most physical sciences do not deal with meaning. And yet it appears that psi may require an explicit bridge between the physical and the psychological worlds. This is why an adequate theory of psi will be not only cross-disciplinary, but also hierarchical in the sense discussed in the previous chapter.

Mind-Body Separations

The theory of psi should explain phenomena associated with evidence suggesting that something may survive bodily death. These phenomena include apparitions, hauntings, out-of-body experiences (OBE), and near-death experiences (NDE).[6]

Because almost all the evidence for these phenomena comes from uncontrolled, spontaneous cases—and thus was necessarily collected as after-the-fact anecdotes rather than as controlled laboratory results—scientific confidence that such effects are what they appear to be is very poor. That's why they weren't discussed in this book. Despite the profundity, hope, and inspiration of the stories associated with these phenomena, the existing evidence must be balanced against our strong motivations to wish for some form of survival. We need to be especially careful in interpreting the evidence for survival, given what we know about expectation bias, wishful thinking, and self-delusion.

Still, an adequate theory of psi must address these phenomena because one of the most striking things about OBE and NDE states is the possibility that the mode of perception is clairvoyance. If this can be confirmed through future experiments, it would be an exciting advancement. The types of clairvoyance in OBEs and NDEs—if that is what is going on—seem to be more vivid and they last longer than the perceptual experiences reported by even the best remote viewers during "IBE" states (i.e., in-the-body experiences).

Poltergeists and Hauntings

The theory may need to account for poltergeist phenomena, which provide the primary evidence for large-scale mind-matter interaction effects. Discussion of the evidence for these phenomena was beyond the scope of this book, but a few good cases—all spontaneous, uncontrolled events—indicate that the microeffects observed with RNGs may scale up into much more powerful effects under certain conditions. The evidence suggests that movements of objects, ranging from a cup to a desk, are not due to "noisy ghosts," but to a human agent, typically a troubled adolescent.[7] However, for the reasons mentioned above with regard to survival phenomena, scientific confidence that such phenomena involve psi per se is poor.

Preparing for a Theory

In preparing for a theory, one of the first questions is whether it's likely that we'll be able to understand psi through existing physical theories. In the 1980s, Halcomb Noble, deputy director of science news at the *New York Times,* asked this question of Nobel laureate physicist Brian Josephson.

> NOBLE: Are the rigors of hard science or quantum mechanics really compatible with investigations of the paranormal and what the intelligent skeptic always regarded as the quackery of, say, the old professional mind reader?
>
> JOSEPHSON: You ask whether parapsychology lies within the bounds of physical law. My feeling is that to some extent it does, but physical law itself may have to be redefined. It may be that some effects in parapsychology are ordered-state effects of a kind not yet encompassed by physical theory.[8]

If existing theories are not adequate to explain psi, it's useful to think of possible metaphors for how psi might work. One proposal, suggested by Princeton University aerospace engineer Robert Jahn and his colleagues, is that just as a photon is both a particle and a wave, perhaps consciousness too has complementary states.[9] In ordinary states, the mind is more particlelike and is firmly localized in space and time. This is supported by the ordinary subjective experience of being an isolated, independent creature. But in unusual, nonordinary states of awareness, our minds may be more wavelike, and no longer localized in space or time. This is supported by subjective experiences of timelessness, mystical unity, and psi.

As with particle-wave duality, it is not the case that only one or the other description is true, but *both are true at the same time.* The fact that we have trouble thinking in terms of "both" rather than "either-or" says more about the limitations of language than it does about the nature of reality. If our minds have complementary characteristics, then perhaps we can be more particlelike or more wavelike depending on what we wish to be, or what it is suitable to be at the time, or what we are motivated to become.

As another metaphor, we know from nonlocal effects described in quantum mechanics and from Einstein's relativity of time and space that the universe is fundamentally interconnected. Because we are constructed out of the same "stuff" as the rest of the universe, what if we were able to directly *experience* that interconnectedness? Our description of it would probably fall into one of two classes: a mystical experience or a psi experience. In the former case, without maintaining any particular perceptual focus, or perhaps while intending to widen the perceptual field, we might sense a completely undifferentiated connection to all things. Particulars would dissolve into a sparkling loom of patterns and connections, all being influenced by, and influencing, each other. We would "know" the movements and meaning of everything. This is how a typical mystical experience is described.

In the latter case, just as a mother pays constant attention to her baby even when her conscious mind is asleep, perhaps each mind also pays attention to loved ones, or to meaningful events, regardless of where the people or events happen to be. If something important happens to those people, especially life-threatening events, the interconnected part of us "recognizes" that this is useful information and brings it to our awareness. This is reminiscent of a scene from the movie *Star Wars* in which the Jedi knight Obewan Kenobi senses a "disturbance in the force" when an entire planet is cruelly destroyed. We are fully interconnected with all things, *and* we are isolated individuals. Both.

Convergence

Beyond the metaphors, it turns out that some scientific developments in recent years suggest a way of thinking about psi that is also compatible with mainstream scientific models. Four such developments are related to quantum theory. All four run counter to common sense, all four were thought to be theoretically possible but practically untestable, and all four have now been empirically proved. Of principal importance here is that all four *must* also be true to be compatible with what we know about psi.

The first, not surprisingly, is the idea of nonlocality itself. The second is that quantum effects may be important to consciousness and biological organisms. The third is that information can be transmitted without expending energy. And the fourth is that information can be instantaneously transmitted—the actual word that physicists use is "teleported"—from one place to another, independent of distance. Let's examine each in turn.

Nonlocality

In the 1960s, physicist John Bell mathematically demonstrated that according to quantum theory, a pair of particles that were once in contact, but have since moved too far apart to interact, should nevertheless instantaneously behave in ways that are too strongly correlated to be explained by classical statistics. As Bell wrote in 1964, "there must be a mechanism whereby the setting of one measuring device can influence the reading of another instrument, however remote."[10]

What this means is that apparently separate particles would not really be separate after all but remain connected regardless of how far apart they were. This startling prediction—and even more startling confirmation—of a fundamental property of the physical world greatly troubled Einstein, who called the idea "spooky actions at a distance."[11] He proposed several ways to get around the unexpected correlations, including the possibility that quantum theory was incomplete. Einstein imagined that there might be a "local hidden variable" that would account for these apparent interconnections, but experiments in the 1980s convincingly demonstrated that, as described in an article in *Science,* "Local hidden variables theory is dead."[12]

It is tempting to extrapolate that the strange properties revealed by quantum physics intrude into the "macroscopic" world we live in, but conservative physicists warn against this. The reason for their warning, however, has more to do with what they consider to be *reasonable* than with what the theories actually predict. As physicist David Lindley explained:

> From a strictly mathematical point of view, light waves or photons that travel backwards in time are just as legitimate as those that travel forwards. From a practical point of view, however, we disregard the backwards solutions. . . . It's not that such a thing is intrinsically impossible, but that the conditions to create it cannot realistically be achieved.[13]

But remember, as we saw in chapter 14, what is considered reasonable and realistic is driven almost entirely by *expectations,* not by the mathematics underlying the theories. When we try to make sense of what a mathematical theory like quantum mechanics means, we are immediately limited by our expectations. If quantum theory predicts, and experiments prove, that the world is in fact nonlocal, then what prevents us from imagining that psi phenomena are our experiences of that connectedness? It is not the *theory* that prevents us from doing so, but our ability to imagine a *connection* from the theory to experience.[14] The usual objections that "there is no evidence" for psi have been discussed at length in earlier chapters, so there is no need to argue this point.

Interpretation of existing theories may change when viewed in the light of psi and nonlocality. For example, in the late 1980s, neuroscientist Benjamin Libet conducted an experiment in which he asked his subjects to flex a finger at the instant of their decision.[15] He monitored their brain waves to see if the instant that the decision was made would be reflected by a change in brain waves. On average, the volunteers took about a fifth of a second to flex their finger after they mentally decided to do so, an expected time lag for the brain to activate the neuromuscular system. But according to their brain waves, their brains also displayed neural activity about a third of a second *before* they were even aware that they had decided to move their finger!

Libet interpreted this result as evidence that our sense of free will in deciding what we do may be unconsciously determined *before* we are consciously aware of the decision. If mental intention, which is connected to our most intimate sense of personal expression, actually does begin in a part of the brain that is outside our conscious reach, then perhaps *all* our behavior is completely determined by processes outside our control. This has led to the idea that most of "free will" is an illusion, and that despite the persuasive power of the illusion, we are really something like animated zombies, with "someone" or "something" else controlling the strings (i.e., we reflexively react to changes in the environment much as an amoeba reflexively reacts to touch or heat).

Another interpretation, however, is that the act of mental intention really is controlled by our conscious self. Perhaps the third of a second anticipation of this intention observed by Libet is equivalent to the presentiment "presponse" observed in the experiments described in chapter 7. That is, if we don't disregard the "backwards solutions" of our physical theories and allow for the possibility of signals traveling backward in time, then what Libet saw may be the brain's presponse to its own decision taking place a third of a second *in the future*.

Note that this interpretation does not automatically imply a dualistic separation of brain and mind. It could equally suggest that the usual assumptions of strict time-synchronization between the conscious mind and the brain may be too restrictive. Perhaps nonlocality in space-time simply blurs the meaning of "now" such that in certain experiments "now" seems to be in the future or in the past.

Quantum Biology

Recently proposed physical theories of consciousness suggest that quantum processes in the brain may be responsible for some of the more puzzling aspects of consciousness, including the unitary sense of self, the sense of free will, and nonalgorithmic "intuitive" insights. Proposed in various forms by a number of prominent scientists, among them physicist David Bohm, physiologist Karl Pribram, Cambridge University Nobel laureate Brian Josephson, Oxford University mathematician Sir Roger Penrose, and neuroscientist Benjamin Libet, these theories predict that quantum properties in biological systems may give rise to nonlocal, field-like processes associated with consciousness.[16]

For example, Brian Josephson and physicist Fotini Pallikari-Viras from the University of Athens have suggested that biological systems may take advantage of quantum effects in unexpected ways.[17] This is because living organisms differ from the "dead" matter studied in pure physics experiments. I argue that as a result of this difference, especially an organism's ability to adapt to a changing environment and its ability to assign "meaning" to what might otherwise appear to be random processes, organisms can make *use* of non-locality. Josephson and Pallikari-Viras specifically refer to examples of remote influences and connections as suggested "by experiments on phenomena such as telepathy."

Until recently, consciousness researchers generally dismissed proposals such as Josephson's and earlier models by physicist Evan Harris Walker[18] because it was thought that nothing in the hot, sticky world of neurons could provide the quantum stability (technically called "coherence") necessary to support nonlocal effects. Quantum effects are usually observed only in the artificially created worlds of the extremely small and cold, so it was thought that nonlocal interactions could not exist in the relatively large, hot brain.

But in the 1970s, nanometer-sized cylindrical structures dubbed "cytoskeletal microtubules" were unexpectedly discovered in brain neurons. For decades no one had any idea what purpose these tiny structures had. Then in 1994, anesthesiologist Stuart Hameroff, from the University of Arizona, proposed that the microtubules could be a possible site for quantum effects in the brain.[19] They were about the right size to sustain quantum coherence, and Hameroff was intrigued by the possible relationships between consciousness and these previously overlooked microtubules.

As an anesthesiologist, Hameroff was in the business of making people become unconscious, yet very little is known about what goes on in the mysterious transition between conscious self-awareness and unconsciousness. Hameroff was also struck by how some of the more puzzling aspects of consciousness resembled equally puzzling aspects of quantum properties. For example, the "unitary sense of self" resembles the properties of quantum coherence and nonlocality; nondeterministic free will resembles quantum indeterminacy; intuitive reasoning resembles quantum computing; and differences and transitions between pre-, sub-, and nonconscious processes resemble how quantum possibilities become hard realities.[20]

Hameroff's proposal has attracted great interest and critical discussion. If it turns out that he is even partly correct, or if his proposal merely helps others think about how quantum processes in the nervous system may be related to consciousness, it opens the theoretical door for explaining how nonlocal effects may manifest in consciousness. And if turns out that nonlocality *does* play a role in the workings of the brain, then something like "quantum telepathy" would no longer be such a strange prospect.

Information Without Energy

Scientists and engineers have long assumed that a minimum energy requirement is associated with transmitting information from one place to another. Consider the information represented by words on a piece of paper. To get the information from one place to another, one can, say, scan the paper into a fax machine. This transforms the word-symbols into electrical signals, which are then transmitted to another fax machine, where the information is reconstructed. It takes energy to scan, transform, send, and reconstruct this information.

But IBM physicist Rolf Landauer has shown in a recent article in *Science* that, based on the ideas of quantum theory, there are in fact no theoretical minimum energy requirements for transmitting a bit of information.[21] This is an important development. Given that small amounts of information can precipitate huge reactions in biological systems, nonlocal biological effects, even if infinitesimally small, could conceivably affect other biological systems at a distance. For example, the words "You have won a million dollars" require very little energy to transmit, but the *meaning* of those words is sufficient to cause a huge emotional response. Thus, the physical energy requirements of sending meaningful "signals" are no longer a barrier in considering how information may be obtained from a distance.[22]

Quantum Teleportation

Other studies looking at applications of quantum nonlocality have shown that it is possible to transmit information instantaneously by what is being called "quantum teleportation." In an article on this phenomenon in *Science* entitled "To Send Data, Physicists Resort to Quantum Voodoo,"[23] science writer Gary Taubes wrote:

> . . . there is an area of physics that holds a vague resemblance to voodoo. It involves one of the weirdest of quantum-mechanical paradoxes, in which two particles can be created simultaneously with their internal quantum states. . . . Quantum mechanics dictates that until a particular state is actually measured, it has no value at all. But when a measurement is made on one entangled particle, its partner instantly takes on the opposite value, even if it happens to be halfway across the universe at the time.

IBM fellow Charles Bennett, one of the developers of this quantum teleportation method, coined the comparison to voodoo. The odd property of quantum "entanglement" (an aspect of nonlocality) establishes a connection between two particles in such a way that the "quantum essence of the particle" can be passed from one to the other, as Bennett poetically expressed it, "like a curse passing from a lock of hair back to its original owner." As quantum researchers have recently shown, this quantum voodoo can be put to work in ways that are less mysterious but no less spooky. The quantum state of a particle can be "teleported" to another location, along with all the information it embodies.[24]

This is no mere theorizing. It has been demonstrated by physicists at the University of Innsbruck in Austria. According to physicist Paul Kwiat, from the Los Alamos National Laboratory, in referring to the ability to instantaneously transmit this information, "It's well . . . it's been teleported. It's completely true that you can't access that information by any known measurements. But the mathematics insist that 'it really is there.'" The link to psi is that biological systems are exquisitely sensitive to certain kinds of information. Perhaps biological systems can both send and access teleported information, in which case we would suddenly have a scientifically acceptable (but still fundamentally mysterious) way to both perceive and influence objects at a distance.

The Future Theory

As some of the stranger aspects of quantum mechanics are clarified and tested, we're finding that our understanding of the physical world is becoming more compatible with psi. An adequate theory of psi, however, will almost certainly not be quantum theory as it is presently understood. Instead, existing quantum theory will ultimately be seen as a special case of how *nonliving* matter behaves under certain circumstances. Living systems may require an altogether new theory. Quantum theory says nothing about higher-level concepts such as *meaning* and *purpose*, yet real-world, "raw" psi phenomena seem to be intimately related to these concepts.

Quantum interconnectedness does tell us that perfectly ordinary "dead matter" operates in remarkable ways that violate our commonsense notions of how the world works. Given that we have only recently glimpsed the strange properties of dead matter, we have every reason to believe that even more remarkable properties of "conscious matter" remain to be discovered. As physicist Nick Herbert said, "I think that Bell's theorem [of nonlocality] is remarkable. I hope I am alive when the first real theory of mind begins to surface. I think it will make Bell's theorem look like 5-finger arithmetic."[25]

In considering the possibility that psi may be one of the major discoveries of the twenty-first century, Halcomb Noble of the *New York Times* wrote:

> "No one understands quantum mechanics," says Nobel Laureate Richard P. Feynman. Its effects are "impossible, absolutely impossible" to explain based on human experience. It may be equally true of ESP. It may exist. It may be important to human and physical behavior. Yet it may not be explainable until long after its discovery.[26]

In other words, if something is real, it can be put to use even if we don't understand it very well. If this were not so, virtually none of the technologies and medical remedies we now take for granted would exist. We've seen that government agencies, business, and medicine are already applying psi, and that it is being explored by high-technology companies. But where is it going? What are the future implications of psi?

CHAPTER 17

Implications

The only solid piece of scientific truth about which I feel totally confident is that we are profoundly ignorant about nature. . . . It is this sudden confrontation with the depth and scope of ignorance that represents the most significant contribution of twentieth-century science to the human intellect.

LEWIS THOMAS, *THE MEDUSA AND THE SNAIL*

What difference does it make if psi is real? Now that psi researchers have resolved a century of skeptical doubts through thousands of replicated laboratory studies, and now that the evolving scientific worldview is becoming increasingly compatible with psi, what's next? What is the future of psi? What does it imply about who and what we are?

We don't know yet. Our most sophisticated scientific theories about the way the world works have not caught up yet with these phenomena. In fact, if tomorrow someone accidentally stumbled upon a satisfactory scientific theory for psi, we might not even recognize it as such. But it's not too soon to ponder the broader implications. The experimental and scholarly research have provided a few hints about what might be going on, and from those hints we can speculate on psi's significance.

One possibility is that psi will eventually be accepted by the scientific community as little more than a curiosity, a psychological reflection of the quantum interconnectedness of the universe. Possibly it will go no further than that.

As discussed in the Introduction, however, when earth-shattering ideas move from Stage 1, "it's impossible," to Stage 2, "it's real, but too weak to be important," Stage 3 often follows. This is when the consequences of "it's real" begin to dawn on a new generation of scientists who did not have to struggle through the blinders of past prejudices.

I believe it's unlikely that Stage 2 will be the end of the story. Science has already proved that psi can be effectively studied through conventional methods, and I strongly suspect that breakthroughs in understanding are far more dependent on society's *willingness* to take these phenomena seriously than on any inherent limitations in our ability to study psi. Such breakthroughs could usher in an era of science and technology so startlingly new that, from today's perspective, it would look like pure magic. Psi-based manipulations of the fundamental properties of space, time, matter, and energy would lead to unimaginable revisions of reality. This is not a wild speculation, but a virtual certainty.

What we ultimately learn about psi, and what we do with it, depends primarily on public interest. If enough people demand that resources flow into psi research, then it is a good bet that we will learn quite a lot in fairly short order, with significant progress measured in years or decades. If people are ambivalent about what might be learned about psi, then the research will remain on the fringe and progress will continue to be measured in half-centuries or centuries. It's that simple.

What Psi Implies . . .

After a century of slowly accumulating scientific evidence, we now know that some aspects of psychic phenomena are real. The importance of this discovery lies somewhere between an interesting oddity and an earth-shattering revolution. At a minimum, genuine psi suggests that what science presently knows about the nature of the universe is seriously incomplete, that the capabilities and limitations of human potential have been vastly underestimated, that beliefs about the strict separation of objective and subjective are almost certainly incorrect, and that some "miracles" previously attributed to religious or supernatural sources may instead be caused by extraordinary capabilities of human consciousness.

Together, these statements suggest that on the "implication scale" of mere oddity to revolutionary, we are probably dealing with revolutionary. If even one of these statements were widely accepted, it would cause huge reverberations in science, technology, society, and theology. Let's consider why in more detail.

. . . for Physics

> **It is almost an absurd prejudice to suppose that existence can only be physical. As a matter of fact, the only form of existence of which we have immediate knowledge is psychic [i.e., in the mind]. We might as well say, on the contrary, that physical existence is a mere inference, since we know of matter only in so far as we perceive psychic images mediated by the senses.**
>
> Carl Jung

Physicists who have retained some humility in the face of nature's mysteries are interested in psi because it implies that we have completely overlooked fundamental properties of space, time, energy, and information. Specifically, psi suggests that the conventional boundaries of space and time can be transcended by the ephemeral concept of "the mind." Theoretical concepts like advanced waves, time symmetry, and nonlocality—all of which were thought of at one time as mere mathematical curiosities—may actually exist *and* be directly experienced.

Overlooking fundamental properties is much more serious than it sounds. Many basic scientific models and experimental techniques are anchored on assumptions that the fundamentals are in fact fundamental. If they start slipping or crumbling, this threatens centuries of cherished and fairly accurate theories of how things work. No wonder some scientists resist psi so vigorously! While it is probably true that some fundamentals will have to be revised, it is also quite clear that the revised worldview will *not* change what we already know as much as recast it in a different light.

It may be that just as we were shocked to learn at the dawn of the twentieth century that matter and energy were essentially the same, perhaps at the dawn of the twenty-first century we are in the midst of discovering that mind and matter are essentially the same. Something like this is already present in the philosophical assumptions of materialistic and transcendental monism, and in Eastern philosophies, but perhaps a new "complementary monism" may evolve. This would allow mind and matter to arise out of a common ground, enjoy intimate interactions with each other, and retain a certain autonomy as well.

Psi effects on random-number generators suggest a particularly perilous heresy for physics: quantum theory may not be complete. The RNG experiments indicate, as physicist Helmut Schmidt wrote, that

> the outcome of quantum jumps, which quantum theory attributes to nothing but chance, can be influenced by a person's mental effort. This implies that quantum theory is wrong when experimentally applied to systems that include human subjects. It remains to be seen whether the quantum formalism can be modified to include psi effects, and perhaps even to clarify the still somewhat puzzling role of the human observer in the theory.[1]

Quantum theory has been one of the most successful physical theories in history, but like any theory, it is an approximation of the world, not the world itself. If ten years from now psi research convincingly demonstrates that quantum theory is merely a special case of a more comprehensive theory, then quantum physicists may be shocked, but it will come as no great surprise to historians and philosophers of science.

Incidentally, clairvoyance is normally thought of as the ability to perceive across vast distances. We might imagine a future "Clairvoyant Space Corps" tasked with exploring distant galaxies. Likewise, we normally think of precognition and retrocognition as seeing across vast gulfs of time, and may envision teams of Indiana Jones–like "time historians" who explore ancient and future civilizations. We also imagine that mind-matter interaction effects may someday be used to push atoms around, operate psychic garage-door openers, and operate wheelchairs.

But it is equally possible that clairvoyance can allow us to see across infinitesimally *tiny* distances, that precognition can allow us to perceive infinitesimally *brief* times, and that mind-matter effects can allow us to push entire *planets* around. These extremes may seem outlandish, but given that we know almost nothing about the limits of psi, setting any imaginative limits at this point would be a big mistake.

And besides, there already is a fascinating bit of evidence that clairvoyance can be used to see the infinitesimally tiny. A recent article by physicist Stephen Phillips provides evidence that a century ago two clairvoyants used psi to examine atomic and subatomic states. Their descriptions didn't make much sense at the time, or for many decades afterward. But now, their descriptions bear a remarkable resemblance to the quark model of particle physics and to superstring theory.[2] Perhaps the next great advancements in our explorations of space, time, and energy will be through psi-enhanced techniques.

. . . for Biology

Psi raises numerous questions whose answers biologists can't even guess at yet: How does psi information "get into" a living organism? Are there secret senses we have overlooked? What are the limits of distant mental interactions on living systems? Is psi an "invisible" carrier of information among living systems? Is psi interconnectedness related to the unitary sense of self in human beings, and the occasional oneness felt by groups engaged in the same activity?

Does psi imply the existence of an even larger unity among human beings, among all sentient creatures, or among all life? Does it serve any evolutionary purpose? Is it an ability, a talent, a throwback to a more primitive sense, or a glimpse of our future? Or is it "merely" a biological reflection of the nonlocal nature of physical matter?

. . . for Psychology

Psi offers a wealth of tantalizing hints about the nature of perception, memory, and communication. How much of the substantial lore about psi experiences can be attributed to psi rather than to more prosaic explanations? Is hidden or unconscious psi actually more prevalent than we have thought, and if so, what role does it play in ordinary human behavior? If we are not as separate from one another as commonly believed, does psi play a role in the behavior of groups, crowds, and society?

Psychological interest in psi is also related to the observation that "magical thinking" lies close beneath the veneer of the sophisticated modern mind. Magical thinking refers to an organic worldview permeated with meaning and deep, living interconnections. In contrast, much of modern science has supported a worldview permeated with "nothing but" meaningless isolation.

Clinical psychologists know that the feeling of being fundamentally alone quickly leads to anxiety, declining health, and depression. To maintain mental and physical health, not only as individuals but as societies, we must believe and act as though we are living in a world that does have deep meaning and personal value. Psi supports the concept of a deeply interconnected "conscious universe," not merely as a psychological coping mechanism, but as *reality*. As science shifts toward a worldview that supports rather than denies our deepest psychological needs, we can expect significant beneficial consequences for society's mental health.

. . . for Sociology

We know that local pollution of the air, land, and sea spreads out and affects the global ecology in many ways. The field-consciousness studies mentioned in chapter 10 suggest, as farfetched as it may seem, that there may be a mental analogy to environmental ecology—something like an "ecology of thought" that invisibly interweaves through the fabric of society. This suggests that disruptive, scattered, or violent thoughts may pollute the social fabric in ways that extend far beyond local influences. That is, a single individual harboring malevolent thoughts can directly affect those around him because of his destructive or antisocial behavior. But his *intent* may also spread out and indirectly "infect" and disrupt others at a distance. Those disruptions may in turn spread out, like a multiplying "psi virus," until the infection encircles the globe. Perhaps periods of widespread madness, such as wars, are indicators of mass-mind infections.

There may also be the equivalent of a mass-mind "immune system" that helps fight off psi viruses—brief, shining moments when intensely nurturing thoughts from a single individual, or groups of like-minded individuals, may spread out and literally heal the world-mind. Perhaps periods of widespread lucidity, like the period preceding the fall of the Berlin Wall, are indicators of mass-mind healings. In general, the field-consciousness studies suggest that thoughts may be less ephemeral or private than we normally believe. One wonders what role the "mind of the world" plays in shaping the evolution of global interconnectedness.

. . . for Philosophy

Psi addresses the core of many age-old philosophical questions, especially the "mind-body" debate and the nature of free will versus determinism. Psi research empirically explores questions such as, What is the role of the mind in the physical world? What is the nature of the objective versus the subjective? Is the mind caused or causal? Is the mind fundamentally different than matter, or are they the same?

As mentioned in chapter 15, a quick pass over contemporary ideas about the nature of mind reveals that no current approach is entirely satisfactory. Orthodox materialism helped spawn the nonsensical notion that the mind is a meaningless illusion. A line of thought called functionalism argues that it does not matter what mind is made of, all that matters is what it *does*. This is a nice pragmatic approach, but it doesn't help us understand what the mind actually *is*. Dualism is haunted by the specter of a disembodied mind. Other philosophical approaches are equally ambiguous, which is why the nature of the mind has remained a hot topic of debate for several thousand years.

David Chalmers, a philosopher at the University of California, Santa Cruz, has provided a pithy summary of four common approaches used to explain consciousness. Chalmers says that they either "explain something else, or deny the phenomenon, or simply declare victory, or find a neatly ambiguous metaphor which sounds for an instant as if it might bridge mind and matter."[3]

While consciousness is still a complete mystery, each approach to understanding it has offered a glimpse at what might be going on. The behaviorist camp showed that the mind is less private than previously supposed, because much of its inner workings can be inferred through careful observation of behavior. The functionalists have proposed that some aspects of mental functioning can be embodied in different ways, for example, as fancy computer programs. Neuroscientists have demonstrated that much of the mind's information-processing capabilities can be understood as patterns of activity in the brain. The dualists have pointed out that no comprehensive model of the mind can leave out subjective experience. "Identity theorists" have suggested that the workings of the brain and the mind are probably linked in some nonseparable way.

A recent twist on identity theory, called "naturalistic panpsychism" by philosopher Michael Lockwood, suggests that a fundamental property of the universe may be a self-reflective sense of "what it's likeness."[4] This would allow a materialistic universe to contain the strange property of subjective mind because "what it's likeness" is built into the same fabric as everything else. We are aware that we are constructions of matter and energy because awareness is fundamental to matter and energy.

What psi offers to the puzzle about consciousness is the observation that information can be obtained in ways that bypass the ordinary sensory system altogether, and there may be ways of directly influencing the outer world by mental means alone. At first, it may seem that by accepting the existence of psi we must immediately reject some of the strictly materialistic and mechanistic proposals about the nature of consciousness, but this is not so. With a concept like naturalistic panpsychism, everything proposed by a hard-core, materialistic neuroscience is still perfectly compatible with psi. All that is needed is the additional assumption that some aspects of a fully interconnected universe can be directly experienced.

In any case, any future philosophical understanding of consciousness that even presumes to be comprehensive must include the sorts of interconnectedness suggested by psi.

. . . for Religion

Much of the wonder and awe of traditional religions comes from stories of miracles, which are used as dramatic illustrations of divine power. From a parapsychological point of view, the great religious scriptures are encyclopedic repositories of stories about psi effects—telepathy, clairvoyance, precognition, mental healing, and mind-matter interactions. For some people, the scientific confirmation that psi is genuine may strengthen their religious faith, because if psi-like miracles are true even by secular standards, then perhaps other messages in the scriptures may be true as well. For others, the scientific study of psi is blasphemous because it "tests God." This latter opinion reflects a widespread belief that some things should not be studied because there are some things we just shouldn't know.

Psi *may* support the idea that there is something more to mind than just the mind-body system. In particular, a mind that is less tightly bounded in space or time than expected by traditional scientific models might be able to communicate with persons from the past or future. If so, when a medium claims to be in contact with a departed spirit, perhaps he is actually in contact with someone who is alive in the *past*. From the "departed" person's perspective, she may find herself communicating with someone from the future, although it is not clear that she would know that.

While it is by no means clear that psi implies anything at all about actual *survival* of consciousness after bodily death, it does imply that genuine communications transcending time may be possible. From that viewpoint, long-departed Grandma Rosie is still very much alive and can be contacted, but she is alive "then" while we are alive "now." Could the occasional passing thought about a strangely familiar but unknown person be a glimmering from an ancestor or descendant?

Future Applications

What does the future hold? To answer this question, we must plunge into a speculation unfettered by present practical and theoretical limitations. This is risky, given humanity's notoriously poor track record in predicting the course of the future. Invariably, the Law of Unintended Consequences combined with Murphy's Law conspires to make forecasting (to say nothing of precognition) a precarious business.

Medicine

Nevertheless, with our speculation hat firmly in place, we envision that future experiments will continue to confirm that distant mental healing is not only real, but is clinically useful in treating certain physical and mental illnesses. It's unlikely in the short term that we'll see doctors routinely prescribing distant-mental-healing treatments, but we probably will see subgroups of mainstream medical and psychiatric associations showing increasing interest in the therapeutic effects of prayer and psi-based medical diagnosis.

Over the longer term, the practice of medicine is going to change radically, completely independent of psi. Economic pressures combined with rapidly advancing conventional technologies assure this. But once we understand more of the factors underlying distant-healing effects, and how to enhance psi-based diagnostic methods, it is likely that we will see new specialties forming within medicine. Such physicians may be trained in what might be called techno-shamanism, an exotic, yet rigorously schooled combination of ancient magical principles and future technologies.

Technology

Developing psi-based technologies in the short term will use the same rationale behind the extensive selection procedures used with jet fighter pilots. That is, we do not randomly select people off the street and expect them to be able to fly fighter jets. But this is not to say that a jet couldn't be built that almost anyone could fly. Likewise, psi-based technologies may eventually work for just about anyone, but the prototypes will require expensive, custom-made devices operated by highly selected individuals.

For economic reasons, any viable psi-based technology will have to do things that can't be done by ordinary means. Building a psychic garage-door opener may be fun, but it will not replace electronic remote controls. On the other hand, using a technology-enhanced telepathic communication system to "call" a friend in a distant spacecraft, or someone in a deeply submerged submarine, does make sense, and these are the applications likely to show up first.

We may be surprised to learn that psi applications will accidentally crop up in the development of atomic-sized devices, known as nanotechnology.

At that scale, even minuscule mind-matter interaction effects may have huge consequences, and what may initially be perceived as extraordinary encounters with Murphy's Law (if anything can go wrong, it will) during development of these tiny devices may eventually be understood as the device's inadvertent responses to the developer's thoughts and wishes.

Military and Intelligence

Military and intelligence communities will continue to use psi because it occasionally provides useful information. There will be predictable stretches, however, when by necessity (to avoid embarrassment) officials will vigorously deny that anyone is interested in psi applications. The same is true for psychic detective work.[5] Public openness about this topic depends entirely on the mood of the times. If it is widely acknowledged that there are some valid aspects to psi, and if the tabloid media limit their absurd stories linking psi to every form of nonsense, then the large underground of police and government agencies who already use psi, or wish to explore its use, will emerge.

Business and Politics

Psi has the potential to enhance decision making in critical or time-sensitive business and political arenas. Combined with the best available information, even the briefest flash about future possibilities can redirect a decision that would have led to devastating losses and instead turn it into giant profits. Likewise, a violent conflict may be deftly turned into sustained peace with just the right bit of additional information. But projecting to a time where psi-refined intuition is routinely used to enhance decision-making raises a curious problem: if too many people begin to accurately peek at their possible futures, and they change their behaviors as a result, the causal loops established between the future and the past may agitate the future from a few likely outcomes into a completely undetermined probabilistic mush.

By analogy, recall that in the early days of uncontrolled computer-based stock market trading, tens of thousands of independent, simple, mathematically aided buy/sell decisions innocently conspired one day to crash the stock market. Similarly, lots of independent, simple glimpses of the future may one day innocently crash the future. It's not clear what it means to "crash the future," but it doesn't sound very good.

On the other hand, a society that consciously uses precognitive information to guide the future is one that is realizing true freedom. That is, the acts of billions of people seeing into their own futures, and acting on those visions, may result in fracturing undesirable, "fated" destinies set in motion long ago. This would allow us to create the future as we wish, rather than blindly follow a predetermined course through our ignorance.

Postscript

Throughout this book I've deemphasized my personal role in conducting psi research. This was intentional, because the scientific case for psi rests not on what an individual claims, or even what one laboratory claims, but upon the replicated findings of dozens of scientists from around the world. That's the main message here.

But while I consider myself to be a fairly conventional scientist, with traditional academic degrees in traditional disciplines from ordinary universities, and I use well-established scientific methods in my research, I admit that something about psi is far from ordinary. As I write this, my lab is only one of two full-time academic psi research labs in the United States. And there are only a handful of labs like this in the entire world. Why is this?

Certainly a big part of the answer is that psi threatens the very core assumptions of science, and it is not easy raising funds to challenge a powerful status quo. But perhaps there's something else different about psi research, something that touches people in unusually deep ways. This "deep touch" manifests in ways that would probably not appeal to most scientists. For example, on Monday, I'm accused of blasphemy by fundamentalists, who imagine that psi threatens their faith in revealed religious doctrine. On Tuesday, I'm accused of religious cultism by militant atheists, who imagine that psi threatens their faith in revealed scientific wisdom. On Wednesday, I am stalked by paranoid schizophrenics who insist that I get the FBI to stop controlling their thoughts.

On Thursday, I submit research grants that are rejected because the referees are unaware that there is any legitimate evidence for psi. On Friday, I get a huge pile of correspondence from students requesting copies of everything I've ever written. On Saturday, I take calls from scientists who want to collaborate on research as long as I can guarantee that no one will discover their secret interest. On Sunday, I rest, and try to think of ways to get the paranoid schizophrenics to start stalking the fundamentalists instead of me.

Psi is like an enigmatic tree blending into an enchanted forest. Overly critical skeptics cannot see the forest because they are too busy cutting down the trees. And overly credulous enthusiasts cannot see the trees because they are in awe of the forest. Everybody else is busy with daily life, but they wonder about strange tales told about trees in the forest, and every so often they are stunned to find one of those trees in their backyard. Meanwhile, a few of us have been attempting to blaze a trail through the forest, admiring and studying the trees along the way, hopefully without getting lost in the process. Why?

One answer was contained in a talk on psi research that I gave at Bell Labs in the mid-1980s. I was at the podium in a large auditorium, preparing my slides while people filed into the room. I didn't pay much attention to the audience until I had my slides in order; then I looked up to see if we were ready to begin. A hush settled over the crowd, and I was surprised to see that all three hundred seats were filled, with more people sitting in the aisles and standing in the doorway.

This was most unusual for a technical seminar at Bell Labs. Such talks usually attract a few dozen people, not a capacity crowd. I assumed that the large attendance was due to the exotic topic, so I opened my talk by saying, "I know what you're thinking. You're thinking, why are we hearing a talk on psychic phenomena, of all things, at a Bell Labs technical seminar?"

The audience never heard past, "I know what you're thinking . . . ," because these words triggered an explosion of laughter. I had overlooked the humor, and during the five minutes it took for the audience to settle down, I briefly considered a career as a stand-up comic. At the end of the talk, I asked if there were any questions, and two dozen hands shot up. An hour later, long over schedule, we were evicted from the auditorium so that another meeting could begin. I was not surprised by the audience's enthusiasm, because I had found in many talks that scientists are first shocked, then doubtful, then fascinated to learn that much of what they *thought* they knew about psi research is simply wrong.

At these seminars, one of the first questions I'm asked is, "How did you become interested in this topic?" This question is fair enough, because all presentations are colored by the speaker's background and motivations. People typically assume that I must have had some sort of life-transforming experience that compelled me to study psi phenomena. This is not the case. I was reared in an agnostic, artistic environment. No one in my family had ever reported anything even vaguely psychic, and I don't recall that it was ever a topic of conversation when I was growing up. So I usually reply by saying that I think psi is one of the most curious and challenging scientific topics I know. In fact, the challenge is so immense, and the implications are so astonishing, that this topic is a creative scientist's dream.

But for some reason, on that day as I looked out over the sea of faces at Bell Labs, I thought of a different answer. I said, "That's a good question, but I'll answer it with another question: Given this overflowing crowd, I am clearly not the only one interested in this topic. In fact, given the number of people present at this talk, why doesn't Bell Labs have an entire *department* devoted to exploring psi applications? There are future, multibillion-dollar technologies here waiting to be discovered."

I paused for effect, then answered my own question. "Most of you, and something like 70 percent of the general population, already believes that there is something interesting here, something worthy of serious investigation." To prove my point, I asked, "If one day management suddenly decided to support psi research on healing applications, or new technologies, or enhanced decision making, how many of you would like to be involved?" About half of the audience tentatively raised their hands. This was not a cross section of the general public, but a select group of highly trained scientists and technologists. So I replied, "Well, there's your answer. I am no more interested in this than anyone else. I just decided that it was the most interesting topic around, and it has the potential for some amazing practical implications."

My original interest in psi was probably fired by the hundreds of science-fiction stories, myths, and folktales I read as a child. One day I read a book describing some scientific tests of ESP, and even though it was a skeptical book, I was impressed that something taken for granted in science fiction and folktales could be studied using scientific methods. Many years later, I was struck by the peculiar fact that through two decades of my formal education, the topic of psychic phenomena was uniformly ignored. It wasn't the case that psi was mentioned and dismissed, but rather in my experience the topic was never even mentioned. It almost seemed as though there were unstated agreements among teachers that certain topics were taboo, and that was that.

In graduate school at the University of Illinois, I tried a few psi experiments with the help of a sympathetic mathematics professor. Some of those experiments produced results that seemed to challenge conventional wisdom, but I soon learned that there weren't many jobs that actually paid you to challenge conventional wisdom. So for pragmatic reasons, after I graduated I got a job at Bell Labs and became involved in more conventional research. This was an enjoyable time, but I never forgot that behind all the popular nonsense about "amazing psychic powers," and beyond the false controversies sustained by debunkers, the world was more astonishing than the theories of mainstream science were ready to admit. At least in principle.

At Bell Labs, I amused myself by reading sections of the voluminous literature of parapsychology. One day, I decided to see whether I could replicate the mind-matter interaction effects reported by physicist Helmut Schmidt. I wrote to him and described my interests, and he kindly loaned me one of his random-number generators. For about a year, I ran experiments with Schmidt's device at Bell Labs, using myself and my colleagues as test subjects. After several dozen studies, I convinced myself that I was able to replicate what Schmidt and others had reported.

But becoming convinced was not an easy task. Despite my feelings in graduate school that scientific theories did not fully describe the world, it took years to reconcile what I was seeing in those experiments at Bell Labs with what my formal education had led me to expect. I hadn't realized just how deeply I had accepted the assumptions of conventional science until I came face-to-face with experimental results that "shouldn't be." At times I became a bit frightened when I started to think about the implications. At other times the cognitive dissonance was so strong that I set aside the experiments for months and spent time on less heady things, like playing the fiddle in a bluegrass band. Then one day, as I was complaining to a friend about the results of the experiments, I said, "I just can't *imagine* how this can be!" My friend calmly replied, "Well, it sounds like you're limiting yourself."

Something went "click," and I realized that the difficulties in reconciling "magic" with science were caused entirely by my prior beliefs. The moment I imagined that some aspects of science would simply expand to accommodate psi, and that most established scientific principles would remain the same as before (although perhaps footnoted that they were special cases), suddenly there was nothing left to reconcile.

It's simple in retrospect, but it shows the incredible power of being schooled to think in certain ways. This is why I still have great sympathy for scientists who never had to seriously think about the scientific worldview as an approximation of certain limited, selected features of the world. The fact is that many very interesting facets of the world are just left out of science altogether. That doesn't mean they don't exist, but it does mean that the scientific worldview is far from complete.

So I continued to conduct experiments, then began to present the results at the annual conferences of the Parapsychological Association. After a few years, I was persistent enough to find positions where I could work full time on psi research. After a decade of working in conventional science and technology, and eight years of close involvement with psi researchers, I learned that contrary to what is often portrayed about this realm, psi research is an exemplar of leading-edge science at its best. It requires critical-thinking skills, rigorous attention to detail, creative use of technologies, and development of new analytical methods, like any other empirical science. Like mathematics it demands an intuitive grasp of aesthetics; like philosophy it requires an appreciation of metaphysics; like sociology it encourages an understanding of the social context of science. It also requires a deep appreciation of established scientific models combined with humility and a sense of humor in the face of experimental results that don't fit with what they teach in school.

I believe that as more scientists become aware of the evidence, innovative corporations will increasingly pour resources into psi applications. There is no doubt that whoever develops psi-based practical applications first will become the leaders of twenty-first-century high technology. The tide of industrial interest has already turned in Asia, and Europe is close behind. The United States lags the rest of the world in this regard.

Future generations will undoubtedly look back upon the twentieth century with a certain poignancy. Our progeny will shake their heads with disbelief over the arrogance we displayed in our meager understanding of nature. It took three hundred years of hard-won scientific advances merely to verify the existence of something that people had been experiencing for millennia.

At the turn of the twentieth century, imaginative scientists were slowly becoming aware of radical new theories on the horizon about space, time, matter, and energy. Some sensed, correctly, that developments such as relativity and quantum theory would radically alter our understanding of reality itself. Almost a century later, the impact of those discoveries is still reverberating throughout science, technology, and society.

As the twenty-first century dawns, astounding new visions of reality are stirring.

Notes

Introduction

1. Crick 1994; Dennett 1991.
2. Sagan 1995, 302; Prasad and Stevenson 1968.
3. Radin and Nelson 1989; Jahn and Dunne 1986; Child 1985; Utts 1991a.
4. Hyman 1985b; Jahn 1982.
5. Rao and Palmer 1987.
6. Bem and Honorton 1994.
7. Stapp 1994.
8. U.S. Library of Congress 1983.
9. Palmer 1985.
10. *Chronicle of Higher Education*, September 14, 1988, p. A5.
11. Office of Technology Assessment 1989.
12. Utts 1996a, 3.
13. Hyman 1996, 57.
14. Atkinson et al. 1990.
15. McCrone 1993, 29.
16. E.g., Neimark 1996; Brown 1996; Begley 1996.
17. In November 1995, the CIA's confirmation that the U.S. government had supported a classified program of psi research for twenty years was front-page news around the world.
18. Barber 1961.

Chapter 1

1. Truzzi 1987, 13.
2. Mitchell 1979, 3.
3. Rosenthal and Rosnow 1984, 7–8.
4. Cited in Machlup and Mansfield 1983, 13.
5. Kaplan 1964, 27.
6. Feyerabend 1975.

7. Whyte 1950, 104–5.
8. Wilber 1977, 33.

Chapter 2

1. Gittelson and Torbet 1987, 216.
2. As told to the author by "Fred" in 1990.
3. Nichols 1966, 113.
4. Targ 1996, 81–82.
5. A drawing is used instead of the actual photograph to conceal the detail available in spy-satellite photos.
6. As witnessed by the author in 1988.
7. In a double-action revolver, when the trigger is pulled the hammer is cocked, the cylinder revolves, and the hammer falls on the next chamber, all in one motion.
8. As told to the author in 1995.
9. Doyle 1884.
10. Gittelson and Torbet 1987, 260.

Chapter 3

1. In special cases like binomial proportions, the width of the confidence interval is determined by the theoretical variance and not by the variability of repeated observations.
2. Strictly speaking, chance is not a condition per se, but the general statistical concepts still hold.
3. Repeatable, replicable, and reproducible will be treated as synonyms, even though they have slightly different connotations.
4. Epstein 1980, 790.
5. Bozarth and Roberts 1972, 774–75.
6. Collins 1985, 40.
7. Neuliep and Crandall 1991.
8. Augustine 1982, 125–26.
9. Polanyi 1961.
10. Rosenthal and Rosnow 1984, 10.
11. Barber 1976.
12. Begley 1996.
13. In Herbert 1985, 29.
14. In Jahn 1981, 99.
15. Coover 1917, 82.
16. Utts 1991a, 1991b.
17. Kennedy and Uphoff 1939.
18. Berger 1989.
19. Honorton 1985; Utts 1986, 1988.
20. Tversky and Kahneman 1971.
21. Utts 1996b.
22. Hyman 1985a, 71.
23. Barns et al. 1964.
24. Hansel 1980, 298.

Chapter 4

1. Bangert-Drowns 1986; Rosenthal 1978, 1990, 1991; Rosnow and Rosenthal 1989.
2. Mullen and Rosenthal 1985, 2.
3. Wachter 1988.
4. Cooper and Rosenthal 1980.
5. Akers 1985; Wilson and Rachman 1983.
6. Chow 1987.
7. Iyengar and Greenhouse 1988.
8. Reprinted in Mann 1990.
9. Hedges 1987, 443.
10. Hedges 1987, 453.
11. Rosenfeld 1975; Wohl et al. 1984.
12. Wohl et al. 1984, p. S5.
13. Wohl et al. 1984, p. S9.
14. Quoted in LeShan 1973, 467.

Chapter 5

1. Gittelson and Torbet 1987, 160–61.
2. The Society for Psychical Research, established in London in 1882, attracted an exceptionally competent international body of scientists interested in what was then called psychical research. The presidents of the society included three Nobel laureates, ten fellows of the Royal Society, one prime minister, the great British scientists Sir William Crookes, Sir Oliver Lodge, and Sir J. J. Thomson, discoverer of the electron, and academic luminaries from Harvard University including William James, William McDougal, and Gardner Murphy.
3. Parapsychologists can be placed into three general classes: scientists, scholars, and therapists. Scientists are primarily interested in experimental and theoretical issues; scholars are interested in historical and philosophical issues; and therapists are interested in mental-health issues. This book concentrates on the scientific aspects of psi, but it is worth noting that there is a branch of parapsychology concerned with helping people understand and deal with genuine psychic phenomena. In "normal" people, psychic experiences can sometimes be disturbing because our culture asserts that only crazy people report such things. As a result, many perfectly well adjusted, high-functioning people are reluctant to tell anyone about their psychic experiences for fear of ridicule.
4. Gurney, Myers, and Podmore 1970 (orig. published 1886).
5. Ullman, Krippner, and Vaughan 1973, 13.
6. Barrett, Gurney, and Myers 1883.
7. Sinclair 1962 (orig. published 1930).
8. Sinclair 1962, ix.
9. Sinclair 1962, 225.
10. Prasad and Stevenson 1968; Rhine 1964.
11. Painted in 1917, oil on canvas, 59-1/2 by 50-3/4 inches; Museum of Modern Art Collection, New York; Curt Valentin bequest.
12. Ullman, Krippner, and Vaughan 1973, 123.
13. Honorton and Harper 1974.

14. Braud and Braud 1973; Braud, Wood, and Braud 1975; Parker 1975.
15. Honorton 1977, 437.
16. Avant 1965; Metzger 1930.
17. Hyman and Honorton 1986.
18. Harris and Rosenthal 1988a, 1988b, 1988c.
19. Dalton et al. 1996.
20. Honorton et al. 1990, 120–21.
21. Honorton 1983; Hyman 1983.
22. Honorton 1985; Hyman 1985b, 1985c.
23. Broughton 1987.
24. Blackmore 1980.
25. Rosenthal 1979.
26. Hyman and Honorton 1986.
27. Harris and Rosenthal 1988a, 1988b, 1988c; Saunders 1985; Utts 1986.
28. Harris and Rosenthal 1988b, 3; Hyman 1991; Utts 1991a, 1991b.
29. Honorton 1985, table A1.
30. Steering Committee of the Physicians' Health Study Research Group 1988.
31. Hyman and Honorton 1986, 351.
32. Publication dates often lag research activities by a few years, so the autoganzfeld research actually started about the same time that Honorton learned the results of the meta-analysis that was ultimately published in 1985.
33. Bem and Honorton 1994.
34. Bem and Honorton 1994.
35. Honorton and Schechter 1987; Honorton et al. 1990.
36. Hyman and Honorton 1986, 351.
37. Hyman 1991, 392.
38. Figure 5.4 includes all studies where the chance hit rate was 25 percent.
39. Dalton et al. 1996; Radin 1993c.
40. Bierman 1995; Broughton and Alexander 1995; Broughton, Kanthamani, and Khilji 1990; Wezelman and Gerding 1994; Wezelman, Gerding, and Verhoeven, in press; Schlitz and Honorton 1992; Morris et al. 1995. An in-press study was obtained from Professor Adrian Parker (personal correspondence), and one study by Professor Daryl Bem, consisting of twenty-five sessions, was not included in this analysis because a group of twenty-five nonmeditators was predicted to not perform as well as a group of meditators, and it did not (personal correspondence). Thus, that group was considered in this analysis as a control.
41. Quoted in Pratt et al. 1966 (orig. published 1940), 11.

Chapter 6

1. Pratt et al. 1966.
2. Gittelson and Torbet 1987, 99.
3. Ullman, Krippner, and Vaughan 1973, 13.
4. L. E. Rhine 1961, 1981; J. B. Rhine 1977.
5. Kennedy and Uphoff 1939; Pratt et al. 1966, 144.
6. Rosenthal 1978.
7. Huntington 1938; Stuart and Greenwood 1937.
8. Pratt et al. 1966, 42.
9. Camp 1937.

10. Pratt et al. 1966.
11. These hit rates and confidence intervals are calculated from the combined number of hits and trials in each category. Thus for distance experiments, in ten studies there were a total of 164,475 trials, resulting in 35,378 hits, for an overall hit rate of 21.5 percent, and a 95 percent confidence interval of ±0.2 percent.
12. Eysenck 1957.
13. Pratt et al. 1966, 42.
14. Swann 1987.
15. May 1995, 204.
16. Puthoff 1996.
17. Wilhelm 1977.
18. Because there were five targets in the pool, by chance alone the actual targets would receive each possible rank, including first place, 20 percent of the time. Since each target could be selected with equal likelihood, by chance the average *ranking* for a series of viewings, and thus the average *score* for a pool of five targets, was simply 3 (the midpoint between 1 and 5). Because a ranking of 1 was the best possible match, an average rank score that was significantly *less* than 3 would be taken as evidence for genuine remote viewing. Rank-order judging is a conservative technique because it provides no extra credit for an excellent match, such as when the remote viewer has given a near-photographic description of the actual target. Nor does it provide less credit for a close-call match, such as when the judge can just barely guess that one target is a better match than another. Rank-order judging is valid even if the viewer is completely aware of the entire set of five possible targets. As long as the viewer and the judge are both blind to the actual target out of a pool of five, the chance of getting a first-place match is still only one in five, or 20 percent.
19. Targ and Puthoff 1974.
20. Puthoff and Targ 1976; Vallee, Hastings, and Askevold 1976; Hastings and Hurt 1976; Hyman 1985b; Jahn 1982.
21. Tart, Puthoff, and Targ 1979; Targ and Harary 1984.
22. Akers 1984.
23. Utts 1996a.
24. May et al. 1988.
25. This was not known to the general public or to the scientific community, because most of this research was classified until November 1995.
26. Utts 1996a.
27. Utts 1996a.
28. Hyman 1996, 39, 40; emphasis added.
29. Hyman 1996, 55.
30. Jahn and Dunne 1987, 164.
31. Hansen, Utts, and Markwick 1992.
32. Dobyns et al. 1992.
33. Milton 1993, 87–104; Eisenberg and Donderi 1979.
34. The results of an experiment expressed in terms of hit rate can be translated into any other hit-rate *equivalent* provided that the number of trials, the actual hit rate, and the probability of each trial by chance are known. A 50-percent-equivalent hit rate is a convenient figure to use because it can be compared with the results of an intuitively familiar coin-flipping experiment. See Bem and Honorton (1994) for discussion of a 50-percent-equivalent effect size called "pi."

35. Schechter 1984; Stanford 1987.
36. Stanford and Stein 1994.
37. Schmeidler 1943.
38. Lawrence 1993.

Chapter 7

1. Dunne 1927; Eisenbud 1982; Zohar 1983.
2. *Psychic Powers*, 17.
3. *Psychic Powers*, 17, 18.
4. Honorton and Ferrari 1989.
5. The correlation was r (246 df) = 0.081, $p = 0.20$, two-tailed.
6. The correlation was r (246 df) = 0.282, $p = 2$ [tms] 10^{-7}, two-tailed.
7. Klintman 1983, 1984.
8. Schmeidler 1988.
9. Radin 1996 and in press.
10. Andreassi 1989; Bouscein 1992.
11. Radin 1996.
12. Because the nature of this experiment made it necessary to display emotionally shocking pictures to cause an orienting response, the participant population was restricted to mature adults. All volunteers were required to read an informed consent notice, which explained that some disturbing pictures might be shown.
13. This includes all data from all twenty-four volunteers who participated in the first experiment and a follow-up replication study. As of this writing (early 1997), we are collecting data in a new series of presentiment experiments.
14. Professor Bierman measured what is called "skin conductance response," whereas we measured "skin conductance level." The former measures fast-moving changes in electrodermal activity; the latter measures both fast and slower-moving drifts of the baseline level.
15. Data provided courtesy of Professor Bierman; Bierman and Radin 1997.
16. This analysis was based on all available data as of early 1997.
17. Bechara et al. 1997, 1293, emphasis added.
18. Bechara et al. 1997, 1295.

Chapter 8

1. Jahn and Dunne 1986; Heims 1981.
2. Herbert 1993; LeShan and Margenau 1982; Mermin 1985; Glanz 1995; Zukav 1979; Stapp 1993; Shimony 1963.
3. d'Espagnat 1979, 158.
4. E.g., Wigner 1963.
5. Squires 1987.
6. Hall et al. 1977.
7. Smith 1968.
8. Jahn and Dunne 1986.
9. Doyle 1981; Desmond 1984; Radin 1990b; Morris 1983, 1986; Reed 1989.
10. Shneiderman 1987.
11. Hecht and Dussault 1987; Pelegrin 1988.
12. McCarthy 1988.

13. Stiffler 1981.
14. Avizienis, Kopetz, and Laprie 1987.
15. Marks and Kammann 1980.
16. Gamow 1959.
17. Conot 1992; Price 1984.
18. World notes, *Time*, October 9, 1989.
19. Dunn 1989.
20. Dunn 1989.
21. Bailey 1988.
22. Rhine 1944.
23. Murphy 1962; Girden 1962; Girden et al. 1964; Girden and Girden 1985.
24. Radin and Ferrari 1991.
25. They included the *Journal of Parapsychology, European Journal of Parapsychology, Journal of the American Society for Psychical Research, Journal of the Society for Psychical Research, Research Letter* of the University of Utrecht's Parapsychology Laboratory, *Newsletter* of the Parapsychology Foundation, *Proceedings of the Society for Psychical Research, Proceedings of the First International Conference of Parapsychological Studies, Journal of Experimental Psychology, Parapsychological Journal of South Africa,* and a book, *The Algonquin Experiments.*
26. (1) *Automatic recording:* Die faces were automatically recorded onto a permanent medium, for example, photographed onto film. (2) *Independent recording:* Someone other than the experimenter also recorded the data. (3) *Data selection prevented:* This refers to designs with sequential-frame photographic data recordings, or studies in which data were kept in bound record books, or some other method of ensuring that all data were used in the final analysis. (4) *Data double-checked:* Data were manually or automatically double-checked for accuracy. (5) *Witnesses present:* Witnesses were present during data recording to help reduce the possibility of mistakes or fraud. (6) *Control noted:* A control study was mentioned, but no details were published. (7) *Local control:* Control data were obtained under the same conditions as the experiment, using the same subject(s) and the same conditions, but with no specific mental effort applied to the dice. (8) *Protocol control:* The study was designed in such a manner that controls were inherently a part of the experiment (e.g., equal number of throws for each die face). (9) *Calibration control:* A long-term randomness test was conducted, usually immediately before and immediately after an experimental series. (10) *Fixed run lengths:* Optional stopping was ruled out by a prespecified design. (11) *Formal study:* The study used a prespecified methodology, as well as analyses specified in advance of experimentation. (12) *Dice toss method:* By hand, bounced against a back wall, use of a cup or chute, or tossed automatically by machine. (13) *Subject type:* Unselected subjects, selected, experimenter as sole subject, experimenter along with subjects, unselected subjects.
27. Radin and Ferrari 1991.
28. Hyman 1996.
29. Radin and Rebman, in press a, in press b.
30. Jahn and Dunne 1987; May, Humphrey, and Hubbard 1980.
31. Schmidt 1969, 1970, 1975, 1976, 1981, 1987.
32. Nelson and Radin 1987, 1989, 1990; Radin and Nelson 1988, 1989.
33. The yearly average "hit rates" shown for the dice and RNG studies are average effect sizes calculated on a 50-percent-equivalent basis. For example, say that a given experiment involving one thousand trials resulted in a standard normal deviate of

$z = 1.5$. We assume that each trial is one random event, typically a single random bit from a binary RNG. A z of 1.5 translates into a hit rate of 52.4 percent, assuming a chance hit rate of 50 percent and one thousand trials. The average yearly hit rates and 95 percent confidence intervals were calculated based on 50-percent-equivalent hit rates for each study in a given year, and the associated empirical standard errors for those hit rates in that year.

34. This graph excludes the PEAR RNG studies, all of which had quality scores of 12 and above.
35. The correlation was $r = 0.165$, $N = 339$.
36. The correlation was $r = 0.01$, $N = 339$.
37. Dobyns 1996.
38. Nelson et al. 1991; Nelson, Dunne, and Jahn 1984.
39. Jahn, Dobyns, and Dunne 1991.
40. Dunne 1993, 1995.
41. Nelson et al. 1991; Dunne et al. 1994.
42. Rhine 1969.
43. Dunne and Jahn 1992.
44. Schmidt, Morris, and Rudolph 1986.
45. A highly accurate descriptive model of these effects proposed by May, Utts, and Spottiswoode (1995) suggests that mental intention does not literally affect matter in the usual forcelike sense of "affect." Rather, the model argues that individuals take advantage of favorable times in which to interact with the random system.
46. Hansel 1966, 1980.
47. Schmidt, Morris, and Rudolph 1986.
48. Schmidt 1993b, 366.
49. Schmidt and Schlitz 1989; Schmidt 1993a.
50. Stapp 1994.
51. Brown 1994, 14.

Chapter 9

1. Woodward et al. 1992.
2. *Life,* March 1994, p. 54.
3. *Time,* June 24, 1996.
4. Crick 1994.
5. Presuming that A and B are truly isolated from each other, in space or time.
6. Cousins 1989.
7. O'Regan 1987.
8. Gazzaniga 1988, 215.
9. Lindo 1985; Milne 1983; Milne and Aldridge 1981; Sheehan 1978; Gravitz 1979, 1981; Lehman 1978.
10. Abrams and Wilson 1983.
11. Galle and Hadni 1984.
12. Cohen 1985.
13. Plotkin 1980.
14. Putnam, Zahn, and Post 1990; Putnam 1984, 1991.
15. Joyce and Welldon 1965; Loehr 1969; Collip 1969.
16. R. Gardner 1983.

17. Dossey 1993, 242.
18. Solfvin 1984; Benor 1990, 1993; Dossey 1993; Schouten 1993.
19. Schouten 1993, 393.
20. Byrd 1988; Dossey 1993, 180.
21. Dossey 1993, 179–86.
22. Dossey 1993, 186.
23. Barry 1968; Beutler et al. 1988; Braud 1990, 1993; Braud and Schlitz 1983; Brier 1969; Campbell 1968; R. Gardner 1983; Grad 1963, 1965; Grad, Cadoret, and Paul 1961; Haraldsson and Thorsteinsson 1973; Kuang et al. 1986; Miller 1982; Nash 1982, 1984; Nash and Nash 1967; Pleass and Dey 1990; Randall 1970; Richmond 1952; Snel and Hol 1983; Solfvin 1982; Watkins and Watkins 1971.
24. Dossey 1993; Benor 1990, 1993.
25. Tichener 1898; Coover 1913.
26. Otani 1955; Beloff 1974; Morris 1977; Schouten 1976.
27. Tart 1963.
28. Dean 1962, 1966; Barry 1967; Haraldsson 1972.
29. Duane and Behrendt 1965. See also Grinberg-Zylberbaum et al. 1992; May, Targ, and Puthoff 1979; Warren, McDonough, and Don 1992.
30. Braud 1981; Braud and Schlitz 1989, 1991.
31. Braud and Schlitz 1991.
32. Braud and Schlitz 1991; study 15 was from Radin, Taylor, and Braud 1995, conducted at the University of Edinburgh; study 16 was from Wezelman et al. 1996, conducted at the University of Nevada, Las Vegas; study 17 was from Rebman et al. 1996, University of Nevada, Las Vegas. A similar study was reported by Delanoy and Sha (1994), at the University of Edinburgh. See also Rebman et al. 1995.
33. Braud and Schlitz 1991.
34. Poortman 1959; Peterson 1978; Williams 1983; Braud, Shafer, and Andrews 1992; Schlitz and LaBerge 1994; Wiseman and Smith 1994; Wiseman et al. 1995; Wiseman and Schlitz 1996.

Chapter 10

1. Lovelock 1979, 1990.
2. Mansfield and Spiegelman 1996.
3. Sheldrake 1981, 1995.
4. Orme-Johnson, no date; see also Orme-Johnson et al. 1982, 1988.
5. Dillbeck 1990; Dillbeck et al. 1988; Gelderloos et al. 1988.
6. Fales and Markovsky, no date.
7. Forman 1994.
8. Schrödinger 1967.
9. Csikszentmikalyi 1975.
10. von Franz 1992, 50.
11. Grof 1988.
12. Radin, Rebman, and Cross 1996.
13. There were breaks for lunch and rest periods, but relatively high group coherence was maintained throughout the workshop.
14. The probability associated with the correlation was adjusted for the smoothing that occurs by averaging over six half-hour epochs. The actual correlation was $r = 0.69$.

15. Nelson et al. 1996; Blasband 1995; Bierman 1996.
16. Nelson and Mayer 1996.
17. Nelson 1996.
18. Nelson 1996, 9.
19. LeShan 1974, 1982, 1987.
20. Mansfield 1995, 226.
21. Mansfield 1995; Combs and Holland 1990; Peat 1987, 1991.
22. Jung 1978, 342.
23. Lovelock 1979, 1990.

Chapter 11

1. McCrone 1994, 34–38.
2. This is not to say that casino operators and experimenters have the same goals, or that casino games and psi experiments are conducted under exactly the same conditions. In addition, while most gamblers insist that they really do want to win, some fraction of addicted gamblers have strong self-destructive tendencies.
3. Tyminski and Brier 1970, cited in Broderick 1992.
4. We are indebted to Bernice Jaeger for providing these data. Radin and Rebman, in press a.
5. These data also raise privacy issues, which may make them difficult to obtain even if they were available.
6. Abt, Smith, and Christiansen 1985.
7. Of course, the Law of Large Numbers declares that even if there were occasional fluctuations in payout rates, over the long term casino winnings would still be predictably stable, and empirically, of course, they are stable.
8. Tromp 1980, 124; see also Adair 1991.
9. Weaver and Astumain 1990.
10. Wilson et al. 1990.
11. Tromp 1980, 227–28.
12. Braud and Dennis 1989; Persinger 1989a, 1989b; Alonso 1993.
13. A partial list includes: Persinger 1985; Lewicki, Schaut, and Persinger 1987; Persinger and Schaut 1988; Spottiswoode 1990; Gissurarson 1992; Wilkinson and Gauld 1993; Radin 1992, 1993a, 1993b; Persinger and Krippner 1989; Radin, McAlpine, and Cunningham 1994; Adams 1986, 1987; Arango and Persinger 1988; Berger and Persinger 1991; Gearhart and Persinger 1986; Haraldsson and Gissurarson 1987; Makarec and Persinger 1987; Schaut and Persinger 1985.
14. Roney-Dougal and Vogl 1993.
15. Intriguing studies about to be published as this book went to press report new analyses following up on the GMF–psi relationship. A study in our lab found that complex combinations of environmental factors ranging from changes in local weather to solar activity significantly affected psi performance (Radin and Rebman, in press a).
16. Rotton and Kelly 1985; Kelly, Rotton, and Culver 1985–86; Alonso 1993; Lieber 1978.
17. Lieber and Sherin 1972; Templer and Veleber 1980; Tasso and Miller 1976; Snoyman and Holdstock 1980; Weiskott and Tipton 1975; Geller and Shannon 1976; Radin and Rebman 1996.

18. Chapman 1961; Pokorny and Jackimczyk 1974.
19. Frey, Rotton, and Barry 1979.
20. Rotton and Kelly 1985, 302.
21. Kelly, Rotton, and Culver 1985–86, 131.
22. Martin, Kelly, and Saklofske 1992, 794.
23. Wait till the moon is full, 1992.
24. Grice 1993.
25. Harner 1980; Witches and witchcraft, 1990.
26. Guiley 1991, 100.
27. Guiley 1991, 113.
28. Guiley 1991, 142.
29. Oliven 1943.
30. Guiley 1991, 148.
31. Abel 1976; Lieber 1978; Moon madness no myth, 1981.
32. Puharich 1973, 281–89.
33. Becker 1990; Roney-Dougal 1993; Playfair and Hill 1978; Tromp 1980, 123.
34. Bigg 1963; Bell and Defouw 1964, 1966.
35. Michel, Dessler, and Walters 1964; Bell and Defouw 1966; Rassbach and Dessler 1966.
36. Fraser-Smith 1982.
37. Kohmann and Willows 1987.
38. Keshavan et al. 1981.
39. The actual correlation was $r = 0.74$, $t = 1.72$, $p = 0.04$ (one-tail). The data used in these correlations were smoothed twice: once by a moving average transformation that smoothed the daily raw data, and again when averaged by a superposed epoch analysis based on the day of the lunar cycle. Double-smoothing helps reduce measurement noise due to daily and monthly variations, but it also decreases the variation between the individual data points. This in turn inflates the apparent strength of the correlation. All results reported were adjusted to take this inflation into account.
40. The correlation was $r = -0.66$, adjusted $t = 1.52$, $p = 0.07$ (one-tail).
41. The last few lunar cycles are not shown because no jackpots occurred during those months.
42. These odds are calculated by assuming that this is a case of getting four "hits" out of six events, where the probability of each event is $p = 3/29$. The "3" comes from the three days within one day of the full moon, and the "29" from the length of the lunar cycle.
43. These analyses included bivariate spectral analyses, multivariate regression analyses, and models using artificial neural networks and abductive networks.
44. Note that the lunar–GMF data for 1991 to 1994 used in the analysis of casino payouts resulted in a negative relationship. This suggests, not surprisingly, that the lunar–GMF link is more complicated than a simple linear relationship can model. Other geophysical and extraterrestrial factors are probably involved, including the solar rotation cycle.
45. The adjustments mentioned in note 39 were applied to this analysis as well.
46. "6/49" means that one needs to guess correctly six numbers out of forty-nine possibilities to win the lottery.
47. Zilberman 1995.

Chapter 12

1. Harner 1980.
2. Quinn 1984; Keller and Bzdek 1986; Krieger 1986.
3. Bro 1989.
4. Burk 1995.
5. Sun Tzu 1963, cited in Mishlove 1993, 219.
6. Anderson and van Atta 1989; Ebon 1983; Levine, Fenyvesi, and Emerson 1988.
7. Mishlove 1993, 220.
8. Mishlove 1993, 223–24.
9. For a detailed description, see Schnabel 1997.
10. Defense Intelligence Agency.
11. Schnabel 1997, 70–72; McMoneagle 1993.
12. Jack Anderson and Jan Moller columns, *Washington Post,* December 23, 30, 1996, and January 9, 1997.
13. For example, the *Journal of Scientific Exploration* (vol. 10, no. 1, 1996) contained several articles on the U.S. program.
14. Hartman and Secrist 1991.
15. Hartman and Secrist 1991.
16. Forrester 1978.
17. Tabori 1974.
18. Lyons and Truzzi 1990.
19. Radin and Bisaga 1991; Radin 1994a.
20. Radin 1989; Radin 1993b.
21. Radin 1994b.
22. Bell Laboratories, now part of Lucent Technologies, was formerly the research arm of the telecommunications giant AT&T. Before the Bell System was broken up in the mid-1980s, some scientists within Bell Labs enjoyed substantial freedom to conduct small-scale studies on whatever they liked. Few industrial labs can afford such luxuries today.
23. The research and development lab for Contel Corporation, a multibillion-dollar telecommunications company that merged with GTE Corporation in 1990.
24. Radin and Utts 1989; Radin 1982.
25. Radin 1990b.
26. Radin 1989, 1990–91, 1993b.
27. Radin and Bisaga 1991.
28. Harman and Rheingold 1984.
29. Harman and Rheingold 1984.
30. *Asian Wall Street Weekly,* April 8, 1985, p. 18.
31. Dean et al. 1974; Radin 1990a.
32. Quoted in Dean et al. 1974, 196.
33. Bowles, Hynds, and Maxwell 1978, 114.
34. Bowles, Hynds, and Maxwell 1978, 114.
35. Mishlove 1993, 232.
36. Quoted in Gittelson and Torbet 1987, 60.

Chapter 13

1. Churchland 1984, 17.
2. Hansen 1992b, 163. See also Birdsell 1989; Truzzi 1983.

3. Marks 1986, 119.
4. Cited by Honorton 1993.
5. M. Gardner 1983, 60.
6. Quoted in Hyman 1996.
7. Honorton 1993, 211.
8. Boring 1955, 1966.
9. Stevens 1967.
10. Blackmore 1996, v.
11. E.g., *Skeptical Inquirer,* September–October 1995.
12. Ayer 1965.
13. Anderson 1990.
14. Hebb 1951.
15. Price 1955, 359.
16. Price 1955, 360.
17. Hansel 1980, 22.
18. Druckman and Swets 1988.
19. News and comment: Academy helps army be all that it can be, 1987, p. 1502.
20. Palmer, Honorton, and Utts 1989.
21. *Chronicle of Higher Education,* September 14, 1988, p. A10.
22. *Chronicle of Higher Education,* September 14, 1988, p. A10.
23. Griffin 1988.
24. News and comment: Academy helps army be all that it can be, 1987.
25. Druckman and Swets 1988, 206.
26. Harris and Rosenthal 1988a.
27. Harris and Rosenthal 1988a.
28. Collins 1987.
29. Marks 1986.
30. Broad and Wade 1982; Kohn 1988.
31. Begley 1996.
32. Begley 1996.
33. Begley 1996.
34. Harris and Rosenthal 1988a, 1988b, 1988c.
35. Jahn and Dunne 1987.
36. Begley 1996.
37. Jahn, Dobyns, and Dunne 1991.
38. Child 1985.
39. Hansel 1980.
40. Hyman 1984; from a *Nova* program transcript.
41. Alcock 1987, 1981.
42. Alcock 1981, 163.
43. Child 1985.
44. Zusne and Jones 1982.
45. Zusne and Jones 1982, 260–61.
46. Child 1985.
47. Roig, Icochea, and Cuzzucoli 1991.
48. Roig, Icochea, and Cuzzucoli 1991, 160.
49. Alcock 1988, 44; emphasis added.
50. Alcock 1981, 7.
51. Alcock 1981, 191.
52. Hawking 1988, 175.

53. M. Gardner 1983, 239.
54. Wagner and Monet 1979.
55. Greeley 1987a 8; see also 1987b, 1991.

Chapter 14

1. Schacter 1996.
2. Bruner and Postman 1949.
3. Milton 1994, 106.
4. Festinger 1962.
5. George 1995.
6. George 1995.
7. Kuhn 1970, 64.
8. Kuhn 1970; Milton 1994, 110.
9. Dixon 1981.
10. News and comment: Science beyond the pale, 1990.
11. Rosenthal and Rubin 1978.
12. Ross and Lepper 1980.
13. Collins and Pinch 1979, 263.
14. Collins 1987, 2.
15. Griffin 1988.
16. Nisbett and Ross 1980.
17. Peale 1956.
18. Rosenthal and Jacobson 1968.
19. Blackmore 1996.
20. Griffin 1988.
21. Tart 1986, 134.
22. Tart 1986.
23. McClenon 1982; McConnell and Clark 1991; Haraldsson and Houtkooper 1991; Wagner and Monet 1979.

Chapter 15

1. Dennett 1991.
2. Elitzur 1995.
3. Thorndike 1905.
4. There was little distinction between philosopher, mathematician, astronomer, and physicist in scientists of those days.
5. It is rumored that a few decades later, God replied with, "Nietzsche is dead."
6. Harman 1991.
7. Kuhn 1970, 24.
8. James 1956, 327.
9. Skinner 1972.
10. Weinberg 1977, 154.
11. Griffen 1988, 7.
12. Krutch 1956, xi.
13. Churchland 1984.
14. Harman 1988.
15. Smart 1979, 159–70, 53–54.

16. Crick and Koch 1992.
17. News and comment, *Brain/Mind Bulletin*, February 16, 1981, p. 1.
18. Koestler and Smythies 1969.
19. Wilber 1977, 37.
20. Hall 1992; Gleick 1987.
21. Churchland 1986, 265.
22. Herbert 1985.
23. Weiss 1969, 13.
24. Whitehead 1933, 134–35.
25. Sperry 1987.
26. Sheldrake 1981, 1995.
27. Lovelock 1979, 1990.
28. Playfair and Hill 1978; Rosen et al. 1991; Gauquelin 1988.
29. Huxley 1944.
30. Morowitz 1980.
31. Wilber 1977, 31.
32. Engelhardt and Caplan 1987.
33. Jeans 1948, 166, 186.
34. Diekman 1974.
35. Schrödinger 1964, 40.
36. Wigner 1969.
37. Herbert 1985, 249.
38. This graph is based on a computer-aided survey of books listed in the Harvard University libraries. The exponential rise in the publications depicted in the graph probably owes to general trends in book publishing, which reflect rapid advances in all scientific disciplines. The point of interest here is the link between books on consciousness and parapsychology.
39. In this case, the terms "psychical research" and "parapsychology" were both counted since the term "parapsychology" was first popularized with J. B. Rhine's publications in the 1930s.
40. Penrose 1989, 226.
41. Harman 1994.
42. Schrödinger 1964, 40.
43. Quoted in Barnett 1979, 108.
44. Goerner 1994, 54.
45. Herbert 1985, 249.
46. Teilhard de Chardin, quoted in Wilber 1993, 39.
47. Rinpoche 1992, 37, 39.
48. Kelly 1955, 6.
49. Jeans 1937, 122.
50. Eddington 1928.
51. Peat 1987, 77.
52. Goerner 1994.
53. LeShan 1974.
54. Quoted in Weber 1986, 203.
55. Wald 1988.
56. Goerner 1994, 173, 174.
57. Wilber 1993, 65.
58. Broad 1949, 291.

59. Wilber 1984.
60. Weber 1986, 6.
61. Weber 1986, 10.

Chapter 16

1. Bell 1966, 1976, 1987; Cushing and McMullin 1989; Einstein, Podolsky, and Rosen 1935; Leggett 1987.
2. Rohrlich 1983, p. 1252; Aspect, Dalibard, and Roger 1982; Freedman and Clauser 1972.
3. Feinberg 1975; Stokes 1987.
4. Laszlo 1993, 1995.
5. May, Utts, and Spottiswoode 1995.
6. Radin and Rebman 1996; Stevenson and Greyson 1979; Bennett 1939; Hart and Hart 1933; Hart 1956; Roll 1977, 1994; Collins 1948.
7. Roll 1977, 1994.
8. Noble 1988, 179.
9. Jahn and Dunne 1987.
10. Bell 1964.
11. Lindley 1996.
12. Rohrlich 1983, p. 1251.
13. Lindley 1996, 149.
14. Shallis 1982.
15. Libet 1994.
16. Conrad, Home, and Josephson 1988; Josephson and Pallikari-Viras 1991; Penrose 1989, 1994; Bohm 1952, 1986, 1987; Bohm, Hiley, and Kaloyerou 1987; Josephson 1988.
17. Josephson and Pallikari-Viras 1991.
18. Walker 1975.
19. Hameroff 1994.
20. Hameroff 1994, 92.
21. Landauer 1996.
22. Bussey 1982.
23. Taubes 1996.
24. Bennett et al. 1993; Taubes 1996, 504.
25. Personal correspondence, January 9, 1997.
26. Noble 1988, 181.

Chapter 17

1. Schmidt 1993b, 367.
2. Phillips 1995.
3. Moreover: Science does it with feeling, 1996, 71–73; Chalmers 1995.
4. Lockwood 1989.
5. Galante 1986; Lyons and Truzzi 1990.

References

Abbreviated titles of journals frequently cited:

EJP: *European Journal of Parapsychology*
JASPR: *Journal of the American Society for Psychical Research*
JP: *Journal of Parapsychology*
JSE: *Journal of Scientific Exploration*
JSPR: *Journal of the Society for Psychical Research*
RIP: *Research in Parapsychology*
SE: *Subtle Energies*

Abel, E. L. 1976. *Moon madness.* Greenwich, CT: Fawcett.

Abrams, D. B., and G. T. Wilson. 1983. Alcohol, sexual arousal, and self-control. *Journal of Personality & Social Psychology* 45:188–98.

Abt, V., J. F. Smith, and E. M. Christiansen. 1985. *The business of risk.* Lawrence: Univ. Press of Kansas.

Adair, R. K. 1991. Constraints on biological effects of weak extremely-low-frequency electromagnetic fields. *Physical Review* A 43:1039–48.

Adams, M. H. 1986. Variability in remote-viewing performance: Possible relationship to the geomagnetic field. In *RIP 1985,* edited by D. H. Weiner and D. I. Radin, 25. Metuchen, NJ: Scarecrow Press.

———. 1987. Persistent temporal relationships of ganzfeld results to geomagnetic activity: Appropriateness of using standard geomagnetic indices. In *RIP 1986,* edited by D. H. Weiner and R. D. Nelson, 78. Metuchen, NJ: Scarecrow Press.

Akers, C. 1984. Methodological criticisms of parapsychology. In *Advances in parapsychological research,* edited by S. Krippner, 4:112–164. Jefferson, NC: McFarland.

———. 1985. Can meta-analysis resolve the ESP controversy? In *A skeptic's handbook of parapsychology,* edited by P. Kurtz. Buffalo, NY: Prometheus Books.

Alcock, J. E. 1981. *Parapsychology: Science or magic? A psychological perspective.* Elmsford, NY: Pergamon Press.

———. 1987. Parapsychology: Science of the anomalous or search for the soul? *Behavioral & Brain Sciences* 10:553–65.

———. 1988. A comprehensive review of major empirical studies in parapsychology involving random event generators or remote viewing. In *Enhancing human performance: Issues, theories, and techniques,* edited by D. Druckman and J. A. Swets. Background Papers, Part VI. Washington, DC: National Academy Press.

Alonso, Y. 1993. Geophysical variables and behavior: LXXII. Barometric pressure, lunar cycle, and traffic accidents. *Perceptual & Motor Skills* 77:371–76.

Anderson, J., and D. van Atta. 1989. CIA secrets and customs agent's firing. *Washington Post* (February 15), D15.
Anderson, P. 1990. On the nature of physical laws. *Physics Today* (December), 9.
Andreassi, J. L. 1989. *Psychophysiology: Human behavior and physiological response.* Hillsdale, NJ: Erlbaum.
Arango, M. A., and M. A. Persinger. 1988. Geophysical variables and behavior: LII. Decreased geomagnetic activity and spontaneous telepathic experiences from the Sidgwick collection. *Perceptual & Motor Skills* 67:907–10.
Asian Wall Street Weekly (April 8, 1985), 18.
Aspect, A., J. Dalibard, and G. Roger. 1982. *Physical Review Letters* 49:1804.
Atkinson, R. L., R. C. Atkinson, E. E. Smith, and D. J. Bem. 1990. *Introduction to psychology*. 10th ed. San Diego: Harcourt, Brace, Jovanovich.
Augustine, N. R. 1982. *Augustine's laws*. New York: American Institute of Aeronautics & Astronautics.
Avant, L. L. 1965. Vision in the ganzfeld. *Psychological Bulletin* 64:246–58.
Avizienis, A., H. Kopetz, and J. C. Laprie, eds. 1987. *Evolution of fault-tolerant computing*. Vienna, Austria: Springer-Verlag.
Ayer, A. J. 1965. Chance. *Scientific American* (October), 44–54.
Bailey, L. W. 1988. Robogod: The divine machine. *Artifex* 7 (Fall): 15–24.
Bangert-Drowns, R. L. 1986. Review of developments in meta-analytic method. *Psychological Bulletin* 99:388–99.
Barber, B. 1961. Resistance by scientists to scientific discoveries. *Science* 134:596–602.
Barber, T. X. 1976. *Pitfalls in human research: Ten pivotal points*. Elmsford, NY: Pergamon Press.
Barnett, L. 1979. *The universe and Dr. Einstein*. Rev. ed. New York: Bantam.
Barns, V. E., et al. 1964. Confirmation of the existence of the omega-minus hyperon. *Physics Letters* 12:134–36.
Barrett, W. F., E. Gurney, and F. W. H. Myers. 1883. First report on thought-reading. *Proceedings of the Society for Psychical Research* 1:13–42. London: Trübner.
Barry, J. 1967. Telepathy and plethysmography. *Revue Metapsychique* 6:56–74.
———. 1968. General and comparative study of the psychokinetic effect on a fungus culture. *JP* 32:237–43.
Bechara, A., H. Damasio, D. Tranel, and A. R. Damasio. 1997. Deciding advantageously before knowing the advantageous strategy. *Science* 275 (February 28): 1293–1295.
Becker, R. O. 1990. *Cross currents: The promise of electromedicine, the perils of electropollution*. Los Angeles: Tarcher.
Begley, S. 1996. Science on the fringe. *Newsweek* (July 8).
Bell, B., and R. J. Defouw. 1964. Concerning a lunar modulation of geomagnetic activity. *Journal of Geophysical Research* 69: 3169–74.
———. 1966. On the lunar modulation of geomagnetic activity, 1884–1931 and 1932–1959. *Journal of Geophysical Research* 71: 4599–4602.
Bell, J. S. 1964. On the Einstein-Podolsky-Rosen paradox. *Physics* 1:195–200.
———. 1966. On the problem of hidden variables in quantum mechanics. *Review of Modern Physics* 38:447–52.
———. 1976. Einstein-Podolsky-Rosen experiments. *Proceedings of the Symposium on Frontier Problems in High Energy Physics*, 33–45. Pisa, Italy: Pisa.
———. 1987. *Speakable and unspeakable in quantum mechanics*. Cambridge: Cambridge Univ. Press.

Beloff, J. 1974. ESP: The search for a physiological index. *JSPR* 47:401–20.
Bem, D. J., and C. Honorton. 1994. Does psi exist? Replicable evidence for an anomalous process of information transfer. *Psychological Bulletin* 115:4–18.
Bennett, C. H., G. Brassard, C. Crepeau, R. Jozsa, A. Peres, and W. Wootters. 1993. Teleporting an unknown quantum state via dual classical and EPR channels. *Physics Review Letters* 70:1895–99.
Bennett, E. 1939. *Apparitions and haunted houses.* London: Faber & Faber.
Benor, D. J. 1990. Survey of spiritual healing research. *Complementary Medical Research* 4:9–33.
———. 1993. *Healing research.* Munich: Helix Verlag GmbH.
Berger, R. E. 1989. A critical examination of the Blackmore psi experiments. *JASPR* 83: 123–44.
Berger, R. E., and M. A. Persinger. 1991. Geophysical variables and behavior: LXVII. Quieter annual geomagnetic activity and larger effect size for experimental psi (ESP) studies over six decades. *Perceptual & Motor Skills* 73:1219–23.
Beutler, J. J., J. T. M. Attevelt, et al. 1988. Paranormal healing and hypertension. *British Medical Journal* 296:1491–94.
Bierman, D. J. 1995. The Amsterdam Ganzfeld Series III and IV: Target clip emotionality, effect sizes and openness. In *Proceedings of Presented Papers,* 38th Annual Parapsychological Association Convention, edited by N. L. Zingrone, 27–37. Fairhaven, MA: Parapsychological Association.
———. 1996. Exploring correlations between local emotional and global emotional events and the behavior of a random number generator. *JSE* 10:363–74.
Bierman, D. J., and D. I. Radin. 1997. Anomalous anticipatory response on randomized future conditions. *Perceptual & Motor Skills* 84:689–90.
Bigg, E. K. 1963. Lunar and planetary influences on geomagnetic disturbances. *Journal of Geophysical Research* 68 (13): 4099–4104.
Birdsell, P. G. 1989. *How magicians relate the occult to modern magic: An investigation and study.* Simi Valley, CA: Silver Dawn Media.
Blackmore, S. 1996. Reply to "Do you believe in psychic phenomena?" *The Times Higher Education Supplement* (April 5), p. v.
Blackmore, S. J. 1980. The extent of selective reporting of ESP ganzfeld studies. *EJP* 3:213–19.
Blasband, R. 1995. The ordering of random events by emotional expression. Presentation to the 14th Annual Meeting of the Society for Scientific Exploration, Huntington Beach, CA, June 15–17.
Bohm, D. J. 1952. A suggested interpretation of the quantum theory in terms of "hidden" variables, I and II. *Physical Review* 85:166–93.
———. 1986. A new theory of the relationship of mind and matter. *JASPR* 80:113–36.
———. 1987. *Unfolding meaning.* New York: Ark.
Bohm, D. J., B. J. Hiley, and P. N. Kaloyerou. 1987. An ontological basis for the quantum theory. *Physics Reports* 144:322–75.
Boring, E. G. 1955. The present status of parapsychology. *American Scientist* 43:108–16.
———. 1966. Paranormal phenomena: Evidence, specification, and chance. Introduction to C. E. M. Hansel's *ESP: A scientific evaluation.* New York: Scribner.
Bouscein, W. 1992. *Electrodermal activity.* New York: Plenum Press.
Bowles, N., F. Hynds, and J. Maxwell. 1978. *Psi search.* San Francisco: Harper & Row.
Bozarth, J. D., and R. R. Roberts. 1972. Signifying significant significance. *American Psychologist* 27:774–75.

Braud, W. G. 1981. Psi performance and autonomic nervous system activity. *JASPR* 75:1–35.

———. 1990. Distant mental influence of rate of hemolysis of human red blood cells. *JASPR* 84:1–24.

———. 1993. On the use of living target systems in distant mental influence research. In *Psi research methodology: A re-examination,* edited by L. Coly. New York: Parapsychology Foundation.

Braud, W. G., and L. W. Braud. 1973. Preliminary explorations of psi-conducive states. Progressive muscular relaxation. *JASPR* 67:27–46.

Braud, W. G., and S. P. Dennis. 1989. Geophysical variables and behavior: LVIII. Autonomic activity, hemolysis, and biological psychokinesis: Possible relationships with geomagnetic field activity. *Perceptual & Motor Skills* 68:1243–54.

Braud, W. G., and M. J. Schlitz. 1983. Psychokinetic influence on electrodermal activity *JP* 47:95–119.

———. 1989. A methodology for the objective study of transpersonal imagery. *JSE* 3:43–63.

———. 1991. Consciousness interactions with remote biological systems: Anomalous intentionality effects. *SE* 2 (1): 1–46.

Braud, W. G., D. Shafer, and C. S. Andrews. 1996. Further studies of autonomic detection of remote staring: Replications, new control procedures, and personality correlates. In *RIP 1992,* edited by E. W. Cook, 1–6. Lanham, MD: Scarecrow Press.

Braud, W. G., R. Wood, and L. W. Braud. 1975. Free-response GESP performance during an experimental hypnagogic state induced by visual and acoustic ganzfeld techniques. A replication and extension. *JASPR* 69:105–13.

Brier, R. 1969. PK on a bio-electrical system. *JP* 33:187–205.

Bro, H. H. 1989. *A seer out of season.* New York: Penguin, New American Library.

Broad, C. D. 1949. *Philosophy* 24:291–309.

Broad, W., and N. Wade. 1982. Betrayers of the truth: Fraud and deceit in the halls of science. New York: Simon & Schuster.

Broderick, D. 1992. *The lotto effect.* Hawthorn, Victoria, Australia: Hudson.

Broughton, R. S. 1987. Publication policy and the JP. *JP* 51:21–32.

Broughton, R. S., and C. Alexander. 1995. Autoganzfeld II: The first 100 sessions. In *Proceedings of Presented Papers,* 38th Annual Parapsychological Association Convention, edited by N. L. Zingrone, 53–61. Fairhaven, MA: Parapsychological Association.

Broughton, R. S., H. Kanthamani, and A. Khilji. 1990. Assessing the PRL success model on an independent ganzfeld data base. In *RIP 1989,* edited by L. Henkel and J. Palmer, 32–35. Metuchen, NJ: Scarecrow Press.

Brown, C. 1996. They laughed at Galileo too. *New York Times Magazine* (August 11).

Brown, J. 1994. Martial arts students influence the past. *New Scientist* (August 27), 14.

Bruner, J. S., and Leo Postman. 1949. On the perception of incongruity: A problem. *Journal of Personality* 18:206–23.

Burk, D. L. 1995. Intuitive medical diagnosis of musculoskeletal diseases with radiographic correlation: Case report, literature review and research protocol design. In *Proceedings of Presented Papers,* 38th Annual Parapsychological Association Convention, edited by N. L. Zingrone, 62–64. Fairhaven, MA: Parapsychological Association.

Bussey, P. J. 1982. Super-luminal communication in Einstein-Podolsky-Rosen experiments. *Physics Letters* A 90:9–12.
Byrd, R. C. 1988. Positive therapeutic effects of intercessory prayer in a coronary care population. *Southern Medical Journal* 81 (7): 826–29.
Camp, B. H. 1937. (Statement in Notes Section.) *JP* 1:305.
Campbell, A. 1968. Treatment of tumours by PK. *JSPR* 46:428.
Chalmers, D. J. 1995. The puzzle of conscious experience. *Scientific American* (December).
Chapman, L. J. 1961. A search for lunacy. *Journal of Nervous & Mental Disease* 132:171–74.
Child, I. L. 1985. Psychology and anomalous observations: The question of ESP in dreams. *American Psychologist* 40:1219–30.
Chow, S. L. 1987. Meta-analysis of pragmatic and theoretical research: A critique. *Journal of Psychology* 121:259–71.
Chronicle of Higher Education (September 14, 1988), p. A5.
Churchland, P. M. 1984. *Matter and consciousness: A contemporary introduction to the philosophy of mind.* Cambridge: MIT Press, Bradford Books.
Churchland, P. S. 1986. *Neurophilosophy: Towards a unified science of the mind/brain.* Cambridge: MIT Press.
Cohen, S. I. 1985. Psychosomatic death: Voodoo death in a modern perspective. *Integrative Psychiatry* 3:46–51.
Collins, B. A. 1948. *The Cheltenham ghost.* London: Psychic Press.
Collins, H. H. 1985. *Changing order: Replication and induction in scientific practice.* Beverly Hills, CA: Sage.
———. 1987. Scientific knowledge and scientific criticism. *Parapsychology Review* 18.
Collins, H. M., and T. J. Pinch. 1979. The construction of the paranormal: Nothing unscientific is happening. *Sociological Review Monograph* 27:237–70.
———. 1982. *Frames of meaning: The social construction of extraordinary science.* Boston: Routledge & Kegan Paul.
Collip, P. J. 1969. The efficacy of prayer: A triple blind study. *Medical Times* 97 (5): 201–4.
Combs, A., and M. Holland. 1990. *Synchronicity: Science, myth, and the trickster.* New York: Paragon House.
Communicating with trits, not bits. 1996. *Science News* 150 (July 13): 31.
Conot, R. 1992. *Thomas A. Edison: A streak of luck.* New York: Da Capo.
Conrad, M., D. Home, and B. D. Josephson. 1988. Beyond quantum theory: A realist psycho-biological interpretation of the quantum theory. In *Microphysical reality and quantum formalism,* edited by G. Tarozzi, A. van der Merwe, and F. Selleri, vol. 1, 285–93. Dordrecht: Kluwer Academic.
Cooper, H. M., and R. Rosenthal. 1980. Statistical versus traditional procedures for summarizing research findings. *Psychological Bulletin* 87:442–49.
Coover, J. E. 1913. The feeling of being stared at. *American Journal of Psychology* 24:57–75.
———. 1917. *Experiments in psychical research at Leland Stanford Junior University.* Stanford, CA: Stanford Univ. Press.
Cousins, N. 1989. Belief becomes biology. *Advances* 6 (3): 20–29.
Crick, F. H. C. 1994. *The astonishing hypothesis: The scientific search for the soul.* London: Simon & Simon.
Crick, F. H. C., and C. Koch. 1992. The problem of consciousness. *Scientific American* (September).

Csikszentmikalyi, M. 1975. *Beyond boredom and anxiety.* San Francisco: Jossey-Bass.
Cushing, J. T., and E. McMullin. 1989. *Philosophical consequences of quantum theory: Reflections on Bell's theorem.* Notre Dame, IN: Univ. of Notre Dame Press.
Dalton, K. S., R. L. Morris, D. Delanoy, D. I. Radin, and R. Wiseman. 1996. Security measures in an automated ganzfeld system. *JP* 60 (2): 120–148.
Dean, D. 1962. The plethysmograph as an indicator of ESP. *JSPR* 41:351–53.
———. 1966. Plethysmograph recordings as ESP responses. *International Journal of Neuropsychiatry* 2:439–46.
Dean, D., J. Mihalasky, S. Ostrander, and L. Schroeder. 1974. *Executive ESP.* Englewood Cliffs, NJ: Prentice-Hall.
Delanoy, D., and S. Sha. 1994. Cognitive and physiological psi responses to remote positive and neutral emotional states. In *Proceedings of Presented Papers,* 37th Annual Parapsychological Association Convention, edited by D. J. Bierman, 128–37. Fairhaven, MA: Parapsychological Association.
Dennett, D. C. 1991. *Consciousness explained.* New York: Little, Brown.
Desmond, J. 1984. Computer crashes: A case of mind over matter? *Computerworld* (June 11), 1–8.
d'Espagnat, B. 1979. The quantum theory and reality. *Scientific American* (November), 158–81.
Diekman, A. J. 1974. The meaning of everything. In *The nature of human consciousness,* edited by R. E. Ornstein, 317–26. New York: Viking.
Dillbeck, M. C. 1990. Test of a field theory of consciousness and social change: Time series analysis of participation in the TM-Sidhi program and reduction of violent death in the U.S. *Social Indicators Research* 22:399–418.
Dillbeck, M. C., C. B. Banus, C. Polanzi, and G. S. Landrith III. 1988. Test of a field model of consciousness and social change: The transcendental meditation and TM-Sidhi program and decreased urban crime. *Journal of Mind & Behavior* 9 (4): 457–86.
Dixon, N. 1981. *Preconscious processing.* New York: Wiley.
Dobyns, Y. H. 1996. Selection versus influence revisited: New methods and conclusions. *JSE* 10 (2): 253–68.
Dobyns, Y. H., B. J. Dunne, R. G. Jahn, and R. D. Nelson. 1992. Response to Hansen, Utts and Markwick: Statistical and methodological problems of the PEAR remote viewing [*sic*] experiments. *JP* 56 (2): 115–46.
Dossey, L. 1993. *Healing words.* San Francisco: HarperSanFrancisco.
Doyle, A. C. 1884. J. Habakuk Jephson's Statement. *Cornhill Magazine* (January). Reprinted in *The Conan Doyle stories* (1956). London: Murray.
Doyle, E. A. 1981. How parts fail. *IEEE Spectrum* (October), 36–43.
Do you believe in psychic phenomena? Are they likely to be able to explain consciousness? 1996. *The Times Higher Education Supplement* (April 5), p. v.
Druckman, D., and J. A. Swets, eds. 1988. *Enhancing human performance: Issues, theories, and techniques.* Washington, DC: National Academy Press.
Duane, T. D., and R. Behrendt. 1965. Extrasensory electroencephalographic induction between identical twins. *Science* 150:367.
Dunn, R. 1989. Computer electrocutes chess player who beat it! *Weekly World News* (March 14).
Dunne, B., and R. Jahn. 1992. Experiments in remote human/machine interaction. *JSE* 6:311–32.

Dunne, B. J. 1993. Co-operator experiments with an REG device. In *Cultivating consciousness: Enhancing human potential, wellness, and healing,* edited by K. R. Rao, 149–63. Westport, CT: Praeger.

———. 1995. Gender differences in engineering anomalies experiments. *Technical Note PEAR 95005*. Princeton Engineering Anomalies Research Laboratory, Princeton Univ. School of Engineering/Applied Science.

Dunne, B. J., Y. H. Dobyns, R. G. Jahn, and R. D. Nelson. 1994. Series position effects in random event generator experiments, with appendix by Angela Thompson. *JSE* 8:197–215.

Dunne, B. J., R. D. Nelson, and R. G. Jahn. 1988. Operator-related anomalies in a random mechanical cascade. *JSE* 2:155–80.

Dunne, J. W. 1927. *An experiment with time*. New York: Macmillan.

Ebon, M. 1983. *Psychic warfare*. New York: McGraw-Hill.

Eddington, A. 1928. *The nature of the physical world*. Cambridge: Cambridge Univ. Press.

Edge, H. L., R. L. Morris, J. H. Rush, and J. Palmer. 1986. *Foundations of parapsychology*. New York: Routledge & Kegan Paul.

Einstein, A., B. Podolsky, and N. Rosen. 1935. Can quantum-mechanical description of physical reality be considered complete? *Physical Review* 47:777–80.

Eisenberg, H., and D. C. Donderi. 1979. Telepathic transfer of emotional information in humans. *Journal of Psychology* 103:19–43.

Eisenbud, J. 1982. *Paranormal foreknowledge*. New York: Human Sciences Press.

Elitzur, A. 1995. Consciousness can no more be ignored: Reflections on Moody's dialog with zombies. *Journal of Consciousness Studies* 2:353–58.

Elworthy, F. T. 1958. *The evil eye*. New York: Julian Press. Originally published by John Murray, London, in 1895.

Engelhardt, H. T., and A. L. Caplan. 1987. *Scientific controversies: Case studies in the resolution and closure of disputes in science and technology*. New York: Cambridge Univ. Press.

Epstein, S. 1980. The stability of behavior. II. Implications for psychological research. *American Psychologist* 35:790–806.

Evans, C. H. 1984. Empirical truth and progress in science. *New Scientist* (January 26), 43–45.

Eysenck, H. J. 1957. *Sense and nonsense in psychology*. New York: Penguin.

———. 1966. Personality and extra-sensory perception. *JSPR* 44:55–71.

Fales, E., and B. Markovsky. Evaluating heterodox theories. Undated, unpublished manuscript.

Feinberg, G. 1975. Precognition—A memory of things future. In *Quantum physics and parapsychology*, edited by L. Oteri, 54–73. New York: Parapsychology Foundation.

Festinger, L. 1962. *A theory of cognitive dissonance*. Stanford, CA: Stanford Univ. Press.

Feyerabend, P. 1975. *Against method: Outline of an anarchistic theory of knowledge*. London: New Left Books.

Forman, R. K. C. 1994. "Of capsules and carts": Mysticism, language and the via negativa. *Journal of Consciousness Studies* 1:38–49.

Forrester, L. 1978. *Fly for your life*. 2d ed. New York: Bantam.

Fraser-Smith, A. C. 1982. Is there an increase of geomagnetic activity preceding total lunar eclipses? *Journal of Geophysical Research* 87 (A2): 895–98.

Freedman S. J., and J. F. Clauser. 1972. Experimental test of local hidden-variable theories. *Physical Review Letters* 28:938–41.

Frey, J., J. Rotton, and T. Barry. 1979. The effects of the full moon on human behavior: Yet another failure to replicate. *Journal of Psychology* 103:159–62.

Galante, M. A. 1986. Psychics: Lawyers using seers to help select juries, find missing children. *National Law Journal* 8 (January 27): 1.

Galle, T. J. R., and J. C. Hadni. 1984. Le pilote et la mort. *Annales Medico-Psychologiques* 142:236–40.

Gamow, G. 1959. The exclusion principle. *Scientific American* 201:74–86.

Gardner, M. 1983. *The whys of a philosophical scrivener.* New York: Quill.

Gardner, R. 1983. Miracles of healing in Anglo-Celtic Northumbria as recorded by the Venerable Bede and his contemporaries: A reappraisal in the light of twentieth century experience. *British Medical Journal* 287:1927–33.

Gauquelin, M. 1988. Is there a Mars effect? *JSE* 2 (1): 29–52.

Gazzaniga, M. S. 1988. *Mind matters: How mind and brain interact to create our conscious lives.* Boston: Houghton Mifflin.

Gearhart, L., and M. A. Persinger. 1986. Geophysical variables and behavior: XXXIII. Onsets of historical and contemporary poltergeist episodes occurred with sudden increases in geomagnetic activity. *Perceptual & Motor Skills* 62:463–66.

Gelderloos, P., M. J. Frid, P. H. Goddard, X. Xue, and S. A. Lliger. 1988. Creating world peace through the collective practice of the Maharishi Technology of the Unified Field: Improved U.S.–Soviet relations. *Social Science Perspectives Journal* 2 (4): 80–94.

Geller, S. H., and H. W. Shannon. 1976. The moon, weather, and mental hospital contacts: Confirmation and explanation of the Transylvanian effect. *Journal of Psychiatric Nursing & Mental Health Services* 14:13–17.

George, L. 1995. *Crimes of perception: An encyclopedia of heresies and heretics.* New York: Paragon House.

Gilovich, T. 1991. *How we know what isn't so: The fallibility of human reason in everyday life.* New York: Free Press.

Girden, E. 1962. A review of psychokinesis (PK). *Psychological Bulletin* 59:353–88.

Girden, E., and E. Girden. 1985. Psychokinesis: Fifty years afterward. In *A skeptic's handbook of parapsychology*, edited by P. Kurtz, 129–146. Buffalo, NY: Prometheus Books.

Girden, E., G. Murphy, J. Beloff, A. Flew, J. H. Rush, G. Schmeidler, and R. H. Thouless. 1964. A discussion of "A review of psychokinesis (PK)." *International JP* 6:26–137.

Gissurarson, L. R. 1992. The psychokinesis effect: Geomagnetic influence, age and sex differences. *JSE* 6:157–66.

Gittelson, B., and L. Torbet. 1987. *Intangible evidence.* New York: Simon & Schuster, Fireside Books.

Glanz, J. 1995. Measurements are the only reality, say quantum tests. *Science* 270:1439–40.

Glass, G. 1977. Integrating findings: The meta-analysis of research. *Review of Research in Education* 5:352–79.

Glass, G. V., B. McGaw, and M. L. Smith. 1981. *Meta-analysis in social research.* Beverly Hills, CA: Sage.

Gleick, J. 1987. *Chaos: Making a new science.* New York: Penguin.

Goerner, S. J. 1994. *Chaos and the evolving ecological universe.* Langhorne, PA: Gordon & Breach.

Grad, B. 1963. A telekinetic effect on plant growth. I. *International JP* 5:117–34.

———. 1965. Some biological effects of laying-on of hands: A review of experiments with animals and plants. *JASPR* 59:95–127.

Grad, B., R. J. Cadoret, and G. K. Paul. 1961. The influence of an unorthodox method of treatment on wound healing in mice. *International JP* 3:5–24.

Gravitz, M. A. 1979. Hypnotherapeutic management of epileptic behavior. *American Journal of Clinical Hypnosis* 21:282–84.

———. 1981. The production of warts by suggestion as a cultural phenomenon. *American Journal of Clinical Hypnosis* 23:281–83.

Greeley, A. 1987a. The "impossible": It's happening. *Noetic Sciences Review,* no. 2 (Spring).

———. 1987b. Mysticism goes mainstream. *American Health* 7:47–49.

———. 1991. The paranormal is normal: A sociologist looks at parapsychology. *JASPR* 85:367–74.

Grice, A. 1993. Full-moon madness. *Essence* (August), 67.

Griffen, D. R. 1988. Introduction: The reenchantment of science. In *The Reenchantment of Science,* edited by D. R. Griffen, 1–46. Albany: State Univ. of New York Press, 1988.

Griffin, D. 1988. Intuitive judgment and the evaluation of evidence. In *Enhancing human performance: Issues, theories, and techniques,* edited by D. Druckman and J. A. Swets, Background Papers, Part 1. Washington, DC: National Academy Press.

Grinberg-Zylberbaum, J., M. Delaflor, M. E. S. Arellano, M. A. Guevara, and M. Perez. 1992. Human communication and the electrophysiological activity of the brain. *SE* 3 (3): 25–44.

Grof, S. 1988. *The adventure of self-discovery.* Albany: State Univ. of New York Press.

Guiley, R. E. 1991. *Moonscapes: A celebration of lunar astronomy, magic, legend and lore.* New York: Prentice-Hall.

Gurney, E., F. W. H. Myers, and F. Podmore. 1970. *Phantasms of the living.* Gainesville, FL: Scholar's Facsimiles and Reprints. Originally published in 1886.

Hall, J., C. Kim, B. McElroy, and A. Shimony. 1977. Wave-packet reduction as a medium of communication. *Foundations of Physics* 7:759–67.

Hall, N. 1992. *The* New Scientist *guide to chaos.* London: Penguin.

Hameroff, S. R. 1994. Quantum coherence in microtubules: A neural basis for emergent consciousness? *Journal of Consciousness Studies* 1:91–118.

Hansel, C. E. M. 1966. *ESP: A scientific evaluation.* New York: Scribner.

———. 1980. *ESP and parapsychology: A critical re-evaluation.* Buffalo, NY: Prometheus Books.

Hansen, G. P. 1992a. CSICOP and the skeptics: An overview. *JASPR* 86:19–64.

———. 1992b. Magicians on the paranormal: An essay with a review of three books. *JASPR* 86:151–86.

Hansen, G. P., J. Utts, and B. Markwick. 1992. Critique of the PEAR remote-viewing experiments. *JP* 56 (2): 97–114.

Haraldsson, E. 1972. Vasomotor reactions as indicators of extrasensory perception. Ph.D. dissertation, Univ. of Freiburg, Germany.

Haraldsson, E., and L. R. Gissurarson. 1987. Does geomagnetic activity affect extrasensory perception? *Personality & Individual Differences* 8:745–47.

Haraldsson, E., and J. M. Houtkooper. 1991. Psychic experiences in the multinational human values study: Who reports them? *JASPR* 85:145–66.

Haraldsson, E., and T. Thorsteinsson. 1973. Psychokinetic effects on yeast: An exploratory experiment. In *RIP 1972,* edited by W. G. Roll, R. L. Morris, and J. O. Morris, 20–21. Metuchen, NJ: Scarecrow Press.

Harman, W. 1994. A re-examination of the metaphysical foundations of modern science: Why is it necessary? In *New metaphysical foundations of modern science,* edited by W. Harman and J. Clark, 1–13. Sausalito: CA: Institute of Noetic Sciences.

———. 1988. The transpersonal challenge to the scientific paradigm: The need for a restructuring of science. *ReVision* 11 (2): 13–21.

———. 1991. The epistemological foundations of science reconsidered. *New Ideas in Psychology* 9 (2): 187–95.

Harman, W., and H. Rheingold. 1984. *Higher creativity.* Los Angeles: Tarcher.

Harner, M. 1980. *The way of the shaman.* New York: Bantam.

Harris, M. J., and R. Rosenthal. 1988a. *Interpersonal expectancy effects and human performance research.* Washington, DC: National Academy Press.

———. 1988b. *Postscript to interpersonal expectancy effects and human performance research.* Washington, DC: National Academy Press.

———. 1988c. *Human performance research: An overview.* Washington, DC: National Academy Press.

Hart, H. 1956. Six theories about apparitions. *Proceedings of the Society for Psychical Research* 50:153–239.

Hart, H., and E. B. Hart. 1933. Visions and apparitions collectively reciprocally perceived. *Proceedings of the Society for Psychical Research* 41:205–49.

Hartman, B. O., and G. E. Secrist. 1991. Situational awareness is more than exceptional vision. *Aviation, Space, and Environmental Medicine* (November), 1084–89.

Hastings, A. C., and D. B. Hurt. 1976. A confirmatory remote viewing experiment in a group setting. *Proceedings of the IEEE* 64:1544–45.

Hawking, S. W. 1988. *A brief history of time.* New York: Bantam.

Hebb, D. O. 1951. The role of neurological ideas in psychology. *Journal of Personality* 20:39–55.

Hecht, H., and H. Dussault. 1987. Correlated failures in fault-tolerant computers. *IEEE Transactions on Reliability* R-36:171–75.

Hedges, L. V. 1987. How hard is hard science, how soft is soft science? The empirical cumulativeness of research. *American Psychologist* 42:443–55.

Heims, S. J. 1981. *John von Neumann and Norbert Wiener: From mathematics to the technologies of life and death.* Cambridge: MIT Press.

Herbert, N. 1985. *Quantum reality: Beyond the new physics.* Garden City, NY: Anchor Books.

———. 1993. *Elemental mind: Human consciousness and the new physics.* New York: Dutton.

Honorton, C. 1977. Psi and internal attention states. In *Handbook of parapsychology,* edited by B. B. Wolman, 435–72. New York: Van Nostrand Reinhold.

———. 1983. Response to Hyman's critique of psi ganzfeld studies. In *RIP 1982,* edited by W. G. Roll, J. Beloff, and R. A. White, 23–26. Metuchen, NJ: Scarecrow Press.

———. 1985. Meta-analysis of psi ganzfeld research: A response to Hyman. *JP* 49:51–91.

———. 1993. Rhetoric over substance: The impoverished state of skepticism. *JP* 57:191–214.

Honorton, C., R. E. Berger, M. P. Varvoglis, M. Quant, P. Derr, E. I. Schechter, and D. C. Ferrari. 1990. Psi communication in the ganzfeld: Experiments with an automated testing system and a comparison with a meta-analysis of earlier studies. *JP* 54:99–139.

Honorton, C., and D. C. Ferrari. 1989. Future telling: A meta-analysis of forced-choice precognition experiments, 1935–1987. *JP* 53:281–308.

Honorton, C., and S. Harper. 1974. Psi-mediated imagery and ideation in an experimental procedure for regulating perceptual input. *JASPR* 68:156–68.

Honorton, C., and E. I. Schechter. 1987. Ganzfeld target retrieval with an automated testing system: A model for initial ganzfeld success. In *RIP 1986,* edited by D. B. Weiner and R. D. Nelson, 36–39. Metuchen, NJ: Scarecrow Press.

Huntington, E. V. 1938. Is it chance or ESP? *American Scholar* 7:201–10.

Huxley, A. 1944. *The perennial philosophy.* New York: Harper Colophon.

Hyman, R. 1983. Does the ganzfeld experiment answer the critics' objections? In *RIP 1982,* edited by W. G. Roll, J. Beloff, and R. A. White, 21–23. Metuchen, NJ: Scarecrow Press.

———. 1984. From a transcript of *Nova* program 1101 (January 17). Boston: WGBH Transcripts.

———. 1985a. A critical overview of parapsychology. In *A skeptic's handbook of parapsychology,* edited by P. Kurtz, 1–96. Buffalo, NY: Prometheus Books.

———. 1985b. Parapsychological research: A tutorial review and critical appraisal. *Proceedings of the IEEE* 74:823–49.

———. 1985c. The ganzfeld psi experiment: A critical appraisal. *JP* 49:3–49.

———. 1991. Comment. *Statistical Science* 6:389–92.

———. 1996. Evaluation of a program on anomalous mental phenomena. *JSE* 10:31–58.

Hyman, R., and C. Honorton. 1986. A joint communiqué: The psi ganzfeld controversy. *JP* 50:351–64.

The Independent (October 22, 1903).

Iyengar, S., and J. Greenhouse. 1988. Selection models and the file drawer problem (with discussion). *Statistical Science* 3:109–35.

Jahn, R. G. 1981. *The role of the consciousness in the physical world.* AAAS Selected Symposium 57. Boulder, CO: Westview Press.

———. 1982. The persistent paradox of psychic phenomena: An engineering perspective. *Proceedings of the IEEE* 70:136–70.

Jahn, R. G., Y. H. Dobyns, and B. J. Dunne. 1991. Count population profiles in engineering anomalies experiments. *JSE* 5:205–32.

Jahn R. G., and B. J. Dunne. 1986. On the quantum mechanics of consciousness, with application to anomalous phenomena. *Foundations of Physics* 16:721–72.

———. 1987. *Margins of reality.* New York: Harcourt Brace Jovanovich.

James, W. 1956. *The will to believe and other essays in popular philosophy and human immortality.* New York: Dover.

———. 1961. *The varieties of religious experience.* New York: Collier Books.

———. 1977. *Human immortality.* Boston: Houghton-Mifflin. Originally published in 1898.

———. 1978. Upon the knees of the gods. In *The Signet handbook of parapsychology*, edited by M. Ebon, 43–53. New York: NAL Penguin.

Jeans, J. 1937. *The mysterious universe.* Cambridge: Cambridge Univ. Press.

———. 1948. *The mysterious universe.* New York: Macmillan.

Josephson, B. D. 1988. Limits to the universality of quantum mechanics. *Foundations of Physics* 18:1195–204.

Josephson, B. D., and F. Pallikari-Viras. 1991. Biological utilisation of quantum nonlocality. *Foundations of Physics* 21:197–207.

Joyce, C. R. B., and R. M. C. Welldon. 1965. The objective efficacy of prayer: A double-blind clinical trial. *Journal of Chronic Diseases* 18:367–77.

Jung, C. G. 1960. *The structure and dynamics of the psyche.* Collected Works, vols. 2 and 8. London: Trevor Hull.

———. 1978. *Basic principles of analytic psychology.* Collected Works, vol. 8. Princeton, NJ: Princeton Univ. Press.

Kaplan, A. 1964. *The conduct of inquiry: Methodology for behavioral science.* Scranton, PA: Chandler.

Keller, E., and V. M. Bzdek. 1986. Effects of therapeutic touch on tension headache pain. *Nursing Research* 35 (2): 101–6.

Kelly, G. A. 1955. *The psychology of personal constructs.* Vol. 1: *A theory of personality.* New York: Norton.

Kelly, I. W., J. Rotton, and R. Culver. 1985–86. The moon was full and nothing happened: A review of studies on the moon and human behavior and lunar beliefs. *The Skeptical Inquirer* 10:129–43.

Kennedy, J. L., and H. F. Uphoff. 1939. Experiments on the nature of extrasensory perception: III. The recording error criticism of extra-chance results. *JP* 3:226–45.

Keshavan, M. S., B. N. Gangadhar, R. U. Gautam, V. B. Ajit, and R. L. Kapur. 1981. Convulsive threshold in humans and rats and magnetic field changes: Observations during total solar eclipse. *Neuroscience Letters* 22:205–8.

Kihlstrom, J. F. 1987. The cognitive unconscious. *Science* 237:1445–52.

Klintman, H. 1983. Is there a paranormal (precognitive) influence in certain types of perceptual sequences? Part I. *EJP* 5:19–49.

———. 1984. Is there a paranormal (precognitive) influence in certain types of perceptual sequences? Part II. *EJP* 5:125–40.

Koestler, A., and J. R. Smythies. 1969. *Beyond reductionism: New perspectives in the life sciences.* London: Hutchinson.

Kohmann, K. J., and A. O. D. Willows. 1987. Lunar-modulated geomagnetic orientation by a marine mollusk. *Science* 235:331–34.

Kohn, A. 1988. *False prophets: Fraud and error in science and medicine.* Oxford: Blackwell.

Krieger, D. 1986. *The therapeutic touch. How to use your hands to help or to heal.* New York: Prentice-Hall.

Krutch, J. K. 1956. *The modern temper: A study and a confession.* New York: Harcourt, Brace & World, Harvest Books.

Kuang, Ankun K., et al. 1986. Long-term observation on Qigong in prevention of stroke—Follow-up of 244 hypertensive patients for 18–22 years. *Journal of Traditional Chinese Medicine* 6 (4): 235–38.

Kuhn, T. S. 1970. *The structure of scientific revolutions.* Chicago: Univ. of Chicago Press.

Kurtz, P. 1981. Is parapsychology a science? In *Paranormal Borderlands of Science,* edited by K. Frazier, 5–23. Buffalo, NY: Prometheus Books.

Landauer, R. 1996. Minimal energy requirements in communication. *Science* 272:1914–18.

Laszlo, Irwin. 1993. *The creative cosmos.* Edinburgh, Scotland: Floris Books.

———. 1995. *The interconnected universe: Conceptual foundations of transdisciplinary unified theory.* River Edge, NJ: World Scientific.

Lawrence, T. 1993. Bringing in the sheep: A meta-analysis of sheep/goat experiments. In *Proceedings of Presented Papers,* 36th Annual Parapsychological Association Convention, edited by M. J. Schlitz. Fairhaven, MA: Parapsychological Association.

Leggett, A. J. 1987. Reflections on the quantum measurement paradox. In *Quantum implications,* edited by B. J. Hiley and F. D. Peat, 85–104. London: Routledge & Kegan Paul.

Lehman, R. E. 1978. Brief hypnotherapy of neurodermatitis: A case with four-year followup. *American Journal of Clinical Hypnosis* 21:48–51.

LeShan, L. 1973. What is important about the paranormal? In *The nature of human consciousness: A book of readings,* edited by R. E. Ornstein, 458–467. New York: Viking.

———. 1974. *The medium, the mystic, and the physicist.* New York: Viking.

———. 1982. *Clairvoyant reality.* Wellingborough, Northants: Turnstone.

———. 1987. *The science of the paranormal.* Wellingborough, Northants: Aquarian.

LeShan, L., and H. Margenau. 1982. *Einstein's space and Van Gogh's sky.* New York: Macmillan.

Levine, A., C. Fenyvesi, and S. Emerson. 1988. The twilight zone in Washington. *U.S. News & World Report* (December 5), 24–30.

Lewicki, D. R., G. H. Schaut, and M. A. Persinger. 1987. Geophysical variables and behavior: XLIV. Days of subjective precognitive experiences and the days before the actual events display correlated geomagnetic activity. *Perceptual & Motor Skills* 65:173–74.

Libet, B. 1994. A testable field theory of mind-brain interaction. *Journal of Consciousness Studies* 1 (1): 119–26.

Lieber, A. L. 1978. *The lunar effect: Biological tides and human emotions.* Garden City, NY: Doubleday.

Lieber, A. L., and C. R. Sherin. 1972. Homicides and the lunar cycle. *American Journal of Psychiatry* 129:69–74.

Lindley, D. 1996. *Where does the weirdness go? Why quantum mechanics is strange, but not as strange as you think.* New York: Basic Books.

Lindo, O. R. 1985. U interessante caso di sollievo dalle sofferenze dovute al cancro, ottenuto con ipnoterapia [An interesting case of relief of suffering caused by cancer through hypnotherapy]. *Rivista Internazionale di Psicologia e Ipnosi* 26:287–90.

Lockwood, M. 1989. *Mind, brain and the quantum. The compound "I."* New York: B. Blackwell.

Lodge, Sir O. 1884. An account of some experiments in thought-transference. *Proceedings of the Society for Psychical Research* 2:189–200.

Loehr, F. 1969. *The power of prayer on plants.* New York: Signet.

Lovelock, J. E. 1979. *Gaia, a new look at life on earth.* Oxford: Oxford Univ. Press.

———. 1990. Commentary on the Gaia hypothesis. *Nature* 344:100–102.

Lyons, A., and M. Truzzi. 1990. *The blue sense: Psychic detectives and crime.* New York: Mysterious Press.

Machlup, F., and U. Mansfield. 1983. Cultural diversity in studies of information. In *The study of information: Interdisciplinary messages,* edited by F. Machup and U. Mansfield. New York: Wiley.

Makarec, K., and M. A. Persinger. 1987. Geophysical variables and behavior: XLIII. Negative correlation between accuracy of card-guessing and geomagnetic activity: A case study. *Perceptual & Motor Skills* 65:105–6.

Mann, C. 1990. Meta-analysis in the breech. *Science* 249:476–80.

Mansfield, V. 1995. *Synchronicity, science, and soul-making.* Chicago: Open Court.

Mansfield, V., and J. M. Spiegelman. 1996. On the physics and psychology of the transference as an interactive field. *Journal of Analytical Psychology.* Downloaded from the World Wide Web, *lightlink.com/vic/field.html.*

Marks, D. F. 1986. Investigating the paranormal. *Nature* 320:119–24.

Marks, D. F., and R. Kammann. 1980. *The psychology of the psychic.* Buffalo, NY: Prometheus Books.

Martin, S. J., I. W. Kelly, and D. H. Saklofske. 1992. Suicide and lunar cycles: A critical review over 28 years. *Psychological Reports* 71:787–95.

May, E. C. 1995. AC technical trials: Inspiration for the target entropy concept. In *Proceedings of Presented Papers,* 38th Annual Parapsychological Association Convention, edited by N. L. Zingrone, 193–211. Fairhaven, MA: The Parapsychological Association.

———. 1996. The American Institutes for Research Review of the Department of Defense's STAR GATE program: A commentary. *JSE* 10:89–108.

May, E. C., B. S. Humphrey, and G. S. Hubbard. 1980. *Electronic system perturbation techniques.* SRI International Final Report (September 30).

May, E. C., R. Targ, and H. E. Puthoff. 1979. EEG correlates to remote light flashes under conditions of sensory shielding. In *Mind at large,* edited by C. T. Tart, H. E. Puthoff, and R. Targ, 127–36. New York: Praeger.

May, E. C., J. M. Utts, and S. J. P. Spottiswoode. 1995. Decision augmentation theory: Application to the random number generator database. *JSE* 9:453–88.

May, E. C., J. M. Utts, V. V. Trask, W. W. Luke, T. J. Frivold, and B. S. Humphrey. 1988. Review of the psychoenergetic research conducted at SRI International (1973–1988). SRI International Technical Report (March).

McCarthy, R. L. 1988. Present and future safety challenges of computer control. *Computer Assurance: COMPASS '88* (IEEE Catalog No. 88CH2628–6), 1–7. New York: IEEE.

McClenon, J. 1982. A survey of elite scientists: Their attitudes toward ESP and parapsychology. *JP* 46:127–52.

McConnell, R. A., and T. K. Clark. 1991. National Academy of Sciences' Opinion on Parapsychology. *JASPR* 85:333–66.

McCrone, J. 1993. Roll up for the telepathy test. *New Scientist* (May 15), 29–33.

———. 1994. Psychic powers, what are the odds? *New Scientist* (November 26), 35–38.

McMoneagle, J. 1993. *Mind trek.* Charlottesville, VA: Hampton Roads.

Meehl, P. E., and M. Scriven. 1956. Compatibility of science and ESP. *Science* 123:14–15.

Mermin, D. 1985. Is the moon there when nobody looks? Reality and the quantum theory. *Physics Today* (April), 38–47.

Metzger, W. 1930. Optische Untersuchungen am Ganzfeld: II. Zur Phänomenologie des homogenen Ganzfelds [Optical investigation of the Ganzfeld: II. Toward the phenomenology of the homogeneous Ganzfeld]. *Psychologische Forschung* 13:6–29.

Michel, F. C., A. J. Dessler, and G. K. Walters. 1964. A search for correlation between K_p and the lunar phase. *Journal of Geophysical Research* 69 (19): 4177–81.

Miller, R. N. 1982. Study of remote mental healing. *Medical Hypotheses* 8:481–90.

Milne, G. 1983. Hypnotherapy with migraine. *Australian Journal of Clinical & Experimental Hypnosis* 11 (3): 23–32.

Milne, G., and D. Adlridge. 1981. Remission with autohypnotic relaxation in a case of longstanding hypertension. *Australian Journal of Clinical & Experimental Hypnosis* 9:77–86.

Milton, J. 1993. Ordinary state ESP meta-analysis. In *Proceedings of Presented Papers,* 36th Annual Parapsychological Association, edited by M. J. Schlitz, 87–104. Fairhaven, MA: Parapsychological Association.

Milton, R. 1994. *Forbidden science.* London: Fourth Estate.

Mishlove, J. 1993. *The roots of consciousness.* Tulsa, OK: Council Oak Books.

Mitchell, E. D. 1979. A look at the exceptional. In *Mind at large,* edited by C. T. Tart, H. E. Puthoff, and R. Targ, 1–10. New York: Praeger.

Moon madness no myth. 1981. *Science Digest* (March), 7–8.

Moreover: Science does it with feeling. 1996. *The Economist* (July 20), 71–73.

Morowitz, H. J. 1980. Rediscovering the mind. *Psychology Today* 14 (3): 12.

Morris, R. L. 1977. Parapsychology, biology and anpsi. In *Handbook of parapsychology,* edited by B. B. Wolman, 687–716. New York: Van Nostrand Reinhold.

———. 1983. Applied psi in the context of human-equipment interaction systems. In *Proceedings: Symposium on applications of anomalous phenomena,* 127–58. November 30–December 1, Leesburg, VA.

———. 1986. Psi and human factors: The role of psi in human-equipment interactions. In *Current trends in psi research,* edited by B. Shapin and L. Coly, 1–26. New York: Parapsychology Foundation.

Morris, R. L., K. Dalton, D. Delanoy, and C. Watt. 1995. Comparison of the sender/no sender condition in the ganzfeld. In *Proceedings of Presented Papers,* 38th Annual Parapsychological Association Convention, edited by N. L. Zingrone, 244–59. Fairhaven, MA: Parapsychological Association.

Mullen, B., and R. Rosenthal. 1985. *BASIC meta-analysis procedures and programs.* Hillsdale, NJ: Erlbaum.

Murphy, G. 1962. Report on paper by Edward Girden on psychokinesis. *Psychological Bulletin* 59:520–28.

Nash, C. B. 1982. Psychokinetic control of bacterial growth. *JSPR* 51:217–21.

———. 1984. Test of psychokinetic control of bacterial mutation. *JASPR* 78:145–52.

Nash, C. B., and C. S. Nash. 1967. The effect of paranormally conditioned solution on yeast fermentation. *JP* 31:314.

Neimark, J. 1996. Do the spirits move you? *Psychology Today* 29 (5): 48.

Nelson, R. D. 1996. Wishing for good weather: A natural experiment in group consciousness. *Technical Note PEAR 96001* (June). Princeton Engineering Anomalies Research Laboratory, Princeton Univ. School of Engineering/Applied Science.

Nelson, R. D., G. J. Bradish, Y. H. Dobyns, B. J. Dunne, and R. G. Jahn. 1996. FieldREG anomalies in group situations. *JSE* 10:111–42.

Nelson, R. D., Y. H. Dobyns, B. J. Dunne, and R. G. Jahn. 1991. Analysis of variance of REG experiments: Operator intention, secondary parameters, database structure. *Technical Note PEAR 91004*. Princeton Engineering Anomalies Research Laboratory, Princeton Univ. School of Engineering/Applied Science.

Nelson, R. D., B. J. Dunne, and R. G. Jahn. 1984. An REG experiment with large database capability, III: Operator related anomalies. *Technical Note PEAR 84003* (September). Princeton Engineering Anomalies Research Laboratory, Princeton Univ. School of Engineering/Applied Science.

Nelson, R. D., and E. L. Mayer. 1996 (in preparation). A fieldREG application at the San Francisco Bay Revels.

Nelson, R. D., and D. I. Radin. 1987. When immovable objections meet irresistible evidence. *Behavioral & Brain Sciences* 10:600–601.

———. 1989. Statistically robust anomalous effects: Replication in random event generator experiments. In *RIP 1988*, edited by L. Henckle and R. E. Berger, 23–26. Metuchen, NJ: Scarecrow Press.

———. 1990. Effects of human intention on random event generators: A meta-analysis. *Proceedings & Abstracts of the Eastern Psychological Association Conference*, March 29–April 1, 1990, Philadelphia.

Neuliep, J. W., and R. Crandall. 1991. Editorial bias against replication research. In *Replication research in the social sciences*, edited by J. W. Neuliep, 85–90. Newbury Park, CA: Sage.

News and comment: Academy helps army be all that it can be. 1987. *Science* 238:1502.

News and comment. 1981. *Brain/Mind Bulletin* (February 16), 1.

News and comment: Science beyond the pale. 1990. *Science* 249:14–16.

Nichols, B. 1966. *The powers that be*. New York: St. Martin's Press.

Nisbett, R., and L. Ross. 1980. *Human inference: Strategies and shortcomings of social judgment*. Englewood Cliffs, NJ: Prentice-Hall.

Noble, H. B. 1988. *Next: The coming era in science*. Boston: Little, Brown.

Office of Technology Assessment. 1989. Report of a workshop on experimental parapsychology. *JASPR* 83:317–39.

Oliven, J. F. 1943. Moonlight and nervous disorders: A historical study. *American Journal of Psychiatry* 99:579–84.

O'Regan, B. 1987. *Healing, remission and miracle cures*. Sausalito, CA: Institute of Noetic Sciences.

Orme-Johnson, D. No date. Summary of Research on the Transcendental Meditation Program, Compiled and Edited. Downloaded in January 1997 from the Maharishi International University site on the World Wide Web, *www.maharishi.org/tm/research/summary.html*.

Orme-Johnson, D., M. C. Dillbeck, R. K. Wallace, and G. S. Landrith III. 1982. Intersubject EEG coherence: Is consciousness a field? *International Journal of Neuroscience* 16:203–9.

Orme-Johnson, D. W., C. N. Alexander, J. L. Davies, H. M. Chandler, and W. E. Larimore. 1988. International peace project in the Middle East: The effects of the Maharishi technology of the unified field. *Journal of Conflict Resolution* 32:776–812.

Otani, S. 1955. Relations of mental set and change of skin resistance to ESP score. *JP* 19 (3): 164–70.

Palmer, J. 1985. *An evaluative report on the current status of parapsychology*. U.S. Army Research Institute, European Science Coordination Office, Contract No. DAJA 45-84-M-0405.

Palmer, J. A., C. Honorton, and J. Utts. 1989. Reply to the National Research Council study on parapsychology. *JASPR* 83:31–49.
Parker, A. 1975. Some findings relevant to the change in state hypothesis. In *RIP 1974*, edited by J. D. Morris, W. G. Roll, and R. L. Morris, 40–42. Metuchen, NJ: Scarecrow Press.
Parker, A., A. Frederiksen, and H. Johansson. In press. Towards specifying the recipe for success with the ganzfeld: Replication of the ganzfeld findings using a manual ganzfeld with subjects reporting subjective paranormal experiences. *EJP* 16.
Peale, N. V. 1956. *The power of positive thinking.* Norwalk, CT: C. R. Gibson.
Peat, F. D. 1987. *Synchronicity: The bridge between matter and mind.* New York: Bantam.
———. 1991. *The philosopher's stone: Chaos, synchronicity and the hidden order of the world.* New York: Bantam.
Pelegrin, M. J. 1988. Computers in planes and satellites. In *Proceedings of the IFAC Symposium,* edited by W. D. Ehrenberger, 121–32. Oxford: Pergamon Press.
Penrose, R. 1989. *The emperor's new mind.* Oxford: Oxford Univ. Press.
———. 1994. *Shadows of the mind.* Oxford: Oxford Univ. Press.
Persinger, M. A. 1985. Geophysical variables and behavior: XXX. Intense paranormal experiences occur during days of quiet, global, geomagnetic activity. *Perceptual & Motor Skills* 61:320–22.
———. 1987. Spontaneous telepathic experiences from *Phantasms of the Living* and low global geomagnetic activity. *JASPR* 81:23–36.
———. 1989a. Psi phenomena and temporal lobe activity: The geomagnetic factor. In *RIP 1988,* edited by L. A. Henkel and R. E. Berger, 121–56. Metuchen, NJ: Scarecrow Press.
———. 1989b. Increases in geomagnetic activity and the occurrence of bereavement hallucinations: Evidence for melantin mediated microseizures in the temporal lobe? *Neuroscience Letters* 88:271–74.
Persinger, M. A., and S. Krippner. 1989. Dream ESP experiences and geomagnetic activity. *JASPR* 83:101–16.
Persinger, M. A., and G. B. Schaut. 1988. Geomagnetic factors in subjective telepathic, precognitive and postmortem experiences. *JASPR* 82:217–35.
Peterson, D. M. 1978. Through the looking-glass: An investigation of the faculty of extra-sensory detection of being stared at. Unpublished thesis, Univ. of Edinburgh, Scotland.
Phillips, S. M. 1995. Extrasensory perception of subatomic particles. I. Historical evidence. *JSE* 9:489–525.
Pinch, T. J., and H. M. Collins. 1984. Private science and public knowledge: The Committee for the Scientific Investigation of the [*sic*] Claims of the Paranormal and its use of the literature. *Social Studies of Science* 14:521–46.
Playfair, G. L., and S. Hill. 1978. *The cycles of heaven.* New York: St. Martin's Press.
Pleass, C. M., and D. Dey. 1990. Conditions that appear to favor extrasensory interactions between homo sapiens and microbes. *JSE* 4:213–31.
Plotkin, W. B. 1980. The role of attributions of responsibility in the facilitation of unusual experiential states during alpha training: An analysis of the biofeedback placebo effect. *Journal of Abnormal Psychology* 89:67–78.
Pokorny, A. D., and J. Jackimczyk. 1974. The questionable relationship between homicides and the lunar cycle. *American Journal of Psychiatry* 131:827–29.
Polanyi, M. 1961. *The study of man.* Chicago: Univ. of Chicago Press.
Poortman, J. J. 1959. The feeling of being stared at. *JSPR* 40:4–12.

Popper, K. R., and J. C. Eccles. 1981. *The self and its brain.* New York: Springer International.

Prasad, J., and I. Stevenson. 1968. A survey of spontaneous psychical experiences in school children of Uttar Pradesh, India. *International JP* 10:241–61.

Pratt, J. G., J. B. Rhine, B. M. Smith, C. E. Stuart, and J. A. Greenwood. 1966. *Extra-sensory perception after sixty years.* Boston: Bruce Humphries. Originally published in 1940.

Price, D. J. 1984. Of sealing wax and string. *Natural History* 1:49–56.

Price, G. R. 1955. Science and the supernatural. *Science* 122:359–67.

Psychic Powers. 1987. In *Mysteries of the unknown.* Richmond, VA: Time-Life Books.

Puharich, A. 1973. *Beyond telepathy.* Garden City, NY: Anchor Books.

Puthoff, H. E. 1996. CIA-initiated remote viewing program at Stanford Research Institute. *JSE* 10:63–76.

Puthoff, H. E., and R. Targ. 1976. A perceptual channel for information transfer over kilometer distances: Historical perspective and recent research. *Proceedings of the IEEE* 64:329–54.

Putnam, F. W. 1984. The psychophysiologic investigation of multiple personality disorder. *Psychiatric Clinics of North America* 7 (1): 31–39.

———. 1991. Recent research on multiple personality disorder. *Psychiatric Clinics of North America* 14 (3): 489–502.

Putnam, F. W., T. P. Zahn, and R. M. Post. 1990. Differential autonomic nervous system activity in multiple personality disorder. *Psychiatry Research* 31 (3): 251–60.

Quinn, J. F. 1984. Therapeutic touch as energy exchange: Testing the theory. *Advances in Nursing Science* 6 (2): 42–49.

Radin, D. I. 1982. Experimental attempts to influence pseudorandom number sequences. *JASPR* 76:359–74.

———. 1989. Searching for "signatures" in anomalous human-machine interaction research: A neural network approach. *JSE* 3:185–200.

———. 1990a. Do you have executive ESP? *Leaders* 13 (July): 123–24.

———. 1990b. Testing the plausibility of psi-mediated computer system failures. *JP* 54:1–19.

———. 1990–91. Statistically enhancing psi effects with sequential analysis: A replication and extension. *EJP* 8:98–111.

———. 1992. Beyond belief: Exploring interactions among mind, body and environment. *SE* 2 (3): 1–40.

———. 1993a. Environmental modulation and statistical equilibrium in mind-matter interaction. *SE* 4 (1): 1–30.

———. 1993b. Neural network analyses of consciousness-related patterns in random sequences. *JSE* 7 (4): 355–74.

———. 1993c. *The Video Ganzfeld System.* Technical Report, Koestler Chair of Parapsychology, Dept. of Psychology, Univ. of Edinburgh.

———. 1994a. Hi-tech consciousness. *Retreat Magazine* 5:19–21.

———. 1994b. On complexity and pragmatism. *JSE* 8:523–34.

———. 1996. Unconscious perception of future emotions. *Journal of Consciousness Studies Abstracts,* Tucson II conference, Univ. of Arizona, Tucson.

———. In press. Unconscious perception of future emotions. *JSE.*

Radin, D. I., and G. Bisaga. 1994. Towards a high technology of the mind. In *RIP 1991,* edited by E. W. Cook and D. Delanoy, 24–28. Metuchen, NJ: Scarecrow Press.

Radin, D. I., and D. C. Ferrari. 1991. Effects of consciousness on the fall of dice: A meta-analysis. *JSE* 5:61–84.

Radin, D. I., S. McAlpine, and S. Cunningham. 1994. Geomagnetism and psi in the ganzfeld. *JSPR* 59:352–63.

Radin, D. I., and R. D. Nelson. 1988. Repeatable evidence for anomalous human-machine interactions. In *Paranormal Research,* edited by M. L. Albertson, D. S. Ward, and K. P. Freeman, 306–17. Fort Collins, CO: Rocky Mountain Research Institute.

———. 1989. Evidence for consciousness-related anomalies in random physical systems. *Foundations of Physics* 19:1499–1514.

Radin, D. I., and J. M. Rebman. 1994. Lunar correlates of normal, abnormal and anomalous human behavior. *SE* 5 (3):209–238.

———. 1996. Are phantasms fact or fantasy? A preliminary investigation of apparitions evoked in the laboratory. *JSPR* 61:65–87.

———. In press a. Seeking psi in the casino. *JSPR.*

———. In press b. Towards a complex systems model of psi performance. *SE.*

Radin, D. I., J. M. Rebman, and M. P. Cross. 1996. Anomalous organization of random events by group consciousness: Two exploratory experiments. *JSE* 10:143–68.

Radin, D. I., R. D. Taylor, and W. Braud. 1995. Remote mental influence of human electrodermal activity: A pilot replication. *EJP* 11:19–34.

Radin, D. I., and J. M. Utts. 1989. Experiments investigating the influence of intention on random and pseudorandom events. *JSE* 3:65–79.

Randall, J. L. 1970. An attempt to detect psi effects with protozoa. *JSPR* 45.

Rao, K. R., and J. Palmer. 1987. The anomaly called psi: Recent research and criticism. *Behavioral & Brain Sciences* 10:539–51.

Rassbach, M. E., and A. J. Dessler. 1966. The lunar period, the solar period, and K_p. *Journal of Geophysical Research* 71 (17): 4141–45.

Rebman, J. M., D. I. Radin, R. A. Hapke, and K. Gaughan. 1996. Remote influence of the autonomic nervous system by focused intention. In *Proceedings of Presented Papers,* 39th Annual Parapsychological Association Convention, edited by E. C. May, 133–148. Fairhaven, MA: Parapsychological Association.

Rebman, J. M., R. Wezelman, D. I. Radin, P. Stevens, R. Hapke, and K. Gaughan. 1995. Remote influence of human physiology by a ritual healing technique. *SE* 6: 111–134.

Reed, M. 1989. The quantum transistor. *Byte* (May), 275–81.

Rhine, J. B. 1944. "Mind over matter" or the PK effect. *JASPR* 38:185–201.

———. 1964. *Extra-sensory perception.* Boston: Bruce Humphries.

———. 1969. Position effects in psi test results. *JP* 33.

———. 1977. History of experimental studies. In *Handbook of parapsychology,* edited by B. B. Wolman, 25–47. New York: Van Nostrand Reinhold.

Rhine, L. E. 1961. *Hidden channels of the mind.* New York: Sloane.

———. 1981. *The invisible picture: A study of psychic experiences.* Jefferson, NC: McFarland.

Richmond, N. 1952. Two series of PK tests on paramecia. *JSPR* 36:577–78.

Rinpoche, S. 1992. *The Tibetan book of living and dying.* San Francisco: HarperSanFrancisco.

Rohrlich, F. 1983. Facing quantum mechanical reality. *Science* 221:1251–55.

Roig, M., H. Icochea, and A. Cuzzucoli. 1991. Coverage of parapsychology in introductory psychology textbooks. *Teaching of Psychology* 18 (3): 157–60.

Roll, W. G. 1977. Poltergeists. In *Handbook of parapsychology,* edited by B. B. Wolman, 577–630. New York: Van Nostrand Reinhold.

———. 1994. Are ghosts really poltergeists? In *Proceedings of Presented Papers,* 37th Annual Parapsychological Association Convention, edited by D. J. Bierman, 347–51. Fairhaven, MA: Parapsychological Association.

Roney-Dougal, S. 1993. *Where science and magic meet.* Rev. ed. Rockport, MA: Element.

Roney-Dougal, S. M., and G. Vogl. 1993. Some speculations on the effect of geomagnetism on the pineal gland. *JSPR* 59:1–15.

Rosen, D. H., S. M. Smith, H. L. Huston, and G. Gonzalez. 1991. Empirical study of associations between symbols and their meanings: Evidence of collective unconscious (archetypal) memory. *Journal of Analytical Psychology* 36:211–29.

Rosenfeld, A. H. 1975. The particle data group: Growth and operations. *Annual Review of Nuclear Science* 25:555–99.

Rosenthal, R. 1978. Combining results of independent studies. *Psychological Bulletin* 85:185–93.

———. 1979. The "file drawer problem" and tolerance for null results. *Psychological Bulletin* 86:638–41.

———. 1990. Replication in behavioral research. *Journal of Social Behavior & Personality* 5:1–30.

———. 1991. *Meta-analytic procedures for social research.* Rev. ed. Newbury Park, CA: Sage.

Rosenthal, R., and L. Jacobson. 1968. *Pygmalion in the classroom: Teacher expectation and pupils' intellectual development.* New York: Holt, Rinehart & Winston.

Rosenthal, R., and R. Rosnow. 1984. *Essentials of behavioral research: Methods and data analysis.* New York: McGraw-Hill.

Rosenthal R., and D. B. Rubin. 1978. Interpersonal expectancy effects: The first 345 studies. *Behavioral & Brain Sciences* 3:377–415.

Rosnow, R. L., and R. Rosenthal. 1989. Statistical procedures and the justification of knowledge in psychological science. *American Psychologist* 44:1276–84.

Ross, L., and M. R. Lepper. 1980. The perseverance of beliefs: Empirical and normative considerations. In *Fallible judgment in behavioral research: New directions for methodology of social and behavioral science,* edited by R. A. Shweder. San Francisco: Jossey-Bass.

Rotton, J., and J. Frey. 1985. Air pollution, weather, and violent crimes: Concomitant time-series analysis of archival data. *Journal of Personality & Social Psychology* 49:1207–20.

Rotton, J., and I. W. Kelly. 1985. Much ado about the full moon: A meta-analysis of lunar-lunacy research. *Psychological Bulletin* 97:286–306.

Sagan, C. 1995. *The demon haunted world.* New York: Random House.

Saunders, D. R. 1985. On Hyman's factor analyses. *JP* 49:86–88.

Schacter, D. L. 1996. *Searching for memory.* New York: Basic Books.

Schaut, G. B., and M. A. Persinger. 1985. Geophysical variables and behavior: XXXI. Global geomagnetic activity during spontaneous paranormal experiences: A replication. *Perceptual & Motor Skills* 61:412–14.

Schechter, E. I. 1984. Hypnotic induction vs. control conditions: Illustrating an approach to the evaluation of replicability in parapsychology. *JASPR* 78:1–27.

Schlitz, M. J., and C. Honorton. 1992. Ganzfeld psi performance within an artistically gifted population. *JASPR* 86:83–98.

Schlitz, M. J., and S. LaBerge. 1994. Autonomic detection of remote observation: Two conceptual replications. In *Proceedings of Presented Papers,* 37th Annual Parapsychological Association Convention, edited by D. J. Bierman, 352–60. Fairhaven, MA: Parapsychological Association.

Schlitz, M. J., and R. Wiseman. 1996. Further studies on remote staring. Paper presented at the 15th annual meeting of the Society for Scientific Exploration, Univ. of Virginia, May.

Schmeidler, G. R. 1943. Predicting good and bad scores in a clairvoyance experiment: A preliminary report. *JASPR* 37:103–10.

———. 1988. *Parapsychology and psychology: Matches and mismatches.* Jefferson, NC: McFarland.

Schmidt, H. 1969. Precognition of a quantum process. *JP* 33:99–108.

———. 1970. Mental influence on random events. *New Scientist & Science Journal* 50:757–58.

———. 1975. Toward a mathematical theory of psi. *JASPR* 69:301–19.

———. 1981. PK tests with pre-recorded and pre-inspected seed numbers. *JP* 45:87–98.

———. 1987. The strange properties of psychokinesis. *JSE* 1:103–18.

———. 1993a. New PK tests with an independent observer. *JP* 57:227–40.

———. 1993b. Observation of a psychokinetic effect under highly controlled conditions. *JP* 57:351–72.

Schmidt, H., R. Morris, and L. Rudolph. 1986. Channeling evidence for a PK effect to independent observers. *JP* 50:1–16.

Schmidt, H., and M. J. Schlitz. 1989. A large scale pilot PK experiment with prerecorded random events. In *RIP 1988,* edited by L. A. Henkel and R. E. Berger, 6–10. Metuchen, NJ: Scarecrow Press.

Schnabel, J. 1997. *Remote viewers: The secret history of America's psychic spies.* New York: Dell.

Schouten, S. A. 1976. Autonomic psychophysiological reactions to sensory and emotive stimuli in a psi experiment. *EJP* 1:72–78.

———. 1993. Applied parapsychology: Studies of psychics and healers. *JSE* 7:375–402.

Schrödinger, E. 1964. *My view of the world.* Cambridge: Cambridge Univ. Press.

———. 1967. *What is life?* Cambridge: Cambridge Univ. Press.

Scientific American (January 13, 1906).

Shallis, M. 1982. *On time.* New York: Schocken Books.

Sheehan, D. V. 1978. Influence of psychosocial factors on wart remission. *American Journal of Clinical Hypnosis* 20:160–64.

Sheldrake, R. 1981. A *new science of life: The hypothesis of formative causation.* Los Angeles: Tarcher.

———. 1995. *Seven experiments that could change the world.* New York: Riverhead Books.

Shimony, A. 1963. Role of the observer in quantum theory. *American Journal of Physics* 31:755.

Shneiderman, B. 1987. *Designing the user interface: Strategies for effective human-computer interaction.* Reading, MA: Addison-Wesley.

Sinclair, U. B. 1962. *Mental radio.* Rev. 2d ed. Springfield, IL: Charles C. Thomas. Originally published in 1930.

Skinner, B. F. 1972. *Beyond freedom and dignity.* New York: Vintage Books.

Smart, J. C. C. 1979. Materialism. In *The mind-brain identity theory*, edited by C. V. Borst. London: Macmillan.
Smith, R. 1968. Unpublished ms., MIT.
Snel, F., and P. R. Hol. 1983. Psychokinesis experiments in casein induced amyloidosis of the hamster. *EJP* 5 (1): 51–76.
Snoyman, P., and T. L. Holdstock. 1980. The influence of the sun, moon, climate and economic conditions on crisis incidence. *Journal of Clinical Psychology* 36:884–93.
Solfvin, G. F. 1982. Psi expectancy effects in psychic healing studies with malarial mice. *EJP* 4 (2): 160–97.
Solfvin, J. 1984. Mental healing. In *Advances in parapsychological research*, edited by S. Krippner, 4:31–63. Jefferson, NC: McFarland.
Sperry, R. 1987. Structure and significance of the consciousness revolution. *Journal of Mind & Behavior* 8:37–66.
Spottiswoode, S. J. P. 1990. Geomagnetic activity and anomalous cognition: A preliminary report of new evidence. *SE* 1:65–77.
Squires, E. J. 1987. Many views of one world—An interpretation of quantum theory. *European Journal of Physics* 8:173.
Stanford, R. G. 1987. Ganzfeld and hypnotic-induction procedures in ESP research: Toward understanding their success. In *Advances in parapsychological research*, edited by S. Krippner, 5:39–76. Jefferson, NC: McFarland.
Stanford, R. G., and A. G. Stein. 1994. A meta-analysis of ESP studies contrasting hypnosis and a comparison condition. *JP* 58 (3): 235–70.
Stapp, H. E. 1993. *Mind, matter, and quantum mechanics*. Berlin: Springer-Verlag.
———. 1994. Theoretical model of a purported empirical violation of the predictions of quantum theory. *Physical Review* A 50:18–22.
Steering Committee of the Physicians' Health Study Research Group. 1988. Preliminary report: Findings from the aspirin component of the ongoing Physicians' Health Study. *New England Journal of Medicine* 318:262–64.
Stein, R. 1997. With new findings, neuroscientists have a hunch intuition makes sense. *Washington Post* (February 28), p. A14.
Stevens, S. S. 1967. The market for miracles. *Contemporary Psychology* 12:103.
Stevenson, I., and B. Greyson. 1979. Near-death experiences: Relevance to the question of survival after death. *Journal of the American Medical Association* 242:265–67.
Stiffler, J. 1981. How computers fail. *IEEE Spectrum* (October), 44–46.
Stokes, D. M. 1987. Theoretical parapsychology. In *Advances in parapsychological research*, edited by S. Krippner, 5:77–189. Jefferson, NC: McFarland.
Stuart, C. E., and J. A. Greenwood. 1937. A review of criticisms of the mathematical evaluation of ESP data. *JP* 1:295–304.
Sun Tzu. 1963. *The art of war*. Translated by S. B. Griffith. New York: Oxford Univ. Press.
Swann, I. 1987. *Natural ESP*. New York: Bantam.
Swets, J. A., and R. A. Bjork. 1990. Enhancing human performance: An evaluation of "new age" techniques considered by the U.S. Army. *Psychological Science* 1:85–96.
Tabori, P. 1974. *Crime and the occult*. New York: Taplinger.
Targ, R. 1996. Remote viewing at Stanford Research Institute in the 1970s: A memoir. *JSE* 10 (1): 77–88.
Targ, R., and K. Harary. 1984. *The mind race*. New York: Villard Books.

Targ, R., and H. E. Puthoff. 1974. Information transmission under conditions of sensory shielding. *Nature* 251:602–7.
Tart, C. T. 1963. Physiological correlates of psi cognition. *International JP* 5:375–86.
———. 1986. *Waking up*. Boston: Shambhala.
Tart, C. T., H. E. Puthoff, and R. Targ. 1979. *Mind at large*. New York: Praeger.
Tasso, J., and D. Miller. 1976. The effects of the full moon on human behavior. *Journal of Psychology* 93:81–83.
Taubes, G. 1996. To send data, physicists resort to quantum voodoo. *Science* 274:504–5.
Templer, D. I., and D. M. Veleber. 1980. The moon and madness: A comprehensive perspective. *Journal of Clinical Psychology* 36:865–68.
Thorndike, L. 1905. *The place of magic in the intellectual history of Europe*. New York: Columbia Univ.
Tichener, E. B. 1898. The feeling of being stared at. *Science* 8:895–97.
Tromp, S. W. 1980. *Biometeorology*. London: Heyden.
Truzzi, M. 1983. Reflections on conjuring and psychical research. Unpublished ms.
———. 1987. Zetetic ruminations on skepticism and anomalies in science. *Zetetic Scholar* 12/13:7–20.
Tversky, A., and D. Kahneman. 1971. Belief in the law of small numbers. *Psychological Bulletin* 2:105–10.
Ullman, M., S. Krippner, and A. Vaughan. 1973. *Dream telepathy*. New York: Macmillan.
U.S. Library of Congress. Congressional Research Service. 1983. *Research into "psi" phenomena: Current status and trends of congressional concern*. Compiled by C. H. Dodge.
Utts, J. M. 1986. The ganzfeld debate: A statistician's perspective. *JP* 50:393–402.
———. 1988. Successful replication vs. statistical significance. *JP* 52:305–20.
———. 1991a. Replication and meta-analysis in parapsychology. *Statistical Science* 6:363–82.
———. 1991b. Rejoinder. *Statistical Science* 6:396–403.
———. 1996a. An assessment of the evidence for psychic functioning. *JSE* 10:3–30.
———. 1996b. *Seeing through statistics*. Belmont, CA: Wadsworth.
Vallee, J., A. C. Hastings, and G. Askevold. 1976. Remote viewing through computer conferencing. *Proceedings of the IEEE* 64:1551.
Vasiliev, L. L. 1963. *Experiments in distant influence*. New York: Dutton.
von Bertalanffy, L. 1968. *General system theory*. New York: George Braziller.
von Franz, M. 1992. *Psyche and matter*. Boston: Shambhala.
Wachter, K. 1988. Disturbed by meta-analysis? *Science* 241:1407–8.
Wagner, M. W., and M. Monet. 1979. Attitudes of college professors towards extrasensory perception. *Zetetic Scholar*, no. 5, pp. 7–16.
Wait till the moon is full. 1992. *Berkeley Wellness Letter* (June), 6.
Wald, G. 1988. Cosmology of life and mind. *Los Alamos Science* 16. Los Alamos National Laboratory, New Mexico.
Walker, E. H. 1975. Consciousness and quantum theory. In *Psychic exploration: A challenge for science*, edited by E. D. Mitchell, 544–68. New York: Putnam.
Warren, C., B. E. McDonough, and N. S. Don. 1992. Event-related brain potential changes in a psi task. *JP* 56:1–30.
Watkins, G. K., and A. M. Watkins. 1971. Possible PK influence on the resuscitation of anesthetized mice. *JP* 35 (4): 257–72.

Weaver, J. C., and R. D. Astumain. 1990. The response of living cells to very weak electric fields: The thermal noise limit. *Science* 247:459–62.

Weber, R. 1986. *Dialogues with scientists and sages: The search for unity.* New York: Routledge & Kegan Paul.

Weinberg, S. 1977. *The first three minutes: A modern view of the origin of the universe.* New York: Basic Books.

Weiskott, G., and G. B. Tipton. 1975. Moon phases and state hospital admissions. *Psychological Reports* 37:486.

Weiss, P. A. 1969. The living system: Determinism stratified. In *Beyond reductionism: New perspectives in the life sciences,* edited by A. Koestler and J. R. Smythies. London: Hutchinson.

Wezelman, R., and J. L. F. Gerding. 1994. The set-effect analysis: A post hoc analysis of displacement in Utrecht ganzfeld-data. In *Proceedings of Presented Papers,* 37th Annual Parapsychological Association Convention, edited by D. J. Bierman, 411–416. Fairhaven, MA: Parapsychological Association.

Wezelman, R., J. L. F. Gerding, and I. Verhoeven. In preparation. Eigensender ganzfeld psi: An experiment in practical philosophy.

Wezelman, R., D. I. Radin, J. M. Rebman, and P. Stevens. 1996. An experimental test of magic healing rituals in remote influence of human physiology. In *Proceedings of Presented Papers,* 39th Annual Parapsychological Association Convention, edited by E. C. May, 1–12. Fairhaven, MA: Parapsychological Association.

Whitehead, A. N. 1933. *Science and the modern world.* Cambridge: Cambridge Univ. Press.

Whyte, L. L. 1950. *The next development in man.* New York: Julian Press.

Wigner, E. P. 1963. The problem of measurement. *American Journal of Physics* 31:6.

———. 1969. Are we machines? *Proceedings of the American Philosophical Society* 113 (2): 95–101.

Wilber, K. 1977. *The spectrum of consciousness.* Wheaton, IL: Theosophical Publishing House.

———. 1984. *Quantum questions.* Boulder, CO: Shambhala.

———. 1993. *The spectrum of consciousness.* 20th anniversary ed. Wheaton, IL: Quest Books.

Wilhelm, J. L. 1977. Outlook (column). *Washington Post* (August 7), p. B5.

Wilkinson, H. P., and A. Gauld. 1993. Geomagnetism and anomalous experiences, 1868–1980. *Proceedings of the Society for Psychical Research* 57:275–310.

Williams, L. 1983. Minimal cue perception of the regard of others: The feeling of being stared at. Paper presented at the 10th Annual Conference of the Southeastern Regional Parapsychological Association, West Georgia College, Carrollton, GA, February 11–12.

Wilson, B. W., C. W. Wright, J. E. Morris, R. L. Buschbom, D. P. Brown, D. L. Miller, R. Sommers-Flannigan, and L. E. Anderson. 1990. Evidence of an effect of ELF electromagnetic fields on human pineal gland function. *Journal of Pineal Research* 9:259–69.

Wilson, G. T., and S. J. Rachman. 1983. Meta-analysis and the evaluation of psychotherapy outcome: Limitations and liabilities. *Journal of Consulting & Clinical Psychology* 51:54–64.

Wiseman, R., and M. J. Schlitz. 1996. Experimenter effects and the remote detection of staring. In *Proceedings of Presented Papers,* 39th Annual Parapsychological

Association Convention, edited by E. C. May, 149–158. Fairhaven, MA: Parapsychological Association.
Wiseman, R., and M. D. Smith. 1994. A further look at the detection of unseen gaze. In *Proceedings of Presented Papers,* 37th Annual Parapsychological Association Convention, edited by D. J. Bierman, 352–60. Fairhaven, MA: Parapsychological Association.
Wiseman, R., M. D. Smith, D. Freedman, T. Wasserman, and C. Hurst. 1995. Examining the remote staring effect: Two further experiments. In *Proceedings of Presented Papers,* 38th Annual Parapsychological Association Convention, edited by N. L. Zingrone, 480–90. Fairhaven, MA: Parapsychological Association.
Witches and witchcraft. 1990. In *Mysteries of the Unknown.* Richmond, VA: Time-Life Books.
Wohl, C. G., R. N. Cahn, A. Rittenberg, T. G. Trippe, and G. P. Yost. 1984. Review of particle properties. *Reviews of Modern Physics* 56 (2), pt. 2, pp. S1–S9.
Woodward, K. L., et al. 1992. Talking to God. *Newsweek* (January 6), 39–44.
World Notes. 1989. *Time* (October 9), 57.
Zilberman, M. S. 1995. Public numerical lotteries, an international parapsychological experiment covering a decade: A data analysis of French and Soviet numerical lotteries. *JSPR* 60:149–60.
Zohar, D. 1983. *Through the time barrier: A study in precognition and modern physics.* London: Granada.
Zukav, G. 1979. *The dancing Wu Li masters.* New York: Bantam.
Zusne, L., and W. H. Jones. 1982. *Anomalistic psychology: A study of extraordinary phenomena of behavior and experience.* Hillsdale, NJ: Erlbaum.

Index